Instrumentation and Techniques in
High Energy Physics

Instrumentation and Techniques in High Energy Physics

Editor

Don Lincoln
Fermi National Accelerator Laboratory, USA

World Scientific

NEW JERSEY · LONDON · SINGAPORE · BEIJING · SHANGHAI · HONG KONG · TAIPEI · CHENNAI · TOKYO

Published by

World Scientific Publishing Co. Pte. Ltd.
5 Toh Tuck Link, Singapore 596224
USA office: 27 Warren Street, Suite 401-402, Hackensack, NJ 07601
UK office: 57 Shelton Street, Covent Garden, London WC2H 9HE

Library of Congress Control Number: 2024945452

British Library Cataloguing-in-Publication Data
A catalogue record for this book is available from the British Library.

INSTRUMENTATION AND TECHNIQUES IN HIGH ENERGY PHYSICS

Sponsoring Consortium for
Open Access Publishing in Particle Physics

Open Access funded by SCOAP³
Copyright © 2025 The Editor(s) (if applicable) and The Author(s)
This ebook was published as Open Access through the sponsorship of SCOAP3, licensed under the terms of the Creative Commons Attribution-NonCommercial 4.0 International License (https://creativecommons.org/licenses/by-nc/4.0/), which permits use, sharing, adaptation, distribution and reproduction in any medium or format, as long as you give appropriate credit to the original author(s) and the source, provide a link to the Creative Commons license and is used for non-commercial purposes.

ISBN 978-981-98-0109-1 (hardcover)
ISBN 978-981-98-0110-7 (ebook for institutions)
ISBN 978-981-98-0111-4 (ebook for individuals)

For any available supplementary material, please visit
https://www.worldscientific.com/worldscibooks/10.1142/14049#t=suppl

Desk Editor: Carmen Teo Bin Jie

Typeset by Stallion Press
Email: enquiries@stallionpress.com

Contents

1. Calorimetry 1
   Nural Akchurin

2. Solid State Tracking Detectors 45
   Maurice Garcia-Sciveres

3. Radiation Damage to Organic Scintillators 71
   Sarah Eno

4. RICH Detectors 97
   Sajan Easo

5. Machine Learning for Analysis and Instrumentation in High Energy Physics 125
   Javier M. Duarte and Dylan S. Rankin

6. Jets at Colliders 179
   Simone Marzani

7. Instrumentation and Techniques in High Energy Physics: Liquid Argon Neutrino Detectors 213

   David Caratelli

8. The Super-Kamiokande and Other Detectors: A Case Study of Large Volume Cherenkov Neutrino Detectors 251

   Shunichi Mine

Index 291

Chapter 1

Calorimetry

Nural Akchurin

*Texas Tech University, Advanced Particle Detector Laboratory,
Department of Physics and Astronomy,
Lubbock, TX 79409, USA*

Calorimetry has been an essential part of nearly all high-energy physics experiments for several decades and has witnessed a remarkable evolution in capability and complexity. Relatively simple and coarse detectors have evolved into highly granular, sophisticated devices that deliver maximal information on energy deposition of particles in matter. In this chapter, we cover the fundamentals of energy measurement at high energies, highlight the salient features of a select few calorimeters at the LHC, and review the emerging ideas that hold promise for the future. Advances in calorimetry have been facilitated by impressive progress in high performance electronics, material science, and computing. There is little doubt that the knowledge gained by targeted R&D efforts and the experiences garnered in calorimetry at the LHC will result in unprecedented devices for precise energy, space, and time measurements of all fundamental particles at future colliders and elsewhere.

Keywords: Calorimetry, particle cascades, sampling calorimetry.

1.1 Introduction

Energy measurement of elementary particles at high energies is the domain of calorimetry. Calorimeters produce signals that are proportional to the energy of the impinging charged or neutral particles when these particles completely deposit their energies within them. The types and processes by which they generate these signals vary widely: a flash of light or a collection of electric charge may represent the energy of the initial particle. These

---

2024 © The Author(s). This is an Open Access chapter published by World Scientific Publishing Company, licensed under the terms of the Creative Commons Attribution 4.0 International License (CC BY 4.0). https://doi.org/10.1142/9789819801107_0001

signals may be produced by electromagnetic and hadronic interactions as the initial particle produces a cascade of secondary particles in the calorimeter. In some cases, the calorimeter is a dense crystal (fully active) and in others, comprises alternating layers of absorber and active materials (sampling). In the last few decades, several insightful publications have given us perspectives on the art and science of calorimetry [1–4]. In this chapter, we focus on the fundamentals of electromagnetic and hadronic showers by reviewing the relevant quantities and scaling relationships. To date, hundreds of calorimeters have been built and operated with different levels of capability. While we do not intend to give an exhaustive account of all, we discuss a select few by delving into the phenomena behind their unique features. We conclude this chapter by examining the latest developments that are likely to play a role in the near future.

1.2 Electromagnetic Showers

Electrons above ~ 1 GeV interact with a dense material mostly by *bremsstrahlung* or "breaking radiation." This is the process through which energetic electrons that are accelerated in the Coulomb field of a nucleus ($+Ze$) emit photons. As the electron traverses the medium, it experiences the electric field of the atomic nucleus, incurs a change in its direction, and emits a photon ($e \rightarrow e + \gamma$) as the nucleus recoils, conserving energy and momentum. In this process, the nucleus does not break up and the photons display a $1/E$ spectrum. The atomic electrons play a negligible role in this process because they carry a much smaller charge compared to the nucleus. Photons of similar energies interact with the electric field of the nucleus and create an electron–positron pair ($\gamma \rightarrow e^+ + e^-$).

The interaction involves production of secondary particles in a cascade process, also called a shower, until the energy of secondaries fall below a threshold. When the average energy per particle is below this threshold, the particle production stops, and electrons and positrons lose their energy through ionization (dE/dx), typically below ~ 10 MeV. Other processes, such as Möller and Bhabha scattering and positron annihilation, contribute to the energy loss but to a much smaller extent.

1.2.1 *Radiation length and critical energy*

The characteristic amount of matter traversed by an electron or positron is called the radiation length, X_o. It is defined as the average longitudinal length at which an energetic electron (>1 GeV) loses $1 - 1/e = 0.632$ of its energy through bremsstrahlung and atomic screening effects. The radiation length is written as

$$X_o^{-1} = 4\alpha r_e^2 \frac{N_A}{A} \left\{ Z^2 [L_{\text{rad}} - f(Z)] + Z L'_{\text{rad}} \right\} \quad (1.1)$$

where α is the fine structure constant, r_e is the classical electron radius ($e^2/4\pi\epsilon_0 m_e c^2 \approx 2.82$ fm), N_A is Avogadro's number, and $A = 1$ g mol^{-1}. Note that the coefficient $4\alpha r_e^2 \frac{N_A}{A} = (716.408 \text{ g cm}^{-2})^{-1}$ applies specifically to electrons through the mass term in the expression for the classical electron radius. The function $f(Z)$ is an infinite sum:

$$f(Z) = a^2 \left[(1+a^2)^{-1} + 0.20206 - 0.0369 a^2 + 0.0083 a^4 - 0.002 a^6 + \cdots \right] \quad (1.2)$$

where $a = \alpha Z$ [5]. L_{rad} and L'_{rad} are calculated by Tsai [6, 7] and given in Table 1.1. A simpler expression that is accurate within a few percent is

$$X_o = \frac{716.4 A}{Z(Z+1) \ln(287/\sqrt{Z})} \quad (1.3)$$

in units of g cm^{-2}. It is understood that a high-energy electron will lose the same fraction of its energy in going through 17.6 mm of iron, or 5.6 mm of lead, or 361 mm of water, all representing 1 X_o. Therefore, the radiation length is a useful measure in quantifying depth, essentially independent of material properties (see Fig. 1.1).

One can think of X_o as the longitudinal length at which the number of electromagnetic particles double. For electrons and positrons, this doubling continues until their energy is degraded such that it becomes more likely that they lose energy by dE/dx than by producing new particles, as mentioned earlier. For photons, e^+e^- pair production becomes impossible below the kinematic limit of $E_\gamma = 2 m_e c^2$. Below this ~ 1 MeV limit,

Table 1.1. The values for L_{rad} and L'_{rad} are used to calculate the radiation length of an element using Eq. (1.1).

Element	Z	L_{rad}	L'_{rad}
H	1	5.31	6.144
He	2	4.79	5.621
Li	3	4.74	5.805
Be	4	4.71	5.924
Others	>4	$\ln(184.15 Z^{-1/3})$	$\ln(1194 Z^{-2/3})$

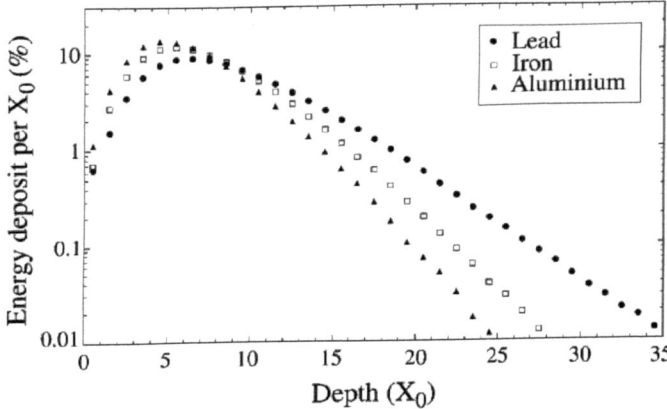

Figure 1.1. The scaling of electromagnetic showers in depth can be observed from EGS4 simulation [8] for different absorbers (Al, Fe, and Pb) for 100 GeV electrons [4]. Depth is expressed in units of radiation lengths.

photoelectric effect and Compton scattering come into play, introducing material dependences that can be significant.

For a mixture of materials, the radiation length can be calculated as

$$X_\circ^{-1} = \sum_i f_i/X_i \qquad (1.4)$$

The f_i is the volume fraction, and X_i is the radiation length (in mm) of the ith component in the mixture. For compounds, the volume fraction should be replaced by the mass fraction, and the radiation length should be expressed in g cm^{-2}.

The critical energy E_c may be identified as the energy at which energy loss through ionization equals energy loss through bremsstrahlung. The Particle Data Group [9] adopts Rossi's definition of critical energy and defines it as the energy at which the ionization loss per radiation length is equal to the electron energy. These two definitions are the same with the approximation $[dE/dx]_{\text{brems}} \approx E/X_\circ$. To account for the density differences in materials, the approximate formulation of the critical energy can be given as $a/(Z+b)^\alpha$. The parameter a equals 710 (gas) or 610 (solid) MeV, and b equals 0.92 (gas) and 1.24 (solid) with $\alpha \approx 1$. For example, the critical energy is 22 MeV for iron, 7.4 MeV for lead, and 83 MeV for water. The largest fractional deviation appears at large Z values.

The Molière radius (ρ_M) is used to characterize the extent of transverse shower size and, unlike the radiation length, does not have a fundamental meaning:

$$\rho_M = E_s \frac{X_o}{E_c} \quad (1.5)$$

The scale energy $E_s = \sqrt{\frac{4\pi}{\alpha}} m_e c^2 = 21.2$ MeV and E_c is based on Rossi's definition. Assuming an infinite material length, a shower is contained at a 90% (99%) level inside a cylinder with 1 (3.5) ρ_M. Beyond this radial size, the Molière radius scaling no longer works properly because the composition effects become significant. The material independence of the Molière radius can be shown by a crude approximation: the radiation length scales as $\frac{A}{Z^2} \approx \frac{1}{Z}$ and the critical energy as $\sim \frac{1}{Z}$. The Molière radius is 16.9 mm for iron, 16.0 mm for lead, and 92 mm for water.

1.2.2 *Longitudinal and transverse shower profile*

The average distance at which energetic photons lose $1 - 1/e = 0.632$ of their energy (mean free path) equals $(9/7)X_o$. The near equality of the two length scales for bremsstrahlung and pair production suggests a comparable significance of these two processes in electromagnetic interactions. In addition, this near-equality renders a well-behaved longitudinal shower profile for electromagnetic showers generated by an incoming particle with energy E_0:

$$\frac{dE}{dt} = E_0 b \left[\frac{(bt)^{a-1} e^{-bt}}{\Gamma(a)} \right] \quad (1.6)$$

where $t = x/X_o$, a and b are free parameters, and $\Gamma(a)$ is the gamma function. The shower maximum occurs at $t_{\max}$

$$t_{\max} = (a-1)/b = \ln y + C_j \quad (j = e, \gamma) \quad (1.7)$$

where $y = E/E_c$, $C_e = -0.5$ for incident electrons, and $C_\gamma = +0.5$ for incident photons. It is evident that the multiplicity of shower particles is maximum at $t_{\max}$ and that the depth of shower maximum logarithmically increases with energy. In order to use Eq. (1.6), one generally uses Eq. (1.7) to find a by assuming $b \approx 0.5$. For some common absorbers, the fitted b values are plotted for incident electrons in the range of $1 \leq E_0 \leq 100$ GeV using EGS4 [8]. It should be stressed that Eq. (1.6) represents the average profile and does not work well in the early part of the shower. The cascade

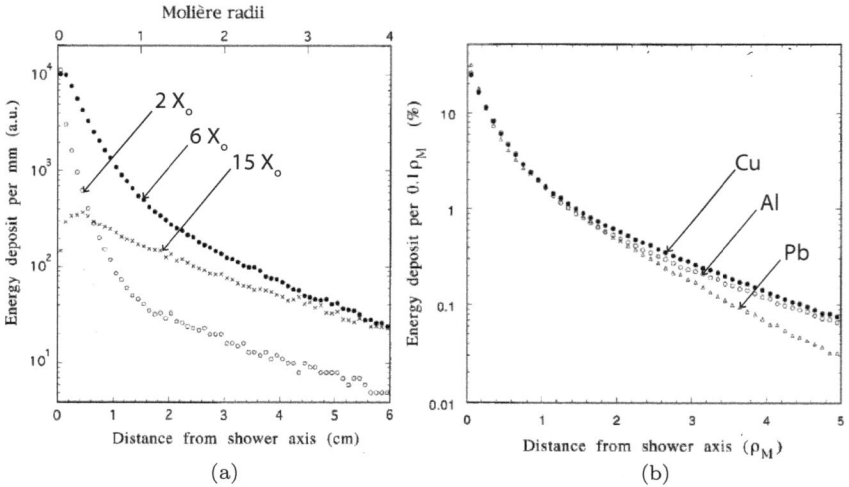

Figure 1.2. (a) The radial profiles of the energy deposited at 2, 6, and 15 radiation lengths of Cu and (b) radial profiles for Al, Cu, and Pb are shown for 100 GeV electrons based on EGS4 [8] simulations from Ref. [4].

process rises rapidly while the gamma function remains relatively flat in this region. The EGS4 should be used when accurate results are needed.

The transverse profile of electromagnetic showers is generally expressed by the sum of two Gaussians. In the earlier stage of the shower development, before the shower maximum, the electrons and positrons travel off-axis by multiple scattering within a narrow cylinder around the shower axis. After the shower maximum, the photons and electrons are produced isotropically. Compton scattering and the photoelectric effect dominate their interactions, and the particles tend to travel farther off-axis in a larger cylinder, expressed by a wider Gaussian (see Fig. 1.2(b)).

We can posit an important property about the energy resolution of an electromagnetic calorimeter by assuming that the number of particles doubles for each radiation length traversed. This process continues until the average particle energy in the shower reaches the critical energy. The average shower particle energy scales as $E/n(t)$ or $E/2^t$, where E is the energy of the incoming particle, n represents the number of shower particles, and t is the number of generations. The number of particles at the shower maximum is $n(t_{\max}) \approx E/E_c = y$, where $t_{\max} \approx \log(y)$. The relationship $n(t_{\max}) \propto E$ indicates a linear response, and n obeys Poisson statistics.

Thus, the fluctuations around the number of particles imply a (stochastic) term in energy resolution that scales as $a_1/\sqrt{E}$. For this basic reason, it is vital to understand the sources of nonlinearities and deviations from $1/\sqrt{E}$ scaling in electromagnetic calorimeters if the data suggest such a behavior. A term that scales as a_2/E can be added to account for noise that is often relevant for precision electromagnetic calorimeters. In addition, there are often energy-independent and uncorrelated effects that degrade the energy resolution, which are represented by an additional (constant) term a_3:

$$\sigma/E = \sqrt{(a_1/\sqrt{E})^2 + (a_2/E)^2 + (a_3)^2} \qquad (1.8)$$

A common way of writing the same is $\sigma/E = a_1/\sqrt{E} \oplus a_2/E \oplus a_3$, where $\oplus$ means square root of sum of squares.

1.3 Two Examples of Electromagnetic Calorimeters

In this section, two typical electromagnetic calorimeters are described, and their performances are analyzed in the light of what has been presented above. The first example is a homogeneous calorimeter (CMS), where all the particle's energy is deposited in the active volume and is used to produce a signal that is proportional to the incoming particle's energy. The second example is a sampling calorimeter (ATLAS), where a dense absorber and active layers are interleaved. This approach allows the high density to fully absorb the particle's energy, while the shower is sampled in the active layers, in which the measurable signal is produced. It should be noted that both of these calorimeters played a crucial role in the discovery of the Higgs boson in 2012 [10, 11] and they have been upgraded in various ways since their original commissioning.

1.3.1 *A fully active electromagnetic calorimeter: CMS ECAL*

Typically, homogeneous electromagnetic calorimeters are constructed using blocks of scintillating high-Z inorganic crystals (e.g., BGO, BaF$_2$, CsI, CsI(Tl), LSO, LYSO, NaI(Tl), and PbWO$_4$), Cherenkov radiators (e.g., PbF$_2$ and lead glass), or noble liquids (e.g., Ar, Kr, and Xe). The best energy resolution is obtained from homogeneous calorimeters, especially if the scintillation light is high. Since they are homogeneous, there are no sampling fluctuations degrading the energy measurement. Some of their parameters are listed in Table 1.2.

Table 1.2. Some high-Z crystal parameters relevant for calorimetry are given. The superscripts for τ_{decay}, $\lambda_{\max}$, and Relative Light Yield (LY) indicate the fast (f) and (s) components of scintillation emission. For more details, see Ref. [9].

	ρ [g/cm^3]	$X_\circ$ [cm]	ρ_M [cm]	dE/dx [MeV/cm]	λ_{int} [cm]	τ_{decay} [ns]	$\lambda_{\max}$ [nm]	n	Relative LY
BGO	7.13	1.12	2.23	9.0	22.8	300	480	2.15	21
BaF$_2$	4.89	2.03	3.10	6.5	30.7	650^s <0.6^f	300^s 220^f	1.50	36^s 4.1^f
CsI	4.51	1.86	3.57	5.6	39.3	30^s 6^f	310	1.95	3.6^s 1.1^f
PbWO$_4$	8.30	0.89	2.00	10.1	20.7	30^s 10^f	425^s 420^f	2.20	0.3^s 0.077^f
NaI(Tl)	3.67	2.59	4.13	4.8	42.9	245	410	1.85	100
PbF$_2$	7.77	0.93	2.21	9.4	21.0	—	—	—	Cherenkov

The CMS electromagnetic calorimeter is based on lead tungstate crystals (PbWO$_4$) [12]. As can be seen from Table 1.2, it is very dense and exhibits a short radiation length and Molière radius, thereby rendering electromagnetic showers more compact compared to the other crystals listed. It also has high dE/dx. The shorter interaction length λ_{int} means that the hadronic showers are also likely to start in the crystals. Compared to NaI(Tl), PbWO$_4$ is considerably less bright, but the photon emission spectrum peaks around 425 nm, well matched to most photodetectors, e.g., the avalanche photodiodes in CMS. The overall structure of the CMS electromagnetic calorimeter is shown in Fig. 1.3.

The barrel section covers the range $|\eta| < 1.479$ and contains 61,200 crystals. The crystals are tapered and mounted such that the cracks between them are not aligned with the particle trajectories coming from the intersection point (3 degrees off in η and ϕ). The crystals are 23 cm ($25.8 X_\circ$) in depth, and the cross-sections are slightly larger than a Molière radius, measuring 2.2×2.2 cm^2 at the front and 2.6×2.6 cm^2 at the rear end. In terms of transverse granularity, this cross-section corresponds to approximately 0.0174×0.0174 in η and ϕ coordinates. The barrel calorimeter weighs 67.4 tons.

The endcap section covers the range $1.479 < |\eta| < 3.0$. Each endcap is divided into 2 halves for mechanical reasons and holds 3,662 crystals. The crystals are pointing 2 to 8 degrees off of projective orientation and are 22 cm ($24.7 X_\circ$) in depth. The cross-section measures 2.86×2.86 cm^2 at the front and 3.0×3.0 cm^2 at the rear end. The endcap calorimeter weighs 24.0 tons.

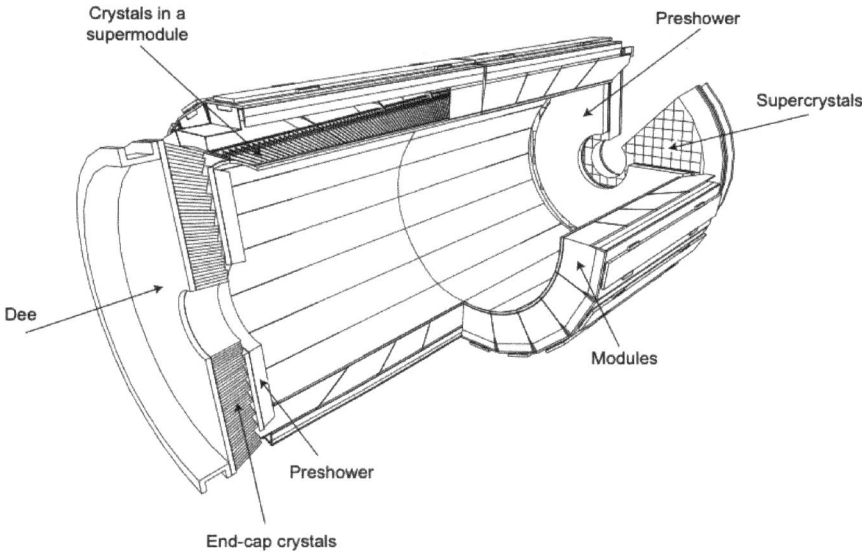

Figure 1.3. Sketch of the CMS electromagnetic calorimeter where the individual PbWO$_4$ crystals are shown in the barrel and endcap regions. The crystals are tapered and nearly projective, pointing to the interaction [12].

Two avalanche photodiodes (APDs) per crystal are used to detect the scintillation light in the barrel section. The vacuum phototriodes (VPTs) are used in the endcaps.

The energy resolution of the CMS electromagnetic calorimeter is given as $3\%/\sqrt{E} \oplus 0.2/E \oplus 0.5\%$ with E in GeV [9]. In what follows, we analyze each of these terms separately. The dominant factor that contributes to the stochastic term of 3% is fluctuation in the photon yield and is given by $\sqrt{F/N_{\text{p.e.}}}$. F is the noise factor that parametrizes fluctuations in the gain process, and it is about 2 for the APDs and 2.5 for the VPTs. $N_{\text{p.e.}}$, the number of photoelectrons generated per GeV, is $\sim$4,500. Thus, the photo-statistical contribution is $\sim$2.1%. The lateral containment of the showers also plays a role at the level of $\sim$1.5% after the signals from 5 × 5 crystals are summed. The contribution increases to $\sim$2% if a 3 × 3 crystal array is used. The stochastic term reaches 2.6–2.9% when these two effects are added in quadrature.

The noise term, 0.2 GeV in this case, scales as $1/E$ and comes predominantly from electronics and digitization noise, with a small contribution from pile-up events. Incoherent noise is set by the square root of the number of detector cells being summed. In a test beam study, the measured noise level for the barrel was 40 MeV/channel after removal of the small channel-to-channel correlated noise. As expected, this procedure resulted in a noise level for the sum of 25 channels that was exactly 5 times the noise in a single channel.

The main factor in the constant term, 0.5%, is the non-uniformity of the longitudinal light collection. In the case of CMS, the crystal quality specification was such that the contribution of this non-uniformity to the constant term was less than 0.3% at the time of construction. This meant a limit on the slope of the longitudinal light collection curve in the region of the shower maximum of ~0.35% per radiation length. With time, radiation damage induces optical absorption, reducing the light yield and introducing sizable non-uniformities in light collection. The CMS collaboration developed an elaborate monitoring system to track and correct for these effects [13]. In addition, the temperature sensitivity of $PbWO_4$ ($-2.5\%/°C$) requires stringent temperature stability: 0.05 °C in the barrel and 0.1 °C in the endcaps. Other smaller contributors come from the inter-calibration errors and longitudinal energy leakage.

1.3.2 A sampling electromagnetic calorimeter: ATLAS ECAL

The ATLAS electromagnetic calorimeter is a sampling calorimeter with alternating layers of lead (Pb) and liquid argon (LAr). While the Pb absorber is efficient in interacting with electrons and photons ($X_o = 5.6$ mm), the LAr layers (2 mm in the barrel and 1.2–2.7 mm in the endcap with $X_o = 14.2$ cm) are responsible for generating measurable electron/ion pairs that drift in the electric field (2 kV for 2 mm gaps in the barrel) and induce a signal (~450 ns drift time), which represents the energy deposition. The calorimeter is arranged in an accordion shape in the barrel and endcap calorimeters (Fig. 1.4).

Both the barrel and endcap calorimeters are longitudinally segmented into three layers. The first layer is about $4.4X_o$ thick and is segmented into strips in the η direction, typically 0.003×0.1 in $\Delta\eta \times \Delta\phi$ in the barrel, so as to provide an event-by-event discrimination between single photon showers and overlapping showers coming from the decay of neutral hadrons in jets. The second layer collects most of the electromagnetic energy in $17X_o$ depth with a granularity of 0.025×0.025. The third and last layer is about

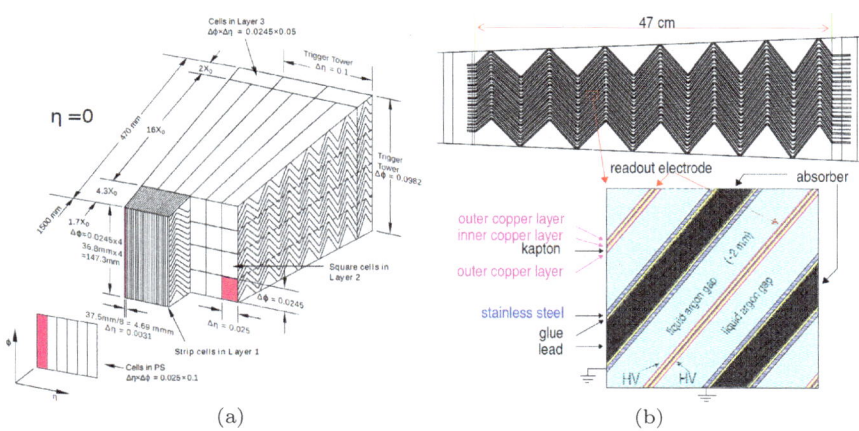

Figure 1.4. (a) Sketch of an ATLAS electromagnetic calorimeter module where the different layers are visible. The granularity in η and ϕ for each of the three layers and for the trigger towers is also shown. (b) A view of a small sector of the barrel calorimeter in a plane transverse to the LHC beams [14].

$2X_o$ thick with a strip granularity of 0.05 × 0.025. This layer is designed to identify energy leakage beyond the calorimeter for high-energy showers. In front of the barrel accordion modules, a 1.1-cm thick LAr pre-sampler layer is used to correct for energy losses upstream of the calorimeter. While this multi-layer structure introduced calibration difficulties, extensive use of Monte Carlo and multivariate analysis techniques helped achieve excellent electromagnetic response linearity and energy resolution for the ATLAS collaboration.

Detector performance was characterized using a test beam over a range of 15–180 GeV. The performance was linear within ±0.1% and the energy resolution was found to be $10\%/\sqrt{E} \oplus 0.2\%$ (E in GeV) [15–17].

1.4 Hadronic Showers

1.4.1 *Structure of hadronic showers*

As Fig. 1.5 illustrates, there are two major components in hadronic showers: an electromagnetic component that comes from the decay of π^os, ηs, and other mesons, and a non-electromagnetic component that encompasses everything else. The non-electromagnetic component includes the energy expended by breaking up nuclei by overcoming the nuclear binding energy of

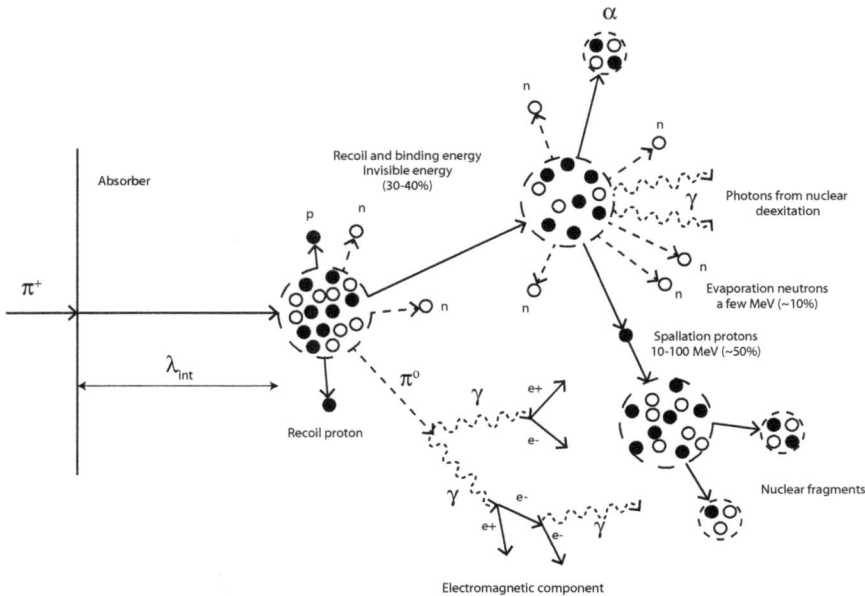

Figure 1.5. A cartoon of a hadronic shower in an absorber such as lead. The percentage numbers indicate the approximate amount of energy carried by each process as a fraction of the non-electromagnetic component.

nucleons that does not result in a measurable signal, e.g., unlike the signals generated by e^- and e^+ tracks in electromagnetic showers. This "invisible" energy may be as large as 40% of the non-electromagnetic component. In addition to the event-by-event fluctuations in energy sharing between the electromagnetic and non-electromagnetic components of the shower, the "invisible" energy within the non-electromagnetic component fluctuates as well. Both fluctuations degrade the precision with which the energy of incoming particles is measured.

The rapid decay ($\sim 10^{-16}$ ns) of π^0s and other mesons to photons in the shower may take place nearly anywhere in the calorimeter. These decays resemble highly localized bursts of electromagnetic energy and are unique to hadron showers, as shown in Figs. 1.6(c) and 1.6(d) [18]. One hadronic shower does not look like another, whereas electromagnetic showers display little variation (Figs. 1.6(a) and 1.6(b)).

For illustration purposes, if one assumes that all available energy goes to meson production in a hadron-initiated shower and, on average, that

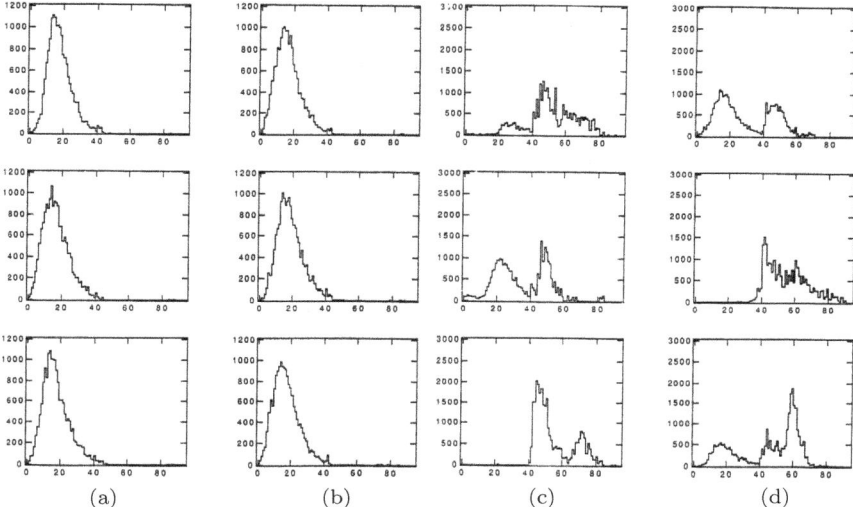

Figure 1.6. Longitudinal energy deposit profiles for randomly selected 170 GeV electron (columns (a) and (b)) and 270 GeV π^- (columns (c) and (d)) showers in a lead/iron/plastic-scintillator calorimeter [18]. The longitudinal profile of electromagnetic showers varies little from shower to shower, as observed in the six events in the left two columns. The sharp peaks in (c) and (d) are due to local electromagnetic showers induced by π^0 decays and occur beyond the first interaction length of the calorimeter. This feature has deep implications for hadronic energy resolution, response linearity, signal distribution, and calibration [19, 20]. The horizontal axes range from 0 to 6 λ_{int}, while the vertical axes are proportional to the deposited energy.

1/3 of the mesons are π^0s in the first interaction, then in the second interaction, if energetically possible, the remaining hadrons produce mesons of which a third will again be π^0s. At the end of the second interaction, the electromagnetic fraction (f_{em}) will thus be $1/3 + 1/3 \times 2/3 = 5/9$. At the end of the third interaction, the electromagnetic fraction will further increase to $5/9 + 1/3 \times 4/9 = 19/27$. For n interactions, f_{em} will simply obey a power law, $1 - (1 - 1/3)^n$. Of course, as depicted in Fig. 1.5, the hadronic interactions are more complicated than purely meson production. In commonly used absorbers, such as iron, copper, or steel, typically $\sim 80\%$ of the non-electromagnetic energy is deposited through nuclear reactions. Apart from the nuclear binding energy losses, which constitute the invisible energy, these reactions produce evaporation protons and neutrons, alphas, and other light ions. The hadron-induced showers also contain about 10% kaons. Vastly different energies and sampling of these particles make hadronic calorimetry somewhat complex.

The average f_{em} in hadronic showers initiated by a hadron with energy E is expressed as

$$\langle f_{\text{em}} \rangle = 1 - (E/E_0)^{k-1} \qquad (1.9)$$

where E_0 is a material dependent scale factor related to the average energy needed to produce one pion in hadronic interactions. The value of E_0 varies from 0.7 GeV for copper to 1.3 GeV for lead in $\pi^{\pm}$-induced reactions. The exponent k defines the energy dependence of f_{em} and is around 0.8 [21, 22]. The connection between Eq. (1.9) and the previously discussed power-law simplification of f_{em} becomes clear by observing that the non-electromagnetic fraction $(1 - 1/3)^n$ can be written as $(1 - f_{\pi^\circ})^n$ and more generally expressed as $(1 - f_{\pi^\circ}) = \langle m \rangle^{(k-1)}$, where $\langle m \rangle$ stands for the average multiplicity per nuclear interactions. It then follows that $\langle f_{\text{em}} \rangle = 1 - \langle m \rangle^{n(k-1)} = 1 - (E/E_0)^{k-1}$.

For proton-induced reactions, $\langle f_{\text{em}} \rangle$ is typically smaller, which is broadly attributed to baryon number conservation. At high energies, an incoming proton is more likely to remain a proton or turn into another baryon (e.g., neutron) after undergoing a hadronic interaction, thereby suppressing the energy fraction going into the electromagnetic component through π° decays. If an incoming particle is a charged pion, a leading charged or neutral pion is produced at the first interaction. If the leading particle is a π°, the electromagnetic fraction is boosted. If, on the other hand, the leading particle is a charged pion, the second interaction may produce a π°. Consequently, for non-compensating calorimeters (see Section 1.4.3), the event-by-event fluctuations are smaller and more symmetrically distributed around the average value for proton-induced showers (Fig. 1.7) [23, 24].

1.4.2 Hadronic shower profiles

The average distance that a high-energy hadron traverses before undergoing a nuclear interaction in a medium is defined as the nuclear interaction length (λ_{int}) and is used as the length scale to describe hadronic showers. The probability that an interaction takes place on a path along z-direction is then $1 - \exp(-z/\lambda_{\text{int}})$. The inelastic interaction cross-section σ is proportional to $A/(N_A \lambda_{\text{int}})$, and in units of g cm^{-2}, $\lambda_{\text{int}} \propto A^{1/3}$ because $\sigma \propto A^{2/3}$. To a good approximation, $\lambda_{\text{int}} = 37.8\,(\text{g cm}^{-2})A^{0.312}$. The

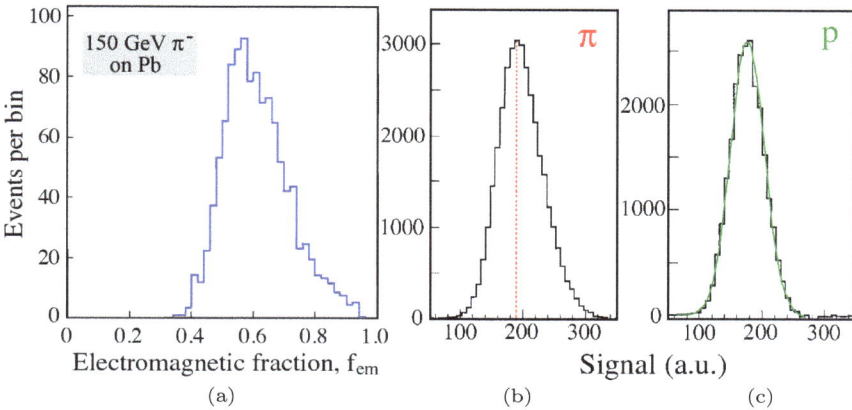

Figure 1.7. (a) Event-by-event fluctuations in the electromagnetic fraction of 150 GeV π^- showers in lead [25]. Signal distributions for 300 GeV pion (b) and proton showers (c) in a copper-based calorimeter [24].

interaction length for iron is 16.8 cm, for lead 17.0 cm, and 84.9 cm for water. Note that the interaction length is generally given for protons. Up to ~100 GeV, the pp cross-sections are larger than πp cross sections by roughly 30%. Thus, the interaction length for pions is longer by the same amount.

Hadron calorimetry requires a better than 99% longitudinal containment if a good energy measurement is desired. In iron, and in materials with similar Z, 99% longitudinal containment requires a thickness ranging from $3.5\lambda_{\text{int}}$ at 10 GeV to $7\lambda_{\text{int}}$ at 100 GeV. Figure 1.8(a) shows 95% and 99% containment depths for pions in iron based on data from the CDHS experiment [26]. The energy containment at the level of 95% in the transverse direction requires a radius of $1.5 - 2\lambda_{\text{int}}$ (Fig. 1.8(b)).

The nuclear interaction length is generally much larger than the radiation length, and the ratio λ_{int}/X_o is proportional to Z. Therefore, it is possible to distinguish electrons from hadrons based on longitudinal shower profile by using high-Z absorbers as part of pre-shower detectors. For example, a thin sheet of lead combined with a scintillation counter makes a very efficient electron/pion discriminator because λ_{int}/X_o is about 30 for lead. A 2-cm ($3.6X_o$) thick block will induce showers for electrons, while hadrons will have only ~11% probability of undergoing an inelastic interaction with it.

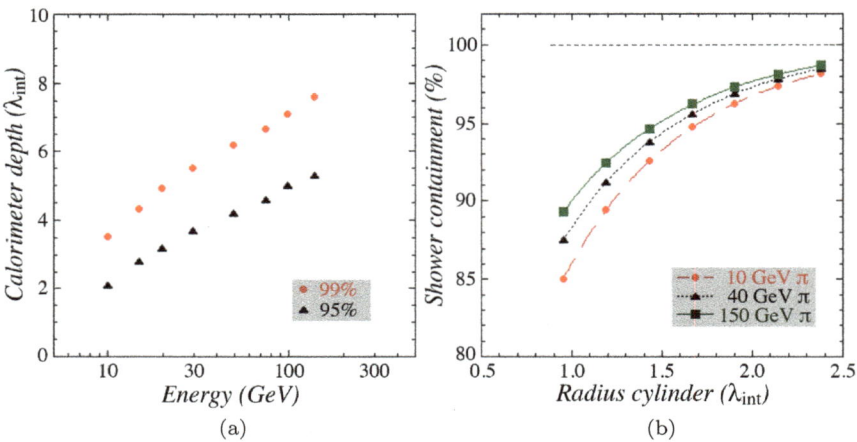

Figure 1.8. (a) The depths needed to contain hadronic showers, on average, at the level of 95% or 99% in iron as a function of the pion energy [26]. Data from Ref. [27] suggest a larger depth requirement for 99% containment. (b) Average transverse containment of pion-induced showers in a lead-based calorimeter as a function of the radius of an infinitely deep cylinder around the shower axis, for three different pion energies [28], shows that the lateral shower leakage fraction decreases with increasing pion energy. Large event-to-event fluctuations in π° production are responsible for the fact that the hadronic energy resolution is more sensitive to the effects of side leakage (Section 1.4.1).

1.4.3 *Calorimeter response and compensation*

Calorimeter response is defined as the average conversion efficiency of the deposited energy into a measurable signal that is normalized to that of electrons. We denote the calorimeter response to the electromagnetic showers as e and to the hadronic shower components as h. The e/h ratio for a calorimeter can be evaluated based on the response measurements to pions and electrons as a function of particle energy E. In practice, one performs an energy scan in as wide an energy range as possible to obtain the π/e ratios and calculates $\langle f_{\mathrm{em}} \rangle$ based on appropriate choices of E_0 and k for each energy using Eq. (1.9):

$$\frac{e}{h} = \frac{1 - \langle f_{\mathrm{em}}(E) \rangle}{\pi/e(E) - \langle f_{\mathrm{em}}(E) \rangle} \qquad (1.10)$$

The e/h value reflects the calorimeter's efficiency to generate signals in response to electromagnetic versus non-electromagnetic energy depositions;

it rarely equals one and is said to be under- ($e/h > 1$) or over-compensating ($e/h < 1$). The vast majority of calorimeters are under-compensating as a consequence of invisible energy. In principle, it is possible to tune the e/h ratio to equal unity through the sampling fraction. The sampling fraction, f_{samp}, is equal to the energy deposited in the active medium divided by the total energy deposited in the calorimeter by a minimum ionizing particle (mip) or to the ratio of the minimum values of (dE/dx) for the active medium and calorimeter. The simplest way of suppressing e is to use a high-Z absorber. The high-Z absorber very effectively suppresses signal generation by low-energy photons because of the strong Z-dependence of photoelectric effect ($\sigma \propto Z^5$). The majority of photoelectrons are captured within the absorber, and only the ones near the surface escape, inducing an observable signal. Enhancing the non-electromagnetic response h, on the other hand, requires hydrogen in the active medium so that recoiling protons from abundant neutrons contribute to the overall hadronic response.

The electromagnetic response linearity with energy is fundamental and dictated by Poisson statistics, as stressed in Section 1.2.1. However, the calorimeter response to hadrons, $\langle f_{\text{em}} \rangle + (1 - \langle f_{\text{em}} \rangle)h/e$, is energy-dependent and non-linear. Event-by-event fluctuations in f_{em} are large and non-Poissonian. If $e/h \neq 1$, these fluctuations dominate the energy resolution and are reflected in asymmetric signal distributions (Fig. 1.7). It is often assumed that the effect of non-compensation on energy resolution is energy-independent and represented by the constant term; however, the effects of fluctuations in f_{em} on energy resolution can be described by a term that is very similar to the one used for the energy dependence of its average value (Eq. (1.9)). This term can be added in quadrature to the $E^{-1/2}$ scaling term that accounts for all Poissonian fluctuations:

$$\frac{\sigma}{E} = \frac{a_1}{\sqrt{E}} \oplus a_2 \left[\left(\frac{E}{E_0}\right)^{l-1} \right] \quad (1.11)$$

where the parameter $a_2 = |1 - h/e|$ is determined by the degree of non-compensation [29]. This approach leads to a somewhat larger stochastic term when the energy resolution is expressed as a linear sum: $a_1/\sqrt{E} + a_2$.

We tend to characterize most calorimeters by the stochastic term in their resolution; however, the $a_1/\sqrt{E}$ parametrization serves a limited purpose at high energies because the resolution is dominated by deviations from this scaling. In addition, statements about energy resolution are only meaningful if they include the effects of nonlinearity, which apply to almost all calorimeter systems currently in use or planned. Signal nonlinearity introduces additional uncertainties in the mean value of the response for jets, which have to be taken into account if one wants to address the precision with which the energy of individual jets can be measured.

1.5 Two Examples of Hadronic Calorimeters

The CMS and ATLAS hadron calorimeters are large, complex, and considerably different from each other in operating principles. After a brief description of each, we focus on their unique features, drawing on the concepts discussed so far.

1.5.1 *Scintillating plates/brass calorimeter: CMS HCAL*

The CMS hadronic calorimeter (HCAL) is a sampling structure consisting of alternating brass and scintillator plates. It is located behind the $PbWO_4$ crystal electromagnetic calorimeter (ECAL) we discussed in Section 1.3.1. The e/h ratios of 2.5 (ECAL) and 1.4 (HCAL) [30] are appreciably different, and the combined calorimeter system presents interesting challenges because of this difference [31].

The barrel HCAL covers the pseudorapidity range $-1.3 < \eta < 1.3$ and consists of 36 identical azimuthal brass wedges ($\Delta\phi = 20°$), which form two half-barrels. Each wedge is further segmented into four azimuthal ($\Delta\phi = 5°$) sectors. The plates are bolted together in a staggered geometric configuration that contains no projective passive material for the full radial extent of a wedge. The interleaved scintillator plates are divided into 16 η sectors, resulting in a segmentation of $(\Delta\eta, \Delta\phi) = (0.087, 0.087)$. The total absorber thickness at 90° is 5.82 λ_{int}. The effective thickness increases with the polar angle to 10.6 λ_{int} at $|\eta| = 1.3$. The ECAL in front adds $\sim 1.1\lambda_{int}$ independent of η [30, 32].

The signal distributions for 5 and 100 GeV π^- test beam particles are displayed in Fig. 1.9. A sizable fraction of pions interact in the ECAL, as the higher and broader signal distributions indicate in Figs. 1.9(a) and 1.9(d). Sharp and narrow minimum ionization peaks, caused by particles that penetrate the ECAL without starting a shower, are also evident. The signals in the HCAL show complementary distributions, i.e., small signals for the early showering particles and larger signals for the ones that penetrate the ECAL.

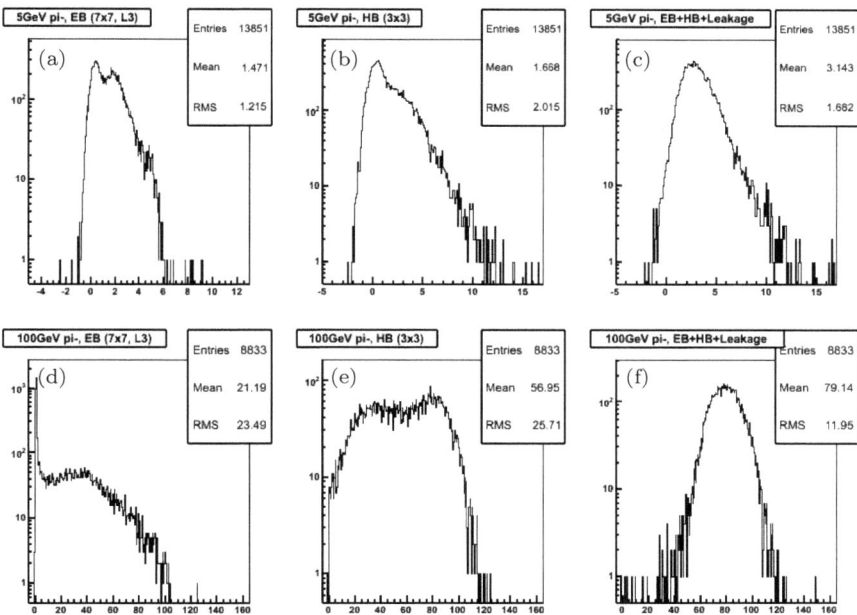

Figure 1.9. The signal distributions for 5 and 100 GeV π^- are shown for the CMS ECAL section (a) and (d), the HCAL section (b) and (e), and the combined system ECAL+HCAL (c) and (f), which also includes longitudinal energy leakage [31, 33].

Figure 1.10 shows the response of the CMS calorimeters to a variety of particles. The data are normalized to the electron response for both sections of the calorimeter. In Fig. 1.10(a), where the calorimeter response is plotted as a function of beam momentum, large differences between the different particles are apparent, especially at low momenta. For example, at 5 GeV, the antiproton response is ∼70% of the electron response, while the responses to charged pions and protons are 62% and 47% of the electron response, respectively. However, a calorimeter responds to available energy, which is different for different particles carrying the same momentum. For pions and kaons, the available energy is their kinetic energy plus their mass. For protons, it is the kinetic energy, and for antiprotons, the available energy for a calorimetric signal equals the kinetic energy plus twice the proton rest mass. In Fig. 1.10(b), the same data are plotted as a function of available energy. One would expect the response to be independent of the hadron

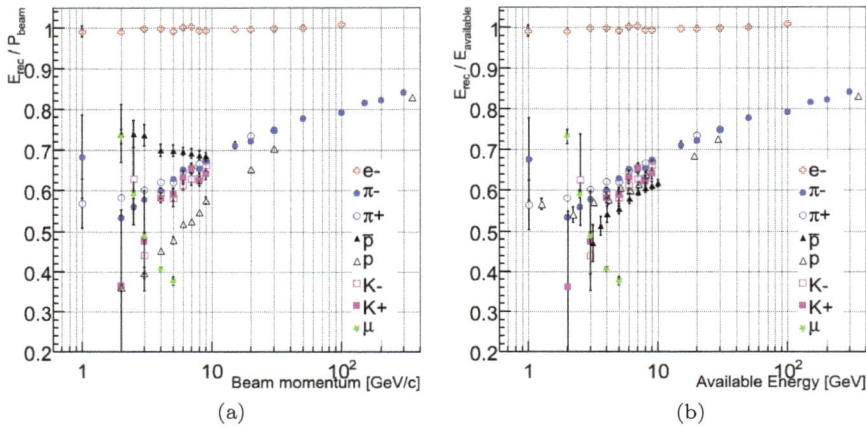

Figure 1.10. (a) The response of the combined CMS ECAL+HCAL calorimeter to different particles is shown as a function of beam momentum or (b) as a function of available energy [31]. The combined response is normalized to that for electrons.

type when the data are represented this way, but differences still remain for reasons we explain next.

The response to π^+ is systematically larger than the π^- response and more so as the energy decreases. This can be understood from the characteristics of the charge exchange reactions, $\pi^+ + n \to \pi^\circ + p$ (I) and $\pi^- + p \to \pi^\circ + n$ (II). In these reactions, a large fraction of the pion energy is carried by the final state π°, which develops electromagnetic showers. Therefore, the calorimeter response to pions interacting this way is close to unity.

Since the target material ($PbWO_4$) contains about 50% more neutrons than protons, the relative effect of reaction (I) is larger than that of reaction (II), and consequently, the calorimeter response to π^+ should be expected to be larger than the π^- response. As noted earlier, the inelastic cross-section for baryon induced interactions is larger than for pions; a larger fraction of the baryons start showering in the ECAL. The effective thickness of the ECAL is thus $1.05\lambda_{int}$ for protons and $0.89\lambda_{int}$ for pions. Since the total cross-sections for protons and antiprotons are about the same, the same holds for the effective ECAL thickness.

Figure 1.11 illustrates how energy is shared between the ECAL and HCAL sections for different hadrons. The fraction of the energy recorded by the ECAL increases from ~25% at the highest energies to ~60% at 2 GeV.

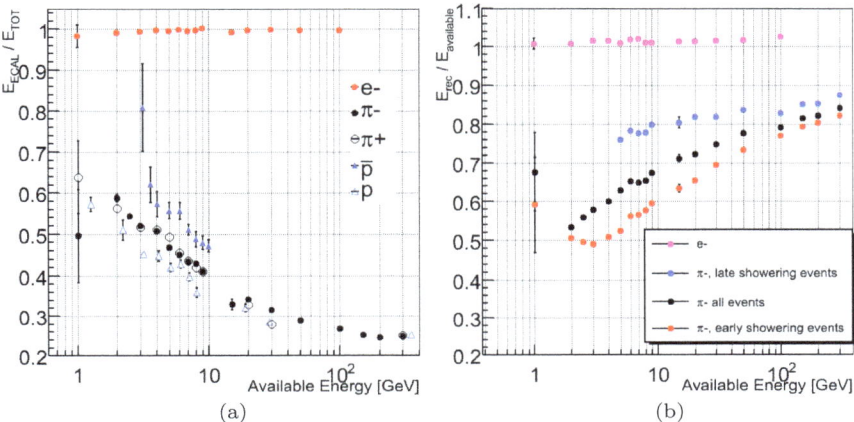

Figure 1.11. (a) The fraction of energy deposited in the CMS ECAL as a function of the available energy for charged pions and (anti)protons. (b) The response to pions as a function of energy, for the CMS barrel calorimeter. The events are subdivided into two samples according to the starting point of the shower, and the response is also shown separately for these two samples [33]. The normalization is based on the response to electrons.

At the same energies, protons deposit on average less than pions in the ECAL, while antiprotons deposit more than pions. Antiprotons start their showers, on average, earlier than pions, and therefore a larger fraction of the energy ends up in the ECAL. It would seem that one could expect the same for proton-induced showers. However, when a proton interacts in the ECAL, the final state should contain 2 baryons, which limits the energy available for π^0s. And, since the ECAL, for all practical purposes, only sees the π^0 component of the showers, this effect suppresses the proton signal in the ECAL despite the fact that protons are more likely to start their showers in the ECAL compared to pions. The baryon number conservation requirement does not limit π^0 production for antiproton induced showers. To first approximation, there is no difference with pion induced showers. The ECAL/HCAL energy sharing properly reflects the difference in interaction length in this case.

The effects described above also explain why the antiproton response is systematically smaller than the pion response (Fig. 1.10(b)). Antiprotons are more likely to start showering in the ECAL compared to pions. Pions deposit, on average, a larger fraction of their energy in the HCAL. Since the e/h value of the HCAL is smaller than that of the ECAL, the pions benefit more from the increased response to the non-electromagnetic shower components.

Another consequence of different e/h ratios for the ECAL and HCAL emerges in energy reconstruction when the showers start early in ECAL or late in HCAL. Figure 1.11(b) shows that the late showers deposit almost no energy in the ECAL, and therefore their response is determined by the more compensating HCAL. Early showers experience the strong (by a factor of 2.5) reduction in the response to the non-electromagnetic shower component deposited in the ECAL. The fact that the discrepancy increases at lower energy reflects the changes in the longitudinal shower profile also observed in the energy sharing plot (Fig. 1.11(a)). The larger the average fraction of the shower energy deposited in the ECAL, the larger the response discrepancy between showers that start in the ECAL and those that do not. CMS applies corrections in reconstructing the energies of hadrons, as described in detail in Ref. [31]. Above 5 GeV, these corrections lead to an energy resolution of the combined system where the stochastic term equals $84.7 \pm 1.6 \%/\sqrt{E}$ (E in GeV) and the constant term is $7.4 \pm 0.8 \%$. The corrected mean response remains constant within 1.3% rms.

1.5.2 Scintillating plates/iron and liquid argon calorimeter: ATLAS HCAL

The ATLAS hadronic calorimeter (TileCal) is a sampling iron/plastic-scintillator detector in the region $|\eta| < 1.7$. It is divided into three cylindrical sections, referred to as the barrel and extended barrel sections. This hadronic calorimeter extends from an inner radius of 2.28 m to an outer radius of 4.25 m. Each section is segmented into 64 azimuthal sections, referred to as modules, covering $\Delta\phi = 2\pi/64 \approx 0.1$. The scintillator plates are oriented perpendicularly to the colliding beam axis and are radially staggered in depth, as shown in Fig. 1.12. By the grouping of wavelength shifting fibers to specific photomultiplier tubes (PMT), modules are segmented in η and in radial depth. In the direction perpendicular to the beam axis, the three radial segments span 1.5, 4.1, and 1.8 λ_{int} in the barrel and 1.5, 2.6, and 3.3 λ_{int} in the extended barrels. The resulting cell dimensions are $\Delta\eta \times \Delta\phi = 0.1 \times 0.1$ (0.1×0.2 in the last segment). This segmentation defines a quasi-projective tower structure, where the deviations from perfect projectivity are small compared to the typical angular extent of hadronic jets. Altogether, TileCal comprises 4,672 readout

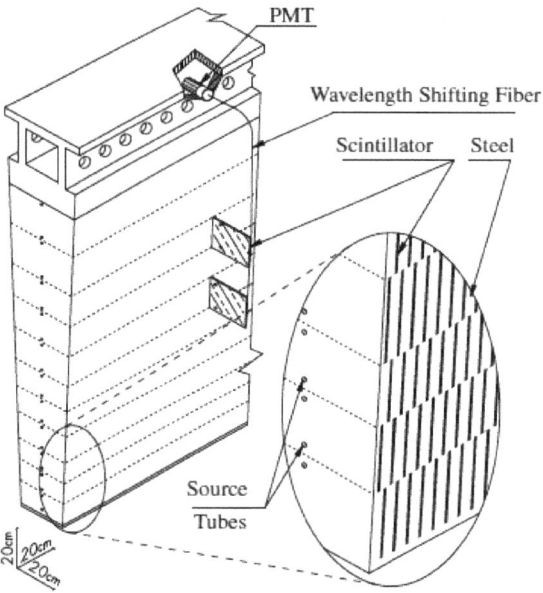

Figure 1.12. The mechanical structure of the ATLAS TileCal module. The plastic scintillator tiles are read out from both sides with wavelength shifting fibers into separate PMTs. The staggered absorber/scintillator and the radioactive source tubes are shown on the right.

cells, each equipped with two PMTs that receive light from opposite sides of every tile; see Refs. [28,34] for detailed description. The endcap and forward hadronic calorimeters ($1.5 \leq \eta \leq 4.9$) are based on LAr technology.

The ATLAS collaboration tested their electromagnetic (Pb/LAr) and hadronic calorimeter systems with low (3–9 GeV) [35] and high momentum (20–350 GeV) [36] particle beams. Both sections were calibrated using electrons, and the shower energy in the calorimeter was determined as the sum of raw signals from these two sections, $E_{\mathrm{raw}} = E_{\mathrm{raw}}(\mathrm{EM}) + E_{\mathrm{raw}}(\mathrm{HAD})$. The $E_{\mathrm{raw}}(\mathrm{EM})$ term was the sum of the energy deposited in the front, middle, and back samples of the electromagnetic section, and the $E_{\mathrm{raw}}(\mathrm{HAD})$ represented the sum of signals from the first and second samples of the hadronic section for low energies. In reconstructing the energy of the event, several conditions were applied: no pre-sampler contribution was added to the electromagnetic signal, and only calorimeter cells with energy depositions larger than twice the standard deviation of the electronic noise

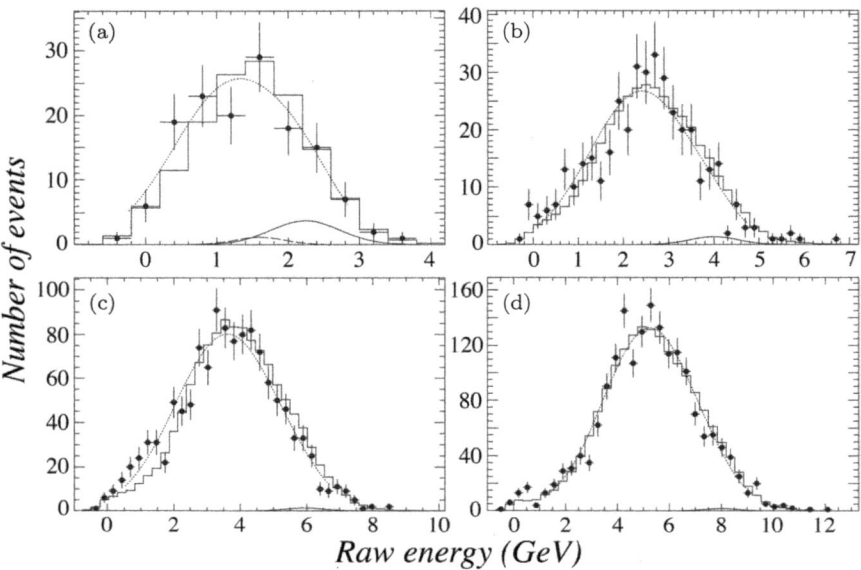

Figure 1.13. Distribution of the reconstructed energy (E_{raw}) for the combined ECAL+HCAL ATLAS system for (a) 3 GeV, (b) 5 GeV, (c) 7 GeV, and (d) 9 GeV pions at $\eta_{\text{beam}} = 0.35$ [35]. The full points represent the measured data. The dashed curves correspond to a fit to the data in a region $\pm 2\sigma$ around the mean value where the electron and muon contaminations in the beam are taken into account. The solid curve represents the expected contribution of the electron contamination in the beam. At 3 GeV, the long-dashed curve shows the expected contribution from the decay muons. The histograms correspond to the prediction of the GEANT4 with the QGSP_BERT physics list [37].

were included in the sum. The total expected electronic noise level was ~160 MeV, and the absolute value of the pedestal shift was less than 2 MeV. No corrections due to shower containment, non-compensation, or dead material were applied. The signal distributions for low energies are shown in Fig. 1.13.

Figure 1.14 reveals a strong signal nonlinearity in response to hadrons as a function of energy. The relative response difference between the measured and simulated data depends on the beam energy and the impact point on the calorimeter (η_{beam}). The simulation overestimates the signal by 5–10% at low momenta (3–9 GeV), while the energy resolution is underestimated by 15% at 3 GeV and 5% at 9 GeV. In general, the agreement is somewhat better at higher energies but degrades at higher η_{beam}.

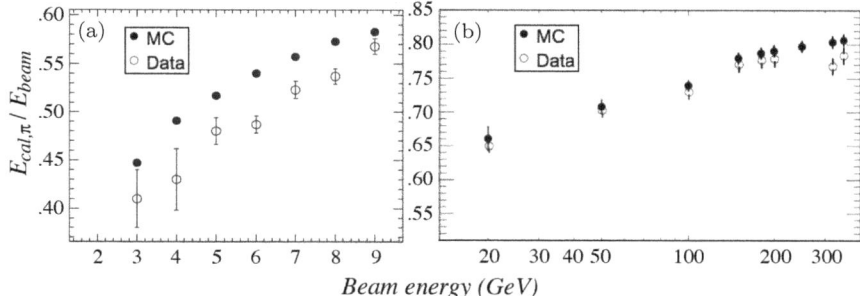

Figure 1.14. The measured E_π/E_{beam} ratio (for $\eta_{\text{beam}} = 0.35$) for the combined ECAL+HCAL ATLAS system (a) at low [35] and (b) high energies [38]. The error bars include statistical as well as systematic errors added in quadrature. The GEANT4 prediction is represented by the black circles.

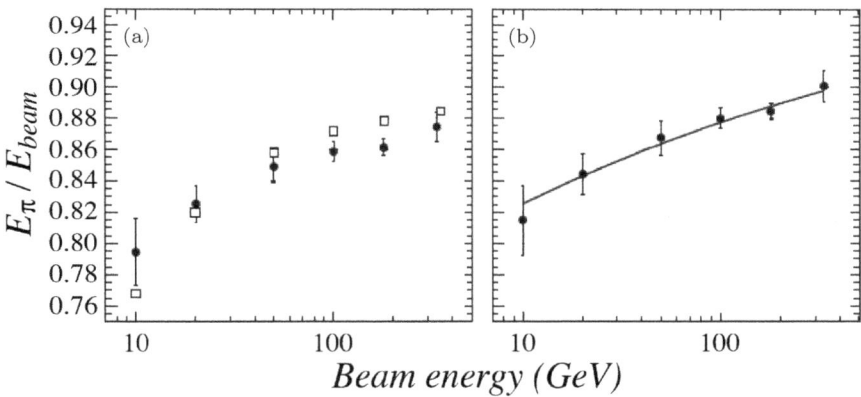

Figure 1.15. The response of TileCal (at $\eta_{\text{beam}} = 0.35$) to pions as a function of pion energy (a) before and (b) after corrections for the effect of shower leakage. The squares represent GEANT4 predictions. The curve shows the result of a fit, $E_\pi = E_{\text{em}}[h/e + \langle f_{\text{em}}\rangle(1-h/e)]^{-1}$ with $e/h = 1.33$, $E_0 = 1$ GeV, and $k = 0.85$ [28].

In Fig. 1.15(a), the responses are plotted without leakage corrections, and in Fig. 1.15(b), with longitudinal and transverse leakage corrections [28]. The response of the TileCal to pions was parametrized as $E(\pi)/E_{\text{beam}} = (1 - f_h) + (\frac{h}{e})f_h$ and fitted to experimental data. When Figs. 1.14 and 1.15 are compared, it becomes clear that most of the signal nonlinearity is induced by the Pb/LAr ECAL calorimeter: in the 20–350

GeV range, ~10% nonlinearity of TileCal alone increases to ~20% in the combined system. At low energies, the nonlinearity is even more dominated by the ECAL.

The TileCal hadronic energy resolution is best described by a stochastic term of $(52.9 \pm 0.9)\%/\sqrt{E}$ (E in GeV) and a constant term of $5.7 \pm 0.2\%$. The noise level is small at all energies and is not considered in this evaluation. There is a good agreement at higher energies between the GEANT4 simulation [37] and measurements.

The response difference between pions and protons, originally observed in Cherenkov calorimeters [24], was also observed in the TileCal beam tests [28]. As the particle energy increases up to 200 GeV, the π/p response ratio decreases; however, the energy resolution for protons is 15–20% better as π^{o} production fluctuates less, event by event.

1.6 Developments and Trends

The idea of compensation ($e/h = 1$) by utilizing signals due to slow neutrons that occur over long times (several 10s of ns) and over large volumes (radius of several 10s of cm) was put forth four decades ago, and it was successfully demonstrated to improve hadronic energy resolution [39, 40]. More than a decade later, the DREAM (Dual-REAdout Module) collaboration [41] showed that event-by-event compensation was also possible by tracking the electromagnetic fraction, f_{em}, if the signals from scintillating and clear fibers were measured simultaneously. Around the same time, the CALICE (CAlorimeter for LInear Collider Experiment) collaboration [42] developed highly granular calorimeters to "image" showers and to be used in combination with precision trackers in support of a particle flow approach for the future collider experiments. The emerging ideas today center around integrating artificial intelligence and machine learning techniques for improved energy, space, and time reconstruction. Simulation results are encouraging; however, experimental confirmation of these results is needed. We discuss these approaches in this section.

At a future e^+e^- collider, likely the next major collider to be built after the LHC, one of the requirements will be to identify hadronically decaying W and Z bosons. An important gain in event rates can be achieved by using the hadronic decay modes of the Z (e.g., in $e^+e^- \rightarrow HZ$), if the hadronically decaying Ws from more abundant processes (e.g., in $e^+e^- \rightarrow W^+W^-$) can be distinguished by the calorimeters. The requirement necessitates measuring 80–90 GeV jets with a resolution of 3–3.5 GeV or achieving a ~30% stochastic term in the hadronic energy resolution.

1.6.1 Dual-readout approach

In the dual-readout approach, one simultaneously measures dE/dx and the Cherenkov light generated in the active components of the calorimeter and determines the f_{em} event by event. Doing this removes the degrading effects of its fluctuation on the energy resolution. These two signals provide complementary information on the shower content: the non-electromagnetic component mostly originates from non-relativistic particles and does not contribute to the generation of the Cherenkov signal. This principle was first experimentally demonstrated by the DREAM Collaboration with a Cu/fiber calorimeter [43]. Scintillating fibers (S) measured dE/dx contribution from all charged particles ($e^{\pm}$, $\pi^{\pm}$, $K^{\pm}$, recoil and spallation p, and nuclear fragments), whereas the quartz/clear plastic fibers (Q) generated the Cherenkov light initiated by the relativistic $e^{\pm}$s coming from the π° decays:

$$S = E \left[f_{em} + \frac{1}{(e/h)_S} (1 - f_{em}) \right] \quad (1.12)$$

$$Q = E \left[f_{em} + \frac{1}{(e/h)_Q} (1 - f_{em}) \right] \quad (1.13)$$

where the e/h ratios for the scintillator and Cherenkov structures of the calorimeter are denoted by the corresponding subscripts. These ratios were $(e/h)_S \approx 1.3$ and $(e/h)_Q \approx 4.8$ for the DREAM module. The Q/S ratio has no energy dependence, as the incoming particle energy E cancels out in the the ratio and the electromagnetic fraction equals:

$$f_{em} = \frac{(h/e)_Q - (Q/S)(h/e)_S}{(Q/S)[1 - (h/e)_S] - [1 - (h/e)_Q]} \quad (1.14)$$

It is convenient to define a parameter χ so that the hadron energy can be simply expressed:

$$E = \frac{S - \chi Q}{1 - \chi} \quad \text{where} \quad \chi = \frac{1 - (h/e)_S}{1 - (h/e)_Q} \quad (1.15)$$

A correlation between the S and Q signals is shown for 100 GeV pions in Fig. 1.16(a). Both axes are in GeV and calibrated by electrons. Several comments are appropriate at this point:

- Since both sections of the calorimeter are calibrated with electrons, the 100 GeV electron data points, if they were plotted, would form a tight cluster at (100 GeV, 100 GeV), indicating $f_{em} = 1$. The larger the electromagnetic component of the hadronic shower, the higher the events

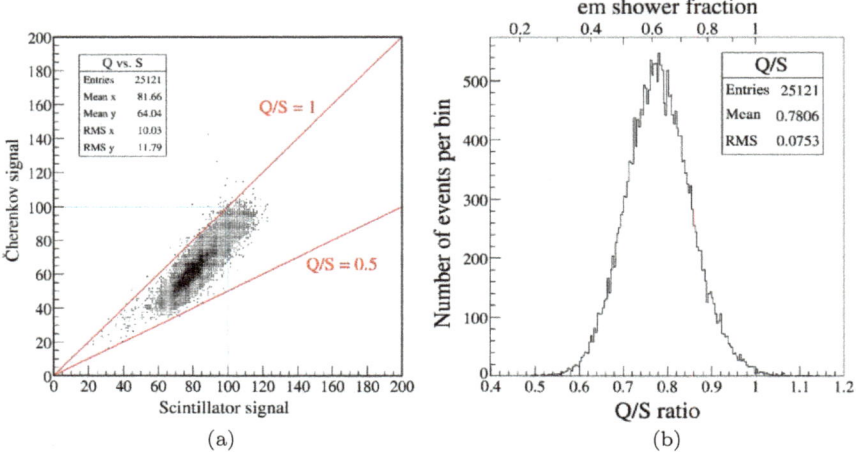

Figure 1.16. (a) Cherenkov (Q) signals versus scintillator (S) signals and (b) the distribution of the Q/S ratio and the f_{em} fraction derived on the basis of Eq. (1.14), for 100 GeV π^-s detected with the DREAM calorimeter [43].

are along the diagonal line $Q/S = 1$. For example, electromagnetic energy leakage out of the calorimeter means that the data points move diagonally down along this line.

- The scintillation signal is always larger than the Cherenkov signal because all charged particles contribute to S, whereas only the relativistic ones, a subset of all charged particles, contribute to Q.
- The ionization signal from a recoil or spallation proton moves the data points along the S-axis. Enriching the hydrogen content of the active material (scintillator) would have the effect of rotating the cluster of events counterclockwise around the $f_{em} = 1$ point. If the signal collection is performed over a long time and in a large volume, the correlation cluster will be entirely vertical, achieving compensation. At this point, the dependence on Q vanishes, implying insensitivity to energy fluctuations between the electromagnetic and non-electromagnetic shower components. Conversely, for example, if the energy leaks out of the calorimeter in the form of neutrons, the data points shift horizontally to the left along the S-axis.
- It is instructive to think of the dual-readout approach as a rotation around the $f_{em} = 1$ point by an angle $\frac{\pi}{2} - \theta$ that displaces the cluster of data points such that their projection onto the S-axis minimizes its

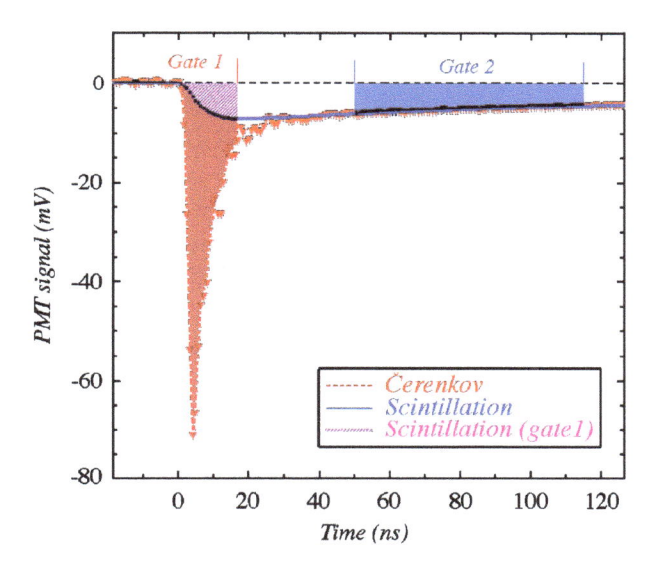

Figure 1.17. The time structure of a typical shower signal measured in the BGO calorimeter equipped with a UV filter. The signals were used to measure the relative contributions of scintillation light (gate 2) and Cherenkov light (gate 1). Note that in this case, with a single readout it is possible to measure the both signals from a single active medium [47].

standard deviation, thus improving energy resolution measured by using the S signals. The slope of the cluster (angle θ with respect to the S-axis) depends only on the two e/h ratios (note $\cot\theta = \chi$ as in Eq. (1.15)). The larger the difference between the two e/h ratios, the more effective is the dual-readout approach [44].

This technique can be applied to dense crystals where both the scintillation and Cherenkov photons are simultaneously generated (Fig. 1.17). The separation of scintillation and Cherenkov photon signals has been shown to be achievable (e.g., $PbWO_4$ and BGO) by exploiting the features in signal time structure and the emission wavelength [45–47]. Other discriminating features such as directionality and polarization [48, 49] of Cherenkov light may also be exploited in the future. The use of crystals is often preferable when exceptionally good electromagnetic energy resolution is desired. This will likely be the case for the heavy flavor and electroweak physics programs at future e^+e^- colliders [50, 51].

In addition to the pulse integrals, the temporal characteristics of pulses, or pulse shapes, carry rich information to be exploited. For instance, the exponential tail of plastic scintillator signals contains the characteristic information expected of non-relativistic neutrons that is absent in the time structure of the Cherenkov signals. The event-by-event contribution of neutrons to the calorimeter signals in this way would further improve the energy resolution beyond the levels made possible by the dual-readout [52]. For example, if rare-earth doped scintillators were introduced in addition to organic scintillating ones in a triple-readout scheme, then it would be possible to track the fluctuations originating from binding energy losses in nuclear break-up by detecting the neutrons of a few MeV. In principle, this can be accomplished by measuring the signal difference between the two scintillating (hydrogenous vs. non-hydrogenous) media on an event-by-event basis. Other variations are also evident: for example, taking advantage of the differences in signal decay times, mixed Cherenkov and (inorganic) scintillation light can be separated by a pulse shape discrimination technique (e.g., Fig. 1.17), reducing the number of multiple readout channels but maintaining the advantage of a multiple-readout scheme.

1.6.2 *Particle flow approach*

The principle concept behind the particle flow approach is the use of a tracker in conjunction with a high-granularity calorimeter. The momenta of charged jet particles are measured with the tracker in a magnetic field, while the energy of the neutral particles is measured with the calorimeter. This approach was successfully demonstrated by the ALEPH collaboration, which measured the mass of hadronically decaying Zs within 6.2 GeV [53]. The CMS collaboration has also adopted a similar approach, appreciably improving missing energy resolution [54].

A challenging aspect of this technique is that the calorimeter has to interpret energy deposits coming neither from charged particles nor from photons as neutral hadrons. But the calorimeter cannot distinguish neutrals from charged, and this ambiguity leads to the so-called "confusion." Proponents argue that highly granular calorimeters with small individual cells, comparable to Molière radius or radiation length throughout the calorimeter, reduce the effect of this confusion [55]. Among several reconstruction algorithms based on particle flow approach, Pandora [56] is commonly used in the context of future lepton collider studies.

The CALICE collaboration has built several high-granularity prototypes and pioneered the development of "imaging" calorimetry. Among them, two high-granularity silicon (SiW-ECAL) [57] and scintillator (ScW-ECAL) [58] electromagnetic calorimeters with tungsten absorbers were investigated. The dense absorber, such as tungsten with small X_o and ρ_M, allows for a compact design that aids in resolving adjacent showers. SiW-ECAL featured nearly 10,000 1 cm^2 silicon pads in 30 active layers (24 X_o). The energy resolution of $16.53\%/\sqrt{E} \oplus 1.07\%$ for electrons was achieved [59] in beam tests at DESY and CERN. An upgraded version with over 15,000 channels was tested with low-energy electrons at DESY [60]. The ScW-ECAL consisted of 30 layers of 3.5 mm thick absorber plates ($20X_o$) with over over 2,000 scintillator channels coupled to wavelength shifting fibers and readout by silicon multipliers. This prototype achieved energy resolution of $12.5\%/\sqrt{E} \oplus 1.2\%$ energy resolution for electrons at Fermilab tests [61].

The CALICE collaboration has been investigating various types of highly granular hadronic calorimeters and has built digital (DHCAL) and analog (AHCAL) prototypes. The DHCAL is based on resistive plate chamber (RPC) technology [62] and operates in a digital mode: it is assumed that the number of hits over a threshold is proportional to the incoming particle's energy. The steel prototype consisted of 350,208 readout channels with transverse granularity of 1×1 cm^2 and longitudinal segmentation of 38 layers. Figure 1.18 shows the electromagnetic and hadronic energy resolutions from beam tests performed at Fermilab [63]. The data points (black squares) are fitted to Eq. (1.8), and the ratio between the simulation and the data (bottom plots) shows an agreement within ~5% for the energies below 20 GeV. The saturation caused by dense electromagnetic showers in this binary readout of 1×1 cm^2 pads clearly degrades the performance to a modest resolution of $(34.6 \pm 0.9)\%/\sqrt{E} \oplus (12.5 \pm 0.3)\%$. This saturation may be mitigated by employing weighting algorithms based on the hit densities [64].

Figure 1.18(b) shows the energy resolution of DHCAL for pions. The black curve represents the fitted Eq. (1.8) for beam energies below 32 GeV, which gives $(51.5 \pm 1.5)\%/\sqrt{E} \oplus (10.6 \pm 0.5)\%$. Above 32 GeV, the resolution degrades for the same reason as in the electromagnetic case. The saturation effects dominate in this binary readout scheme and result in poor performance.

The CALICE collaboration developed a semi-digital hadron calorimeter (SDHCAL) with a three-threshold readout system to address the saturation

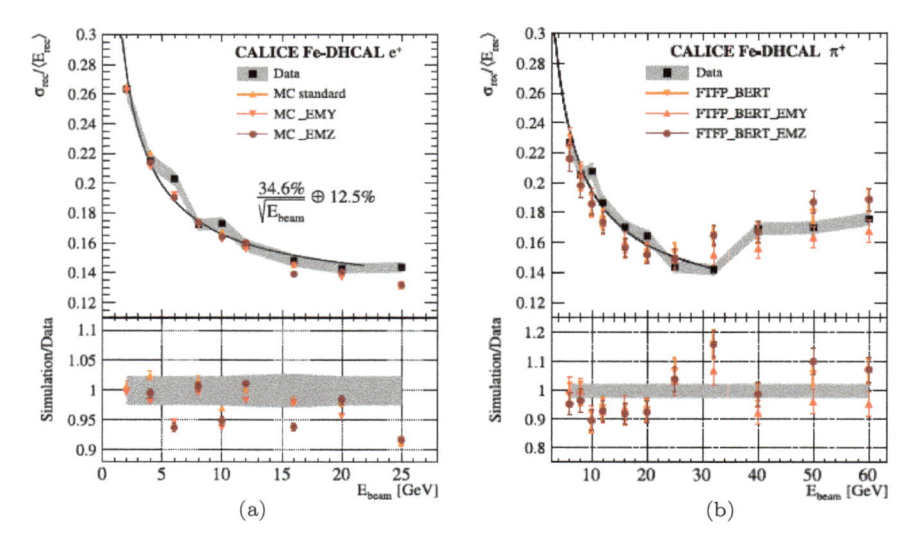

Figure 1.18. (a) The energy resolution for positrons with energies from 2 to 25 GeV and (b) the energy resolution for π^+ with energies from 6 to 60 GeV. The bottom plots show the ratio of the simulations and data. The data points are indicated by black squares, and the black curve represents the fit to Eq. (1.8). The error bands show the systematic and statical uncertainty added in quadrature. The statistical errors are smaller than the size of the markers [63].

effects observed in the binary readout system discussed above [65]. They found that the energy response with a 4–5% deviation from linearity in the 5–80 GeV energy range. The resolution associated with the linearized energy response of the same selected data sample was also estimated in the binary and the multi-threshold modes. The multi-threshold capabilities clearly showed improved resolution at energies above 30 GeV. The energy resolution is 7.7% at 80 GeV, likely due to information provided by the second and third thresholds. For more details, see Refs. [66, 67].

The CMS collaboration has adopted silicon and SiPM-on-tile high granularity technology and is in the process of building a 6-million-channel endcap calorimeter for the HL LHC phase [68–70]. The large number of channels presents challenges in mechanical and electronics integration and data processing. The calibration method relies on a minimum ionizing muon signal being above noise for the lifetime of the detector, a challenge exacerbated by radiation damage of active elements (silicon sensors and scintillating tiles) [71–73]. Figure 1.19 shows a 300 GeV pion track before a

Figure 1.19. An event display illustrating the development of a hadronic shower initiated by a pion of 300 GeV energy starting in the last layers of the electromagnetic section and depositing energy in the hadronic sections [74].

hadronic shower develops in the later layers of the first section (ECAL) and extending into the second and third sections (HCAL) in a beam test [74].

1.6.3 *Emerging ideas*

The capability to image showers with highly granular calorimeters has opened new avenues in energy, or more broadly, information reconstruction. The recognition that the topological structures within the showers encompass valuable information about interaction types lends itself to new ways of energy reconstruction. Figure 1.20 shows the spatial distributions of shower particles generated by a single 30 GeV π^+ in copper: pions and protons display a clear vertex structure, and positrons are produced in pair production following $\pi^°$ decays that are associated with the hadronic vertices. These pions, protons, and positrons make up the fast ($t < 5$ ns) shower components. Neutrons and gammas are slower and spread more widely within the calorimeter. Electrons similarly spread out: some come from the $\pi^°$ decay chain, and many others are from Compton scattering.

Figure 1.21 shows a strong correlation between the invisible energy and the number of hadronic vertices, as in π^+, π^-, and p in Fig. 1.20, produced in inelastic hadronic interactions in the first <5 ns. If the number of vertices can be counted or imaged by a highly granular calorimeter, the invisible fraction of the deposited energy can be estimated with good precision, as simulation studies suggest, leading to remarkably good energy reconstruction using neural networks (see Refs. [75, 76] for details).

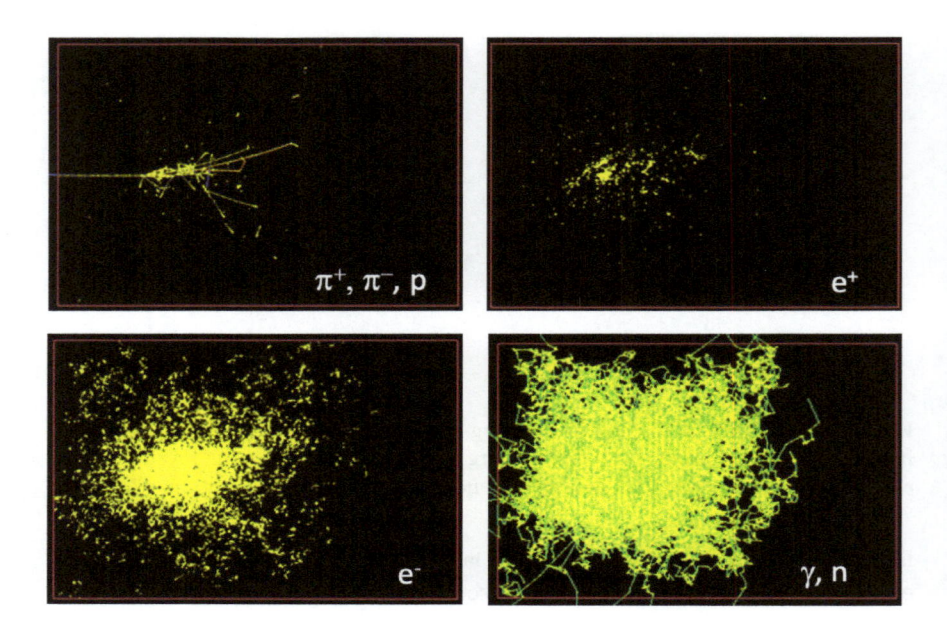

Figure 1.20. Simulated images of shower particles for a single 30 GeV π^+ in a solid copper absorber ($100 \times 100 \times 150$ cm^3): (a) fast π^+, π^-, and protons display distinctive hadronic vertices and tracks. The e^+ and e^- are shown separately. (b) e^+s arise mainly from fast pair production via π^0 decays, whereas (c) e^-s come from the slow component of the shower due to Compton scattering of widely spread γs, as well as counterparts of positrons in the fast component. (d) Slow neutrons and γs from neutron capture spread broadly and form a "fuzzy" image [75].

In addition to the value of the shower image in three-dimensional space, the impact of shower timing using neural networks is also being investigated [75]. Single pion and electron energy distributions in space for various integration times (i.e., the duration of the window of time a signal may be collected and observed as the integration time) are shown in Fig. 1.22 [77]. Each subfigure gives a representation of the spatial distribution of energy at some integration time; the vertical axis represents the radial distance from the shower axis, and the horizontal axis represents the longitudinal depth of the shower. The energy is accumulated up to a 10 ns integration interval, and the number of time slices is treated as a variable to evaluate the impact of timing precision. The time of any simulated energy deposition is recorded as $t = t_{G4} - z/c$, where t_{G4} is the time when the energy is

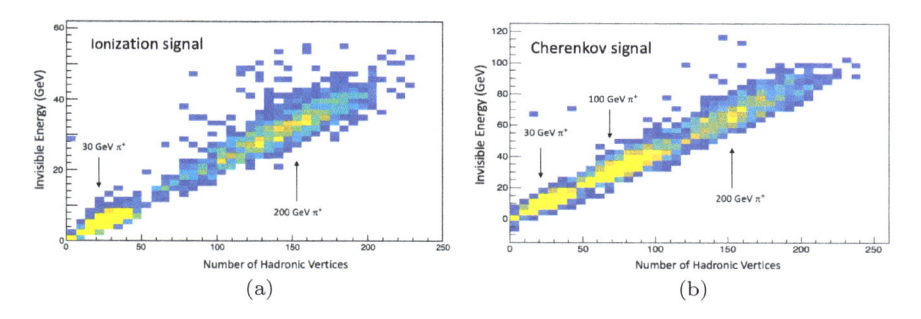

Figure 1.21. Invisible energy *vs* the number of hadronic vertices in a sampling calorimeter: (a) with ionization signal in Cu (17 mm)+Si (3 mm) and (b) with a Cherenkov signal in Cu (17 mm)+Quartz plate (3 mm) for 30, 100, and 200 GeV π^+s. The "invisible energy" in this case is defined as the difference between the beam energy and the simple sum of ionization signal or Cherenkov signal. The "hadronic vertex" is defined as a vertex of hadron–nucleus inelastic interaction excluding neutron–nucleus interaction. The energy scale was calibrated with electrons [77].

deposited as reported by GEANT4 and z/c is the travel time of light in vacuum to cover the longitudinal depth. This form of visualization draws attention to the development of the radial extent of showers at different timescales. Unlike hadron showers, the electron-initiated energy deposits take place promptly without much spatial structure; low-energy deposits from photons are found transversely far from the shower axis. Roughly speaking, by going from long to short integration times, the calorimeter is effectively transversely segmented. This segmentation supplies valuable information for the networks to exploit since the timescales of some hadronic processes are much longer than the electromagnetic processes.

Figure 1.23 illustrates the effect of increasingly better timing measurement on energy resolution [77]. Each data point (solid and open circles) includes several time slices. While the signal is integrated for 10 ns for all data points, the precision with which timing intervals are known is plotted on the horizontal axis. For example, the timing precision of 0.5 ns includes time intervals (0, 0.5 ns), (0, 1 ns), (0, 4 ns), and (0, 10 ns) and is plotted at 0.5 ns. As shorter time slices are included, the energy resolution improves. The information inherent to short time slices, possibly due to low energy protons, contributes additional information to the network. Essentially, protons in shorter timescales seem to stand in for neutrons in longer timescales for enhanced energy measurement. The value of

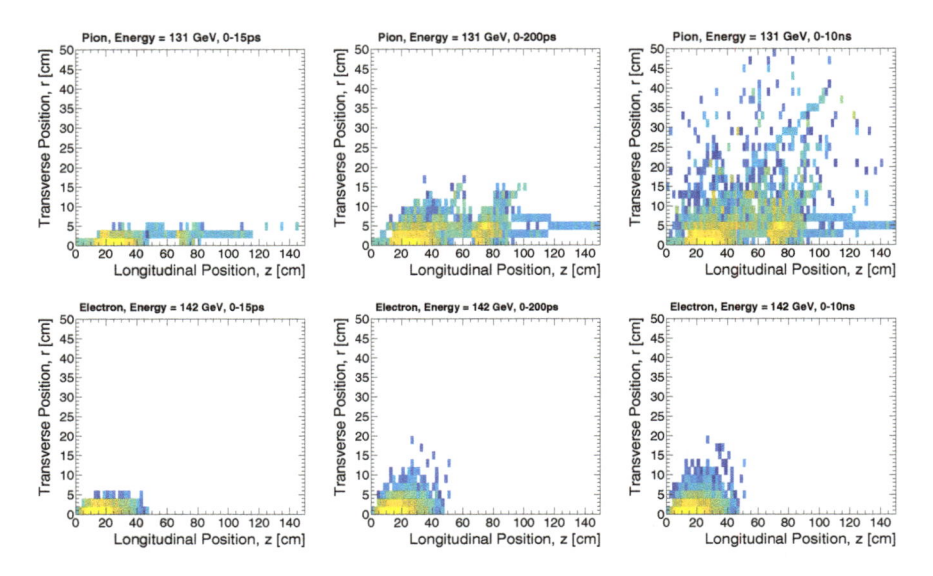

Figure 1.22. The energy deposits due to a single 131 GeV charged pion (top row) and a single 142 GeV electron (bottom row) are shown in $r-z$ coordinates, where the colors indicate deposited energy. As indicated at the top of each plot, the integration times gradually increase from 0–15 ps to 0–10 ns. Time is 'local'; in other words, it is corrected for the travel time, $t = t_{G4} - z/c$, along z-axis for all particles [77].

timing information for enhanced calorimeter performance, apart from the purposes of pile-up mitigation or time-of-flight measurements, requires further research at the level of optimized neural networks and beam tests with similarly optimized calorimeter prototypes.

1.7 Outlook

The field of high-energy calorimetry has undergone significant revitalization in recent decades. Much progress has been made in understanding the fundamentals of calorimetry, and effective techniques have been developed to enhance performance of various aspects. We expect this positive trajectory to continue at an accelerated pace, thanks to the integration of imaging showers, neural networks, and optimized reconstruction algorithms. The inclusion of images and temporal structures of showers provides additional information that reveals previously inaccessible quantities, such as invisible energy, which has been a long-standing challenge specific to hadron calorimetry. It is likely that we will witness significant advances,

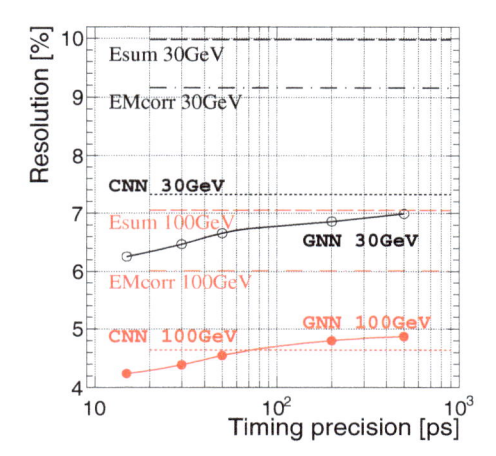

Figure 1.23. The energy resolution (σ/E) for 30 GeV (black) and 100 GeV (red) pions: simple energy sum (Esum), f_{em} corrected energy sum (EMcorr) as in the dual-readout approach, CNN and GNN reconstruction techniques. The horizontal axis indicates the assumed timing precision for the GNN technique. The energy resolutions obtained from different reconstruction techniques are shown for comparison [77].

perhaps achieve similarly precise energy measurements for all fundamental particles as for electrons and photons, before the next collider is built.

It's important to note that this chapter does not delve into topics related to calibration, simulations, sensor technologies, material science, and the effects of external magnetic field and radiation damage. Nor have we discussed inventive calorimeters in non-accelerator experiments. For readers seeking more detailed information, the references provided at the end of this chapter should prove useful. Furthermore, for the latest updates and specific details, we recommend referring to the online documentation provided by the Particle Data Group [9].

References

[1] T. Ferbel. *Experimental Techniques in High-Energy Nuclear and Particle Physics*, 2nd edn. World Scientific, 1991. arXiv: https://www.worldscientific.com/doi/pdf/10.1142/1571, DOI: 10.1142/1571. https://www.worldscientific.com/doi/abs/10.1142/1571.
[2] F. Sauli. *Instrumentation in High Energy Physics*. World Scientific, 1992. arXiv: https://www.worldscientific.com/doi/pdf/10.1142/1356, DOI: 10.1142/1356. https://www.worldscientific.com/doi/abs/10.1142/1356.

[3] C. W. Fabjan and F. Gianotti. Calorimetry for particle physics. *Reviews of Modern Physics*, 75:1243–1286, 2003. DOI: 10.1103/RevModPhys.75.1243.

[4] R. Wigmans. *Calorimetry: Energy Measurement in Particle Physics.* International Series of Monographs on Physics, Vol. 107, 2000, pp. 1–726.

[5] H. Davies, H. A. Bethe, and L. C. Maximon. Theory of bremsstrahlung and pair production. II. Integral cross section for pair production. *Physical Review*, 93:788–795, 1954. DOI: 10.1103/PhysRev.93.788. https://link.aps.org/doi/10.1103/PhysRev.93.788.

[6] Y.-S. Tsai. Pair production and bremsstrahlung of charged leptons. *Reviews of Modern Physics*, 46:815–851, 1974. DOI: 10.1103/RevModPhys.46.815. https://link.aps.org/doi/10.1103/RevModPhys.46.815.

[7] Y.-S. Tsai. Erratum: Pair production and bremsstrahlung of charged leptons. *Reviews of Modern Physics*, 49:421–423, 1977.DOI: 10.1103/RevModPhys.49.421. https://link.aps.org/doi/10.1103/RevModPhys.49.421.

[8] W. R. Nelson, H. Hirayama, and D. W. Rogers. EGS4 code system. Technical Report, Stanford Linear Accelerator Center, Menlo Park, CA (USA), 1985.

[9] R. L. Workman, et al. Review of particle physics. *PTEP*, 2022:083C01, 2022. DOI: 10.1093/ptep/ptac097.

[10] G. Unal on behalf of the ATLAS Collaboration. The role of the LAr calorimeter in the search for $H \to \gamma\gamma$ in ATLAS. *Journal of Physics: Conference Series*, 404(1):012001, 2012. DOI: 10.1088/1742-6596/404/1/012001. https://dx.doi.org/10.1088/1742-6596/404/1/012001.

[11] T. Tabarelli de Fatis on behalf of the CMS Collaboration. Role of the CMS electromagnetic calorimeter in the hunt for the higgs boson in the two-gamma channel. *Journal of Physics: Conference Series*, 404(1):012002, 2012. DOI: 10.1088/1742-6596/404/1/012002. https://dx.doi.org/10.1088/1742-6596/404/1/012002.

[12] CMS Collaboration. Performance and operation of the CMS electromagnetic calorimeter. *Journal of Instrumentation*, 5(03):T03010, 2010. DOI: 10.1088/1748-0221/5/03/T03010. https://dx.doi.org/10.1088/1748-0221/5/03/T03010.

[13] F. N. Tedaldi. Response evolution of the CMS ECAL and R&D studies for electromagnetic calorimetry at the high-luminosity LHC, 2012. *arXiv preprint* arXiv:1211.3885.

[14] ATLAS Liquid-Argon Calorimeter: Technical Design Report. ATLAS, CERN, Geneva, 1996. DOI: 10.17181/CERN.FWRW.FOOQ. https://cds.cern.ch/record/331061.

[15] M. Aharrouche, et al. Energy linearity and resolution of the ATLAS electromagnetic barrel calorimeter in an electron test-beam. *Nuclear Instruments and Methods in Physics Research Section A: Accelerators, Spectrometers, Detectors and Associated Equipment*, 568(2):601–623, 2006. DOI: 10.1016/j.nima.2006.07.053. https://doi.org/10.1016%2Fj.nima.2006.07.053.

[16] G. Aad, et al. Electron and photon energy calibration with the ATLAS detector using LHC Run 1 data. *The European Physical Journal C*, 74(10):3071, 2014.

[17] G. Aad, *et al.* Electron performance measurements with the ATLAS detector using the 2010 LHC proton-proton collision data. *The European Physical Journal C*, 72(3), 2012. DOI: 10.1140/epjc/s10052-012-1909-1. https://doi.org/10.1140%2Fepjc%2Fs10052-012-1909-1.

[18] D. Green, Selected topics in sampling calorimetry. In *4th International Conference on Calorimetry in High-energy Physics*, La Biodola, Italy, 19–25 September 1993.

[19] M. G. Albrow, *et al.* Intercalibration of the longitudinal segments of a calorimeter system. *Nuclear Instruments and Methods in Physics Research Section A: Accelerators, Spectrometers, Detectors and Associated Equipment*, 487:381–395, 2002. DOI: 10.1016/S0168-9002(01)02190-8.

[20] R. Wigmans. On the calibration of segmented calorimeters. In *AIP Conference Proceedings*, 867:90–97, 2006. DOI: 10.1063/1.2396942.

[21] T. Gabriel, *et al.* Energy dependence of hadronic activity. *Nuclear Instruments and Methods in Physics Research Section A: Accelerators, Spectrometers, Detectors and Associated Equipment*, 338:336–347, 1994.

[22] N. Akchurin, *et al.* Beam test results from a fine-sampling quartz fiber calorimeter for electron, photon and hadron detection. *Nuclear Instruments and Methods in Physics Research Section A: Accelerators, Spectrometers, Detectors and Associated Equipment*, 399:202–226, 1997. DOI: 10.1016/S0168-9002(97)00789-4.

[23] T. Gabriel, D. Groom, P. Job, N. Mokhov, G. Stevenson, Energy dependence of hadronic activity. *Nuclear Instruments and Methods in Physics Research Section A: Accelerators, Spectrometers, Detectors and Associated Equipment*, 338(2):336–347, 1994. DOI: https://doi.org/10.1016/0168-9002(94)91317-X. https://www.sciencedirect.com/science/article/pii/016890029491317X.

[24] N. Akchurin, *et al.* On the differences between high-energy proton and pion showers and their signals in a non-compensating calorimeter. *Nuclear Instruments and Methods in Physics Research Section A: Accelerators, Spectrometers, Detectors and Associated Equipment*, 408:380–396, 1998. DOI: 10.1016/S0168-9002(98)00021-7.

[25] D. Acosta, *et al.* Lateral shower profiles in a lead/scintillating fiber calorimeter. *Nuclear Instruments and Methods in Physics Research Section A: Accelerators, Spectrometers, Detectors and Associated Equipment*, 316:184–201, 1992. DOI: 10.1016/0168-9002(92)90901-F.

[26] H. Abramowicz, *et al.* The response and resolution of iron scintillator calorimeter for hadronic and electromagnetic showers between 10-GeV and 140 GeV. *Nuclear Instruments and Methods in Physics Research Section A: Accelerators, Spectrometers, Detectors and Associated Equipment*, 180:429, 1981. DOI: 10.1016/0029-554X(81)90083-5.

[27] P. Adragna, *et al.* Measurement of pion and proton response and longitudinal shower profiles up to 20 nuclear interaction lengths with the ATLAS tile calorimeter. *Nuclear Instruments and Methods in Physics Research Section A: Accelerators, Spectrometers, Detectors and Associated Equipment*, 615:158–181, 2010. DOI: 10.1016/j.nima.2010.01.037.

[28] P. Adragna, et al. Testbeam studies of production modules of the ATLAS tile calorimeter. *Nuclear Instruments and Methods in Physics Research Section A: Accelerators, Spectrometers, Detectors and Associated Equipment*, 606:362–394, 2009. DOI: 10.1016/j.nima.2009.04.009.

[29] D. E. Groom, Energy flow in a hadronic cascade: Application to hadron calorimetry. *Nuclear Instruments and Methods in Physics Research Section A: Accelerators, Spectrometers, Detectors and Associated Equipment*, 572(2):633–653, 2007. DOI: DOI: 10.1016/j.nima.2006.11.070.

[30] S. Abdullin, et al. Design, performance, and calibration of CMS hadron-barrel calorimeter wedges. *European Physical Journal*, C55:159–171, 2008. DOI: 10.1140/epjc/s10052-008-0573-y.

[31] S. Abdullin, et al. The CMS barrel calorimeter response to particle beams from 2-GeV/c to 350-GeV/c. *European Physical Journal*, C60:359–373, 2009. DOI: 10.1140/epjc/s10052-009-0959-5.

[32] S. Abdullin, et al. Design, performance, and calibration of the CMS hadron-outer calorimeter. *European Physical Journal*, C57:653–663, 2008. DOI: 10.1140/epjc/s10052-008-0756-6.

[33] N. Akchurin, et al. The response of CMS combined calorimeters to single hadrons, electrons and muons. CERN-CMS-NOTE-2007-012.

[34] ATLAS-Collaboration. ATLAS Tile Calorimeter: Technical Design Report. CERN-LHCC-96-42, 1996, 347 p.

[35] E. Abat, et al. Study of the response of the ATLAS central calorimeter to pions of energies from 3 to 9 GeV. *Nuclear Instruments and Methods in Physics Research Section A: Accelerators, Spectrometers, Detectors and Associated Equipment*, 607:372–386, 2009. DOI: 10.1016/j.nima.2009.05.158.

[36] E. Abat, et al. Study of energy response and resolution of the atlas barrel calorimeter to hadrons of energies from 20 to 350 gev. *Nuclear Instruments and Methods in Physics Research Section A: Accelerators, Spectrometers, Detectors and Associated Equipment*, 621(1):134–150, 2010. DOI: https://doi.org/10.1016/j.nima.2010.04.054. https://www.sciencedirect.com/science/article/pii/S0168900210009034.

[37] S. Agostinelli, et al. GEANT4: A simulation toolkit. *Nuclear Instruments and Methods in Physics Research Section A: Accelerators, Spectrometers, Detectors and Associated Equipment*, 506:250–303, 2003. DOI: 10.1016/S0168-9002(03)01368-8.

[38] E. Abat, et al. Study of energy response and resolution of the ATLAS barrel calorimeter to hadrons of energies from 20 to 350 GeV. *Nuclear Instruments and Methods in Physics Research Section A: Accelerators, Spectrometers, Detectors and Associated Equipment*, 621:134–150, 2010. DOI: 10.1016/j.nima.2010.04.054.

[39] D. Acosta, et al. Electron, pion and multiparticle detection with a lead/scintillating — fiber calorimeter. *Nuclear Instruments and Methods in Physics Research Section A: Accelerators, Spectrometers, Detectors and Associated Equipment*, 308:481–508, 1991. DOI: 10.1016/0168-9002(91)90062-U.

[40] U. Behrens, *et al.* Test of the ZEUS forward calorimeter prototype. *Nuclear Instruments and Methods in Physics Research Section A: Accelerators, Spectrometers, Detectors and Associated Equipment*, 289:115–138, 1990. DOI: 10.1016/0168-9002(90)90253-3.
[41] DREAM Collaboration. http://www.phys.ttu.edu/~dream/. Accessed: 27 August 2023.
[42] CALICE Collaboration. https://twiki.cern.ch/twiki/bin/view/CALICE/CaliceDetectors. Accessed: 27 August 2023.
[43] N. Akchurin, *et al.* Hadron and jet detection with a dual-readout calorimeter. *Nuclear Instruments and Methods in Physics Research Section A: Accelerators, Spectrometers, Detectors and Associated Equipment*, 537:537–561, 2005. DOI: 10.1016/j.nima.2004.07.285.
[44] D. E. Groom. Degradation of resolution in a homogeneous dual-readout hadronic calorimeter. *Nuclear Instruments and Methods in Physics Research Section A: Accelerators, Spectrometers, Detectors and Associated Equipment*, 705:24–31, 2013. https://doi.org/10.1016/j.nima.2012.12.080. https://www.sciencedirect.com/science/article/pii/S016890021201618X.
[45] N. Akchurin, *et al.* Separation of crystal signals into scintillation and Cherenkov components. *Nuclear Instruments and Methods in Physics Research Section A: Accelerators, Spectrometers, Detectors and Associated Equipment*, 595:359–374, 2008. DOI: 10.1016/j.nima.2008.07.136.
[46] N. Akchurin, *et al.* Dual-readout calorimetry with lead tungstate crystals. *Nuclear Instruments and Methods in Physics Research Section A: Accelerators, Spectrometers, Detectors and Associated Equipment*, 584:273–284, 2008. arXiv:0707.4021, DOI: 10.1016/j.nima.2007.09.035.
[47] N. Akchurin, *et al.* Dual-readout calorimetry with crystal calorimeters. *Nuclear Instruments and Methods in Physics Research Section A: Accelerators, Spectrometers, Detectors and Associated Equipment*, 598:710–721, 2009. DOI: 10.1016/j.nima.2008.10.010.
[48] N. Akchurin. Polarization as a tool in calorimetry. *Physics Procedia* 37:333–339, 2012. *Proceedings of the 2nd International Conference on Technology and Instrumentation in Particle Physics* (TIPP 2011). https://doi.org/10.1016/j.phpro.2012.02.383. http://www.sciencedirect.com/science/article/pii/S1875389212017130.
[49] N. Akchurin, F. Bedeschi, A. Cardini, M. Cascella, G. Ciapetti, D. De Pedis, P. Dimpfl, R. Ferrari, S. Franchino, M. Fraternali, *et al.* Polarization as a tool for dual-readout calorimetry. *Nuclear Instruments and Methods in Physics Research Section A: Accelerators, Spectrometers, Detectors and Associated Equipment*, 638(1):47–54, 2011.
[50] M. Lucchini, L. Pezzotti, G. Polesello, and C. Tully. Particle flow with a hybrid segmented crystal and fiber dual-readout calorimeter. *Journal of Instrumentation*, 17(06):P06008, 2022. DOI: 10.1088/1748-0221/17/06/P06008. https://dx.doi.org/10.1088/1748-0221/17/06/P06008.
[51] M. Aleksa, F. Bedeschi, R. Ferrari, F. Sefkow, and C. G. Tully. Calorimetry at fcc-ee. *The European Physical Journal Plus*, 136(10):1066, 2021. DOI:

10.1140/epjp/s13360-021-02034-2. https://doi.org/10.1140/epjp/s13360-02 1-02034-2.

[52] N. Akchurin, et al. Neutron signals for dual-readout calorimetry. *Nuclear Instruments and Methods in Physics Research Section A: Accelerators, Spectrometers, Detectors and Associated Equipment*, 598:422–431, 2009. DOI: 10.1016/j.nima.2008.09.045.

[53] D. Buskulic, et al. Performance of the ALEPH detector at LEP. *Nuclear Instruments and Methods in Physics Research Section A: Accelerators, Spectrometers, Detectors and Associated Equipment*, 360:481–506, 1995. DOI: 10.1016/0168-9002(95)00138-7.

[54] A. Sirunyan, et al. Particle-flow reconstruction and global event description with the CMS detector. *Journal of Instrumentation*, 12(10):P10003, 2017. DOI: 10.1088/1748-0221/12/10/P10003. https://dx.doi.org/10.1088/1748-0 221/12/10/P10003.

[55] F. Sefkow, A. White, K. Kawagoe, R. Pöschl, and J. Repond. Experimental tests of particle flow calorimetry. *Reviews of Modern Physics*, 88:015003, 2016. DOI: 10.1103/RevModPhys.88.015003. https://link.aps.org/doi/10.1 103/RevModPhys.88.015003.

[56] M. Thomson. Particle flow calorimetry and the pandorapfa algorithm. *Nuclear Instruments and Methods in Physics Research Section A: Accelerators, Spectrometers, Detectors and Associated Equipment*, 611(1):25–40, 2009. https://doi.org/10.1016/j.nima.2009.09.009. https://www.sciencedire ct.com/science/article/pii/S0168900209017264.

[57] J. Repond, et al. Design and electronics commissioning of the physics prototype of a Si-W electromagnetic calorimeter for the International Linear Collider. *Journal of Instrumentation*, 3(08):P08001, 2008. DOI: 10.1088/1748-0221/3/08/P08001. https://dx.doi.org/10.1088/1748-0221/3/08/P08001.

[58] K. Francis, et al. Performance of the first prototype of the CALICE scintillator strip electromagnetic calorimeter. *Nuclear Instruments and Methods in Physics Research Section A: Accelerators, Spectrometers, Detectors and Associated Equipment*, 763:278–289, 2014. https://doi.org/10.1016/j.nima .2014.06.039. https://www.sciencedirect.com/science/article/pii/S01689002 14007621.

[59] C. Adloff, et al. Response of the CALICE Si-W electromagnetic calorimeter physics prototype to electrons. *Nuclear Instruments and Methods in Physics Research Section A: Accelerators, Spectrometers, Detectors and Associated Equipment*, 608(3):372–383, 2009. https://doi.org/10.1016/j.nima.2009.07.0 26. https://www.sciencedirect.com/science/article/pii/S0168900209014673.

[60] V. Boudry. New results of the technological prototype of the calice highly granular silicon tungsten calorimeter. *Nuclear Instruments and Methods in Physics Research Section A: Accelerators, Spectrometers, Detectors and Associated Equipment*, 1051:168185, 2023. https://doi.org/10.1016/j.nima.2 023.168185. https://www.sciencedirect.com/science/article/pii/S016890022 3001754.

[61] J. Repond, et al. Construction and response of a highly granular scintillator-based electromagnetic calorimeter. *Nuclear Instruments and Methods in Physics Research Section A: Accelerators, Spectrometers, Detectors and Associated Equipment*, 887:150–168, 2018. https://doi.org/10.1016/j.nima.2018.01.016. https://www.sciencedirect.com/science/article/pii/S0168900218300305.

[62] G. Drake, J. Repond, D. Underwood, and L. Xia, Resistive plate chambers for hadron calorimetry: Tests with analog readout. *Nuclear Instruments and Methods in Physics Research Section A: Accelerators, Spectrometers, Detectors and Associated Equipment*, 578:88–97, 2007. https://doi.org/10.1016/j.nima.2007.04.160.

[63] M. Chefdeville, et al. Analysis of testbeam data of the highly granular RPC-steel CALICE digital hadron calorimeter and validation of Geant4 Monte Carlo models. *Nuclear Instruments and Methods in Physics Research Section A: Accelerators, Spectrometers, Detectors and Associated Equipment*, 939:89–105, 2019. https://doi.org/10.1016/j.nima.2019.05.013. https://www.sciencedirect.com/science/article/pii/S0168900219306230.

[64] C. Adloff, et al. Hadronic energy resolution of a highly granular scintillator-steel hadron calorimeter using software compensation techniques. *Journal of Instrumentation*, 7(09):P09017, 2012. DOI: 10.1088/1748-0221/7/09/P09017. https://dx.doi.org/10.1088/1748-0221/7/09/P09017.

[65] G. Baulieu, et al. Construction and commissioning of a technological prototype of a high-granularity semi-digital hadronic calorimeter. *Journal of Instrumentation*, 10(10):P10039, 2015. DOI: 10.1088/1748-0221/10/10/P10039. https://dx.doi.org/10.1088/1748-0221/10/10/P10039.

[66] CALICE Collaboration. First results of the CALICE SDHCAL technological prototype. *Journal of Instrumentation*, 11(04):P04001, 2016. DOI: 10.1088/1748-0221/11/04/P04001. https://dx.doi.org/10.1088/1748-0221/11/04/P04001.

[67] C. Neubueser. Comparison of two highly granular hadronic calorimeter concepts, 26 October 2016. DOI: 10.3204/PUBDB-2016-05499. https://cds.cern.ch/record/2637266.

[68] The phase-2 upgrade of the CMS endcap calorimeter. Technical Report, CERN, Geneva, 2017. DOI: 10.17181/CERN.IV8M.1JY2. https://cds.cern.ch/record/2293646.

[69] B. Acar, et al. Response of a CMS HGCAL silicon-pad electromagnetic calorimeter prototype to 20–300 gev positrons. *Journal of Instrumentation*, 17(05):P05022, 2022. DOI: 10.1088/1748-0221/17/05/P05022. https://dx.doi.org/10.1088/1748-0221/17/05/P05022.

[70] B. Acar, et al. Performance of the cms high granularity calorimeter prototype to charged pion beams of 20−300 gev/c, 2023. arXiv:2211.04740.

[71] N. Akchurin, P. Almeida, G. Altopp, M. Alyari, T. Bergauer, E. Brondolin, B. Burkle, W. Frey, Z. Gecse, U. Heintz, N. Hinton, V. Kuryatkov, R. Lipton, M. Mannelli, T. Mengke, P. Paulitsch, T. Peltola, F. Pitters, E. Sicking, E. Spencer, M. Tripathi, M. B. Pinto, J. Voelker, Z. Wang, and R. Yohay.

Charge collection and electrical characterization of neutron irradiated silicon pad detectors for the CMS high granularity calorimeter. *Journal of Instrumentation*, 15(09):P09031, 2020. DOI: 10.1088/1748-0221/15/09/P09031. https://dx.doi.org/10.1088/1748-0221/15/09/P09031.

[72] B. Acar, *et al.* Neutron irradiation and electrical characterisation of the first 8" silicon pad sensor prototypes for the CMS calorimeter endcap upgrade. *Journal of Instrumentation*, 18(08):P08024, 2023. DOI: 10.1088/1748-0221/18/08/P08024. https://dx.doi.org/10.1088/1748-0221/18/08/P08024.

[73] N. Akchurin, G. Altopp, B. Burkle, W. Frey, U. Heintz, N. Hinton, M. Hoeferkamp, Y. Kazhykarim, V. Kuryatkov, T. Mengke, T. Peltola, S. Seidel, E. Spencer, M. Tripathi, and J. Voelker. Modeling of surface damage at the Si/SiO2-interface of irradiated mos-capacitors. *Journal of Instrumentation*, 18(08):P08001, 2023. DOI: 10.1088/1748-0221/18/08/P08001. https://dx.doi.org/10.1088/1748-0221/18/08/P08001.

[74] B. Acar, *et al.* Performance of the CMS high granularity calorimeter prototype to charged pion beams of 20–300 GeV/c. *Journal of Instrumentation*, 18(08):P08014, 2023. DOI: 10.1088/1748-0221/18/08/P08014. https://dx.doi.org/10.1088/1748-0221/18/08/P08014.

[75] N. Akchurin, C. Cowden, J. Damgov, A. Hussain, and S. Kunori. The (un)reasonable effectiveness of neural network in cherenkov calorimetry. *Instruments*, 6(4), 2022. DOI: 10.3390/instruments6040043. https://www.mdpi.com/2410-390X/6/4/43.

[76] Y. Wang, Y. Sun, Z. Liu, S. E. Sarma, M. M. Bronstein, and J. M. Solomon. Dynamic graph cnn for learning on point clouds, 2019. arXiv:1801.07829.

[77] N. Akchurin, C. Cowden, J. Damgov, A. Hussain, and S. Kunori. On the use of neural networks for energy reconstruction in high-granularity calorimeters. *Journal of Instrumentation*, 16(12):P12036, 2021. DOI: 10.1088/1748-0221/16/12/P12036. https://dx.doi.org/10.1088/1748-0221/16/12/P12036.

Chapter 2
Solid State Tracking Detectors

Maurice Garcia-Sciveres

Lawrence Berkeley National Lab
MS 50B-5239, 1 Cyclotron Rd., Berkeley, CA 94720, (510) 486-7354
mgs@lbl.gov
http://physics.lbl.gov/garcia-sciveres

2.1 What Is a Tracking Detector?

In a high energy particle experiment, a beam of particles (electrons, protons, muons, etc.) is focused onto a target and the scattered particles are measured using a detector. The properties of the target can be reconstructed from measurements of the scattered particles. It can be insightful to compare the problem to conventional microscopy where a small object is imaged using scattered light rays. There are important differences from optical microscopy:

1. The target being studied can be the quantum vacuum rather than a solid object, in which case the incident beam of particles is focused onto another incident beam of particles: this is the particle collider configuration.
2. The scattered particles need not be the same as the incident beam particles. Larger numbers and types of particles typically emerge. Energy is conserved, but if the incident beam energy is higher than the rest mass of some particles, those particles may be produced in the collision.
3. There is no lens that can focus scattered particles as done with light. The characteristic ray tracing of optics reconstructs straight line rays from two points: the lens (common to all rays) and the position on the image sensor. The analogous ray tracing for particles is done by measuring two

---

2024 © The Author(s). This is an Open Access chapter published by World Scientific Publishing Company, licensed under the terms of the Creative Commons Attribution 4.0 International License (CC BY 4.0). https://doi.org/10.1142/9789819801107_0002

or more points along the trajectory of each particle and inferring the trajectory from those points. This is called track reconstruction or tracking.

When considering a single particle emerging from a target and traveling in vacuum, measuring two points would suffice to infer the straight line trajectory. However, there are hundreds or thousands of particles emerging at the same time, which introduces ambiguities when connecting the two measured points. In addition, the particles have electric charge, which means they will not travel in a straight line; rather they will follow a helix due to the applied magnetic field. Finally, the particles do not travel in a vacuum, as both accelerator and detector contain significant material. To address these conditions, significantly more than two points per trajectory must be measured. The detector that makes such measurements is call a tracker.

Practical trackers measure points along the trajectories of electrically charged particles only. The interaction cross-section of charged particles with matter peaks at low values of energy loss [1] so that an energetic charged particle traversing material will gradually lose energy along its path. Thus, small amounts of material can be used to extract enough energy for a detection signal from a traversing charged particle with high efficiency, yet without stopping or significantly changing the trajectory of the particle. Neutral particles (gamma rays, neutrons, and neutral kaons), on the other hand, lose energy in larger, discrete interactions, either being completely stopped or deflected, making it unfeasible to sample their trajectories without disturbing them.

2.2 Why Solid State?

Gas, liquid, and solid materials have all been used to build trackers. Earlier trackers all used internal gain, meaning that the energy lost by a traversing particle is amplified though some physical process within the detection medium before being measured by electronic or optical means. Bubble chambers used a superheated liquid as the energy loss medium, in which traversing particles cause nucleation, leading to local boiling along the trajectory, leaving a trail of bubbles that could be photographed. This is a relatively slow process and not compatible with continuous operation. In gaseous detectors, such as drift chambers, a high electric field near a sense electrode leads to avalanche breakdown in the presence of ionization from a traversing particle. This process is much faster than bubble formation but still too slow (the recovery in particular) for contemporary colliders [2]. Furthermore, the position resolution of gaseous trackers is limited by ionization statistics. The density of a gas is low enough that the distance between ionizing interactions of a passing charged particle[1] is Poisson distributed with mean of order $100\,\mu$m, and this limits how well the track position can be known from the ionization [3].

Solid state trackers, which overwhelmingly use silicon as the detection medium, are both faster and more precise than gaseous trackers. The high density of solids results in sufficient ionization that no intrinsic gain is required in order to measure the signal from a traversing particle. This is also thanks to low noise electronics enabled by integrated circuits and high density interconnection methods (wire bonding and bump bonding). The capacitance of interconnects enters into the electronic noise as will be seen later. Detection without intrinsic gain is faster because one does not need to wait for the intrinsic gain process and its recovery to take place, and this enables the very high rate capability of silicon trackers. Ionization statistics no longer limits resolution (for the time being), as the mean free path between energy deposits of minimum ionizing particles in silicon is of order $1\,\mu$m.

While resulting in high ionization which is good for fast, precise detection, the high density of solids is also a liability because it leads to scattering of the particles whose trajectory is being measured. Fundamentally, a tracker must extract energy from traversing particles in order to measure them while at the same time not extracting energy in order to leave their trajectory undisturbed. This is clearly an optimization problem for the thickness and separation of the solid state sensors in a tracker. Early silicon trackers all used $300\,\mu$m thick silicon sensors because this happens to be the standard thickness for commercial silicon wafers and results in a signal that can be readily measured, but contemporary trackers use ever thinner sensors, as other factors enter into the optimization, in particular radiation damage to the sensor.

2.3 Strips, Pixels, and Monolithic Pixels

A tracker design involves the optimization of many parameters and tracker designs have evolved in response to changing parameters. Some parameters are driven by the experiment and science requirements, while others are due to available technology and practical matters, such as cost. In broad strokes, the historical trends go from small to large systems (both size

---

[1] We are mainly concerned with relativistic particles in the so-called minimum ionizing regime [1].

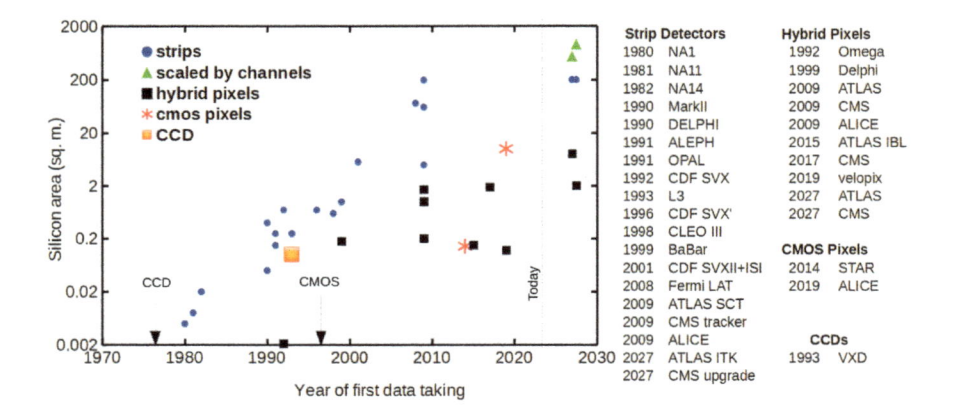

Figure 2.1. Evolution of silicon tracking detectors for particle and nuclear physics, separated into the three main types. All of them are on ground-based experiments except for Fermi-LAT, which is a space mission. The green triangles for strip detectors are duplicate points scaled by channel count rather than area because area is constrained by the existing solenoid magnets the detectors must fit into. The solid square on the horizontal axis (Omega) is just $2\,\mathrm{cm}^2$ (so off scale). The dashed arrow indicates when the CCDs first appeared in consumer digital cameras, while the solid arrow marks when the monolithic CMOS sensors first appeared in webcams. The vertical dotted line shows the time of this writing.

and channel count), from slow to fast, and from less to more radiation tolerant. Figure 2.1 shows the historical area trends for three main types of silicon trackers: strips, hybrid pixels, and monolithic. In all trackers, one can identify a basic unit called a *module* which is replicated many times. A module is like a tile in a tiled floor or wall. A tracker consists of surfaces (which are often cylindrical) tiled with modules. The main difference between the three types of trackers is what the modules are made of.

Silicon Strips: The modules consist of a silicon sensor and readout electronics (Fig. 2.2(a)). The sensor is an array of parallel line channels or *strips* connected to readout at one end. This is the oldest type of tracker and the simplest to build using more or less generic components. The first such detectors were built with discrete electronics fed by one wire per strip, but area scaling and collider configurations were enabled by integrated circuit electronics and automated wire bonding technology. The first custom integrated circuit for particle physics, the Microplex chip, was developed in 1983 precisely to be able to scale the use of silicon strip sensors to larger areas and collider detector geometry. It is worth noting that the first silicon strip sensors were commercially sold in 1969, so the sensor technology was

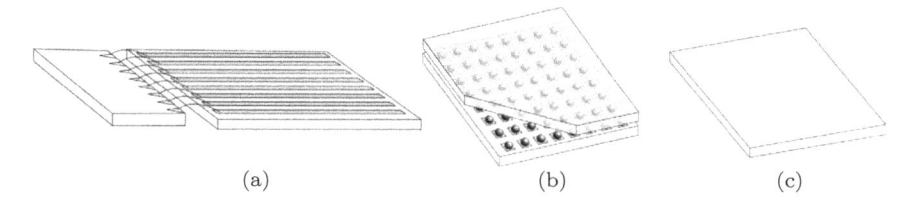

Figure 2.2. Schematic depiction of module types. (a) Strip module with sensor on right and sensor strips connected by wire bonds to the readout electronics on the left. (b) Hybrid pixel module with sensor and readout integrated circuit having matching geometry and connected to each other face-to-face by metal bump bonds. A corner has been cut away to help visualize the assembly. (c) Monolithic module.

mature; it was the development of integrated circuits that made them ideal for particle physics applications.

Hybrid Pixels: The modules consist of a silicon sensor and readout electronics (Fig. 2.2(b)), but the sensor is a matrix of pixels connected to readout electronics via bump bonds (one bump per pixel). Since the sensor and readout integrated circuit must have matched geometries, and because fine pitch bump bonding requires specialized equipment, hybrid pixels are complex to build and require custom designed components. However, a major advantage is that the sensor and readout integrated circuit can be separately optimized using very different fabrication processes. Mature strip sensor technology could be used to produce science grade pixel sensors almost "out of the box." High efficiency at the same time as high readout rate and radiation tolerance could be achieved thanks to the separate optimization.

Monolithic Active Pixel Sensors (MAPSs): A MAPS module can consist of just one MAPS sensor and some interconnects. MAPS are the closest relative of consumer image sensors ubiquitous in every mobile device. While MAPSs are simple to assemble, they are complex to produce, requiring a customized integrated circuit fabrication process. Since integrated circuits are mass produced in large foundries, particle physics-specific customization for producing small volumes is challenging to negotiate. MAPS sensors provide the lowest mass option for particle tracking but have rate and radiation tolerance limitations relative to hybrid pixels. Thus, the highest rate and radiation experiments have not managed to make use of them. Such monolithic technology was investigated very early on [4] (before CMOS imagers replaced CCD imagers in consumer electronics) but was initially disfavored relative to hybrid pixels, which had a mature sensor technology base.

2.4 How Does the Detector Work?

2.4.1 *Hybrid sensor*

A silicon sensor channel (strip or pixel) can be very simply modeled as a parallel plate capacitor in vacuum, in which electric charges magically appear along the trajectory of a particle crossing the gap. A resistor is in parallel to the capacitor and a DC voltage called bias of order 100 V is required across this RC for operation. For silicon sensors, the amount of charge from a passing particle is on average 80 electrons (and 80 positive charges or *holes*) per micron of path length, which for a 300 μm thick sensor traversed perpendicularly is 4 fC. This is the signal. To read out this signal, the charge must be extracted from the sensor channel and turned into a voltage. The capacitance of a silicon strip (pixel) channel is of order 1 pF/cm (100 fF) and the resistor is of order 100 GΩ. So the voltage change on the channel from the added charge is of order 1 mV (40 mV) for a 4 cm strip (pixel). One cannot easily read this voltage that sits on top of the 100 V bias. Instead, the charge is extracted by a readout circuit, which generates a voltage from it.

The sensor is not in reality parallel plates in vacuum but a solid silicon crystal. High resistivity silicon is used, which means it has a low level of doping and hence a low charge carrier concentration. Highly doped implants are used to define the pixels or strips. These implants form PN diode junctions with the high resistivity bulk. For an introduction to and reference on silicon devices, see [5]. Applying a reverse bias voltage to the diodes will grow the depletion region until the entire bulk is depleted of free carriers. The depleted sensor bulk is the vacuum of the parallel plate capacitor model in which charge can magically appear. A traversing high energy particle will lose energy in the silicon crystal and this energy will promote charge carriers from the valence band to the depleted conduction band. The average energy per electron–hole pair promoted to the conduction band in silicon is 3.7 eV. Thus, the 80 e-h pairs per micron correspond to an energy loss of 300 eV per micron by the passing charge particle. This is of course the energy loss by a minimum ionizing charged particle in silicon [1]. As a fun digression, consider what if one really had vacuum instead of silicon. It is still possible to turn energy into charge carriers, but one needs twice the electron mass per electron positron pair created, which would be 80 GeV per micron instead of 300 eV. So other than the 8 orders of magnitude energy difference and a solid lattice to conserve momentum for the lost energy, a depleted sensor is a good analogy for Dirac's negative energy electron sea filling all of space [6].

A depleted semiconductor is almost like an insulator (or vacuum) but not quite. Imperfections in the crystal have different energy levels than pure silicon and can result in carriers within the forbidden region separating the valence band from the conduction band. Combined with carrier diffusion, this leads to promotion of carriers into the conduction band manifesting as a steady "leakage current." The parallel resistor in the simple model was there to simulate this current. Leakage current, also known as dark current in photodiodes, is a major topic in semiconductor detectors. The first thing to note is that it looks exactly like signal: charge carriers appearing in the depleted conduction band. Consider the 100 GΩ resistor under 100 V vias of the simple model. This results in 1 nA of leakage current, or 1000 fC/s, to be compared with the 4 fC signal. This may sound hopeless, but it just means that the readout electronics need to be fast because the signal appears all at once as an AC pulse, while the leakage current is DC. If the electronics can integrate the signal in 1 μs, for example, then the above leakage contribution becomes negligible. However, the story does not end there because 1 nA is the leakage current of a very high quality, brand new sensor channel, but it can increase by orders of magnitude with radiation damage. A common technique in strip detectors is to capacitively couple each strip to the readout (called AC coupling), which preserves the signal while stopping DC current, but this still leaves the fluctuations in the leakage current to contend with because leakage current is a stochastic process. Furthermore, AC coupling adds processing cost, limits the bias voltage that can be applied, and is not practical for pixels. Two main technologies are used to address leakage current for detectors that must withstand high radiation (which is most of them): (1) sensor material engineering and (2) cooling. An international collaboration called RD50 has been perfecting silicon material for tracker applications for decades [7]. Cooling reduces leakage current because, since leakage current depends on diffusion of intrinsic carriers, it scales like the intrinsic carrier concentration, which is proportional to $T^{1.5}e^{-E_g/2kT}$, where T is absolute temperature, E_g is the bandgap (1.12 eV in silicon), and k is Boltzmann's constant. The use of advanced materials and methods in the low mass mechanical design (see Section 2.8) enables cold operation of silicon trackers, ranging from $-20°$C to $+10°$C sensor temperature.

2.4.2 Integrated circuit and readout

The readout integrated circuit (either hybrid or monolithic) is primarily responsible for extracting the signal from the sensor and turning it into communicable information that can be relayed to a data acquisition system, which can be a significant distance away, of order 100 m. The distinct feature of hybrid pixel and strip detectors is that each channel (each individual pixel or strip) is an independent unit with its own electronics, and all channels run in parallel. This is in sharp contrast to consumer image sensors, either CMOS or their CCD ancestors, where all or many pixels on one device are staged onto a common readout channel. The all-in-parallel operation permits trackers to reach the high frame rates required by particle physics experiments, for example, 40 MHz at the LHC. A 40 MHz frame rate is unheard of (and unnecessary) for commercial electronics. If one takes as a figure of merit detector area times frame rate, to cover $1\,\text{m}^2$ of detector at 40 Mfps would require 20 billion mobile phone image sensors since each sensor is about $0.5\,\text{cm}^2$ and capable of 40 fps. Instead, this is done with 2,500 custom integrated circuits in the case of pixels and 10 times fewer in the case of strips. The problem is data volume, as should be clear from the mobile phone comparison. (If everyone on the planet had 10 mobile phones and tried to upload video to the cloud simultaneously for a few years non-stop, this would be a problem.) The central question for tracker readout is: what is the information that must be extracted and how can that be done?

The information content (or information entropy) in a tracker can be calculated [8, 9]. The information entropy, which can be expressed in bits, depends linearly on the number of particles traversing the tracker, which should not be too surprising. But it depends very weakly (logarithmically) on the number channels, and it depends more strongly on the noise performance and timing resolution of the electronics. For example, for High Luminosity LHC pixel detectors (rightmost solid squares in Fig. 2.1), the information content is about 25 bits per particle in each detector layer crossed. This is a measure of the information available. Without lower noise or faster electronics, there is no more useful information to be had. But this does not mean that one will transmit 25 data bits per particle from each detector module to the data acquisition system. This is an information entropy bound, which means that unless information is discarded, it is impossible to move the information off-detector with fewer than 25 bits per particle. But it is actually very challenging to achieve this entropy bound. The High Luminosity LHC pixel detectors will get to within 25%

or so, which is better than ever done before, and with very good reason: the High Luminosity LHC particle rate is extremely large, and moving information out of the pixel detectors (namely, readout cables) is what limits their tracking performance, due to the mass of the cables causing multiple scattering. This illustrates the complexity of tracker design, where apparently disconnected parameters depend on each other, such as the noise target for electronics depending on the amount of data to be moved off detector in two separate ways: first, because it would be useless to generate more information (lower noise = more information) than can be transmitted, and second, because higher precision (lower noise = higher precision) than the multiple scattering smearing due to data cables will not improve tracking performance (Section 2.6).

Viewed as images, the frames from a tracker module will be dark fields with a few bright dots. Transmitting such frames as images would use many bits because of all those dark channels, which are of no interest for tracking yet are preserved in an image. Instead, dark channels are suppressed by applying a threshold (called zero-suppression), so only the bright dots must be read out. The threshold is applied as early as possible, before any digitization. In fact, this threshold discrimination is itself used as the digitization technique. Some detectors simply record which channels are above threshold (also called hits) with no further information, while others store a low precision amplitude value for every hit by counting "Time over Threshold" (ToT).

The readout process is depicted conceptually in Fig. 2.3. Even if the entropy bound were achieved, transmitting all information for every bunch crossing usually results in too much data to handle off-detector. The High Luminosity LHC ATLAS and CMS experiments will see about 10,000 charged particles per beam bunch crossing. If each particle crosses 10 detector layers, that would be 2.5 Mbits per bunch crossing, which leads to 100 Tb/s. Therefore, the trackers need a way to decide which bunch crossings to send off-detector rather than sending all of them. This selection is shown as a trigger signal in Fig. 2.3. Since deciding to keep or discard a bunch crossing takes some time, this means the readout integrated circuit must be a large digital memory to store *all* the data until a trigger arrives. For this reason, tracker integrated circuits are produced in the technology nodes with the smallest transistors that detector builders have access to. Smaller transistors means more memory per unit area to be able to store more hits per unit area. Therefore, Moore's law has been an important enabler of advances in semiconductor trackers.

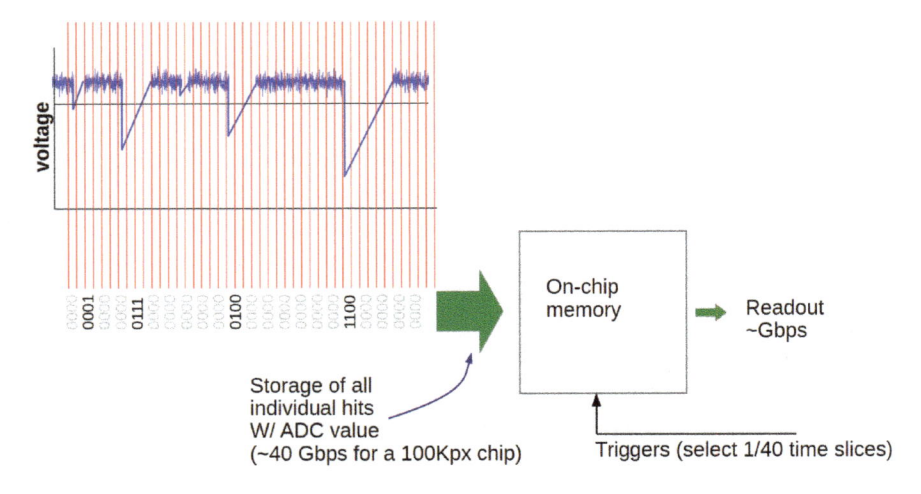

Figure 2.3. Representation of readout integrated circuit data flow with High Luminosity LHC pixel detector values. The waveform shows the analog signal for a single channel, with a noisy baseline and negative-going signal pulses. The horizontal line represents the applied threshold. Only pulses exceeding threshold are further processed, digitized, and stored in on-chip memory. The vertical lines mark the 40 MHz beam bunch crossings, with only one possible digitization per bunch crossing. The trigger signal selects only specific bunch crossings for off-detector readout.

Trigger signals are often supplied from outside the tracker, but a tracker can also be self-triggered, locally deciding which hits to transmit off-detector and which to discard. Hits produced by charged tracks will be correlated from one layer to the next, and the spatial correlation will depend on the track origin and momentum (if there is a magnetic field). Measuring the charge deposition profile in a sensor of finite thickness is sensitive to the incident particle direction but with precision limited by the sensor thickness d, as well as the position precision of the sensor Δx. The angular measurement error will be $\Delta x/d$. Interconnecting two layers some distance apart can greatly increase precision by increasing d far beyond sensor thickness. The CMS experiment is implementing such a direction-sensitive detector with electrically interconnected pairs of sensing layers consisting of one strip and one pixel, which measure correlated hit pairs called stubs. These are 3-D vectors rather than 3-D points in space. The stubs are filtered on-chip, and only those compatible with particle tracks of interest are read out, and combined off-detector to serve as a trigger signal for other subsystems.

Earlier we said the data rate was too high to read out everything without any trigger. However, in the case of the LHCb experiment, the number of particles per beam collision is an order of magnitude less, and additionally the fixed-target detector geometry permits routing cables outside the volume where the particle propagates. Thanks to these features, the LHCb experiment has implemented full triggerless readout of their detectors. This actually simplifies the architecture of the readout chips, as a storage of data while waiting for a trigger is no longer needed. This simplification is offset by the need to move data faster both on-chip and to the DAQ, requiring higher bandwidth circuits. The data must still be filtered off-detector to select events of interest. More sophisticated filtering is possible, as correlations between all detector elements can be exploited but only if enough computing resources can be deployed, which can be a significant cost. Even where triggerless readout is possible, detector design must balance the potential physics advantage from having all data off-detector against the lower cost and processing complexity of filtering data at the source, before it is read out.

Analog Front End and Signal Discrimination: The analog Front End (FE) is the readout circuit element responsible for extracting the signal from the sensor and converting it to a substantial voltage for further processing. It must move an electric charge Q from a significant sensor capacitance C_d, where it only equates to a small voltage $V = Q/C_d$, onto a smaller capacitor C_f where it will result in a larger voltage $V = Q/C_f$. The design and function of integrated circuit amplifiers for silicon detectors are extensive topics and only a superficial overview is given here in order to motivate the different types of devices used; for a more in-depth treatment, see, for example [10]. A plumbing analogy for the FE starts a small amount of water at the bottom of a bathtub, which needs to be moved to a glass in order to measure how much water there is. For the case of the bathtub, this can be accomplished with a pump, with the time taken and power used to transfer the water to the glass depending on the pump characteristics. Furthermore, for unattended operation, one needs a mechanism to empty the glass periodically or continually (through a small hole at the bottom). With a small hole, the water level in the glass will rise as the pump empties the tub and then fall as the glass drains, resulting in a water level vs. time pulse shape depending on how fast the pump pumps and the size of the hole. Silicon detector FEs are most commonly of this kind, where the electrical equivalent of the pump is a high open-loop gain preamplifier, the

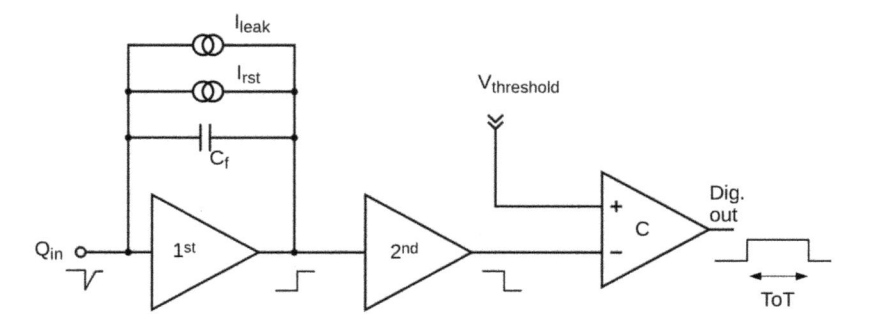

Figure 2.4. Schematic diagram of a typical analog front end (FE). Signal polarities and Time over Threshold (ToT) are indicated along the bottom. The 1st stage or preamp shows a feedback capacitor C_f, a reset current I_{rst}, and a leakage compensating current I_{leak}. The comparator stage, C, shows a threshold voltage adjustment $V_{\text{threshold}}$.

glass is a capacitor placed in the preamp feedback loop (C_f), and the hole is a continuous reset of the capacitor (I_{rst}) provided by a resistor, a current source, or a more complex circuit to control linearity and baseline. These elements can be seen schematically on the left of Fig. 2.4.

Figure 2.4 also shows an additional feedback element, I_{leak}, a 2nd stage preamp, and voltage comparator (C) with an applied threshold. The 2nd stage is conceptually not a separate functional element but typically required to provide additional gain and/or drive the comparator input. The I_{leak} feedback element is functionally important if the sensor is DC connected to the FE input, such that sensor leakage current will flow through the FE. Sensors can also be capacitively coupled to the FE, in which case the no leakage current circuitry is needed, but this adds cost and complexity to the sensors and is typically only possible for strip detectors.

The comparator carries out the pulse height discrimination, distinguishing pulses above a user-chosen threshold, which ideally are only produced by particles hitting the sensor, from electronic noise (or thermal noise) and from leakage current fluctuations (called shot noise). Optimization of the threshold to be more than 99% efficient for real particle hits while having low firing rate in the absence of real hits is a main challenge of analog design.

Thermal noise can be thought of as fluctuations of voltage in an electronic system. At no point in an electronic system is the voltage a perfectly constant value with time, but it wiggles with some RMS noise value. Therefore, if one samples a voltage (or compares two voltages as is done in a comparator), the result depends on exactly at what point in time the sampling happens. If a circuit is very slow (low bandwidth), the wiggles are slow and the result does not change much for small changes in sampling time, but if the bandwidth is high, the change is larger. So a slow amplifier will have less noise, but it can't be made too slow because it has to distinguish bunch crossings. A very important point is that thermal noise is a voltage noise, whereas the signal in particle detectors is a charge. Noise in silicon detectors is typically expressed in units of Equivalent Noise Charge (ENC), also known as input-referred noise. We want to know the value of signal-like charge fluctuation that results in a voltage fluctuation equal to the electronic voltage noise. Crucially, this depends on the sensor capacitance for no more complicated reason than the $Q = CV$ relationship. Thus, thermal noise (or more correctly thermal signal to noise, S/N) scales with sensor capacitance! Another important point is that the ENC depends on the amplifier signal gain and charge collection efficiency. Recall that in a continuous reset FE the glass being filled by the pump has a hole it, so the slower the pump fills it, the lower the maximum level the water will rise to. For a resistive reset, the discharge current is V_{out}/R, while the charging current from the amplifier operation is $(Q_0/\tau)e^{-(t/\tau)}$. The current starts at some value Q_0/τ and asymptotes to zero when the entire signal charge Q_0 has been extracted from the sensor. The integral of this extracted current is Q_0 and the time constant, τ, is a characteristic of the amplifier (how fast the pump pumps in the plumbing analogy). The output voltage is $V_{\text{out}} = QC_f$, where Q is instantaneous charge on the feedback capacitor C_f (analogous to the volume of water in glass next to the bathtub):

$$\frac{dQ}{dt} = \frac{Q_0}{\tau}e^{-(t/\tau)} - \frac{Q}{RC_f} \qquad (2.1)$$

This has a characteristic pulse shape with a rise time given by τ and a fall time given by RC_f. The peak amplitude Q_{peak} occurs when $dQ/dt = 0$, and it can be readily appreciated from Eq. (2.1) that the faster the fall time (the smaller R), the lower Q_{peak} for a given Q_0. All this is to say that the ENC is not only the voltage noise times the detector capacitance, but it is further increased by a factor Q_0/Q_{peak}.

The shot noise arises because charge is quantized. Integrating leakage current for a period of time equates to counting a number of charges and like any counting is subject to a Poisson fluctuation of $\sqrt{N}$ for large N. The larger the N (or the integrated leakage current Q_{leak}), the smaller

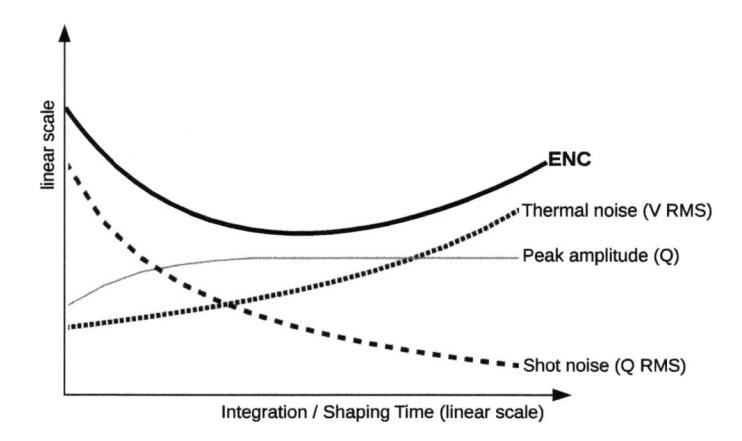

Figure 2.5. Depiction of contributions to Equivalent Noise Charge (ENC) from thermal (voltage) noise and leakage current shot noise. The ENC is the quadrature sum of the thermal noise converted to equivalent charge, including peak amplitude (gain) correction as explained in the text, and the shot noise, which already has units of charge. The peak amplitude decrease on the left of the figure signals that the shaping time is becoming comparable to the amplifier rise time (the hole in the glass is becoming too large in the plumbing analog of the text).

the relative error $\sqrt{N}/N$. Thus, shot noise decreases for longer integration time (larger value of R). Note that it makes no difference if the sensor is AC or DC coupled for shot noise, since the fluctuations are an AC signal. However, since shot noise is already a charge fluctuation, it does not depend on detector capacitance. Unlike shot noise, thermal noise increases with increasing R because the bigger time window allows more voltage wiggles to be sampled. The amplifier shaping (RC_f in this simple case) can thus be optimized to minimize ENC (maximize S/N), as shown in Fig. 2.5.

There are many important conclusions that can be drawn from the simple noise analysis presented. Strip detectors have significant sensor capacitance and leakage current, simply because the area of each strip is relatively large. They need more sophisticated amplifiers with high gain due to the large sensor capacitance and cannot operate at very high bandwidth because of leakage current. Hybrid pixel detectors, on the other hand, have exactly the same signal size as strip detectors but lower capacitance and leakage currents; they are just like extremely small strips. So they can have a simpler, lower gain FE and can operate at higher bandwidths. MAPSs have even smaller capacitance and therefore can use an even simpler FE or no FE at all: the sensor capacitance is analogous to the size of the bathtub in the plumbing example, so when the bathtub becomes the size of a glass, one no longer needs a pump to collect the water into a measuring glass. However, MAPSs also have smaller signal size, and the development of that signal can be slower than in a standalone sensor, so they cannot necessarily exploit the low capacitance to achieve high speed; they are generally significantly slower devices than hybrid pixels. This already hints at the importance of shot noise for MAPSs, since the signal is small, the integration time is relatively long, and shot noise does not care that the detector capacitance is tiny.

For detectors with charge readout, the fact that a comparator follows the FE of silicon detectors is universally exploited to digitize the signal amplitude with the technique to Time over Threshold (ToT), as mentioned in Section 2.4.2. This kind of digitization requires minimal circuitry (just a counter) by making use of the already available comparator output and the beam crossing clock. The comparator output is a digital pulse whose duration is equal to the time that the FE output exceeded the comparator threshold. This time is measured by using that pulse to gate a counter that counts beam crossing clock cycles. Clearly, this only works if the FE output pulses are typically longer than one beam crossing, which tends to be the case because the amplifier rise time is chosen to be just fast enough to distinguish beam crossings, and the fall time must be longer to minimize ENC, as per Fig. 2.5.

We can now return to to the hit discriminating function of the comparator in more detail. The ENC must be small enough (equivalently S/N large enough) that one can choose a threshold which achieves the required 99% hit efficiency with a low rate of noise hits. This is depicted in Fig. 2.6. In the ideal case of the figure, the threshold can be chosen where there is zero response probability, resulting in zero hit rate in the absence of signal and 100% efficiency for signal. In practice, there is typically some overlap between tails of the two distributions. The distributions can't simply be added to determine the minimum probability point suitable for the threshold because they have different normalizations. The particle response function is normalized by the real hit rate impinging on the detector, while the ENC distribution is normalized by the maximum hit rate that the FE can respond to, which depends on FE speed and shaping. The ENC hit rate will rise exponentially as the threshold is decreased, up to the maximum at zero threshold. The operating threshold can be chosen to either reduce noise hit rate or increase signal efficiency. An interesting

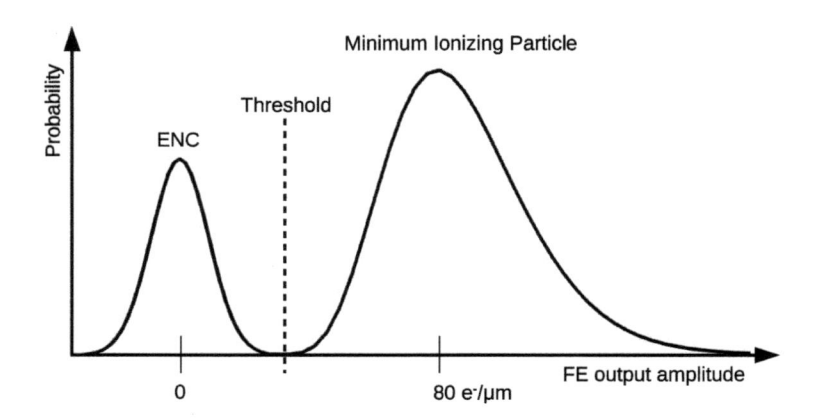

Figure 2.6. Depiction of FE response probability distribution functions and the suggested threshold setting. In the absence of signal, the FE response is given by the ENC, a Gaussian centered at zero (fluctuations can be negative or positive). When a minimum ionizing particle ionizes the detector, a random charge value obeying a Landau distribution with most probably value of $80\,\mathrm{e}^-/\mu\mathrm{m}$ appears at the FE input, leading to a response at the output given by Landau, convoluted with the ENC (since noise is always present whether or not there is a signal). In this ideal example, there is zero response probability between the two distributions, where the threshold is shown.

observation is that for high rate experiments such as those at the High Luminosity LHC, the real particle hit rate will reach values of $3\,\mathrm{GHz/cm^2}$ and therefore very high noise hit rates (in the $\mathrm{MHz/cm^2}$ range) could in principle be tolerated with minimal impact to the data readout. However, experiments typically opt for running with much lower noise hit rates for a variety of reasons, including the ability to perform noise-free calibrations.

A final consideration about the threshold is that it must be the same (when referred to the input) for all channels in a chip and also stable with time so that the hit efficiency is uniform and, in the case of ToT, the charge measurement is uniform. Uniform response of analog circuits is not automatic within integrated circuits. Such uniformity in the static case is typically achieved with a programmable channel-by-channel threshold adjustment. But threshold fluctuations in time can be more challenging to control. For example, changes in chip power consumption can lead to threshold shifts, and processing of hits can affect power consumption. In pixel detectors, this tends to lead to a maximum occupancy beyond which operation may become unstable. For these reasons, the threshold is better thought of as a band of some thickness rather than the thin line in Fig. 2.6.

2.5 Tracker Performance

In the ideal case of a helical trajectory in a magnetic field, the momentum resolution of a tracker with N layers is given by the following Gluckstern formula [11]:

$$\frac{\sigma_{p_T}}{p_T} = \left(\frac{p_T}{0.3|z|} \frac{\sigma_{\text{point}}}{L^2 B} \sqrt{\frac{720}{N+4}}\right) \oplus \left(\frac{\sigma_{p_T}}{p_T}\right)_{\text{MS}} \quad (2.2)$$

where p_T is transverse momentum in GeV/c, L is the radial length in m, B is the magnetic field in T filling the tracker volume, z is the particle electric charge in elementary units, σ_{point} is the resolution of the detector measurements in m, and N is assumed to be large in this approximation. L has the largest effect in the equation, but note that the whole of L has to be filled with magnetic field, and the stored energy in a solenoidal field scales like $B^2 L^2$, so making L very large is not trivial. It may therefore seem that improving point resolution may be the most economical way to enhance tracker performance. However, this has to compete against the second term in quadrature. Point resolution also has to compete against alignment precision, which is not considered in Eq. (2.2).

The second term in Eq. (2.2) is the multiple scattering (MS) contribution for a number of detector layers N. It can be written as

$$\left(\frac{\sigma_{p_T}}{p_T}\right)_{\text{MS}} = \frac{0.0136}{0.3\,\beta\, BL} \sqrt{\frac{(N-1)x/\sin\theta}{X_0}} \sqrt{C_N}, \quad [L] = \text{m},\ [B] = \text{T} \quad (2.3)$$

where L is the radial track length and $(x/\sin\theta)/X_0$ is the total material thickness traversed by a particle incident with polar angle θ with respect to the beam, in units of the radiation length. One radiation length is the thickness of material that reduces the energy of impinging relativistic electrons by $1/e$, or equivalently that pair-converts $e^{-7/9}$ (45%) of impinging energetic gamma rays. C_N is a factor depending on the number of layers: $C_N = 2.5$ for the minimum of three layers to measure a circle and approaches $C_N = 1.33$ for $N \to \infty$ (continuous scattering). Equation (2.3) shows that the low momentum performance of a tracker is limited by its mass. The radiation length of 1 mm thick silicon, copper, and carbon composite, is 1%, 7%, and 0.3%, respectively. State-of-the-art hybrid (monolithic) trackers achieve under 2% (1%) of a radiation length per layer.

2.6 How Does Tracking Work?

Track reconstruction, or tracking, is the process of converting the hit coordinates that were read out from the detector, known as space points, into momentum vectors of the charge particles that produced them. The results of Section 2.5 only apply after tracking has done a good job. Multiple operations are involved, which we group into four main categories: pattern recognition, track fitting, alignment, and simulation. One might here ask why not train an artificial neural network to perform this complex transformation from space points to momentum vectors? This is a very active development area but does not affect the division into these four categories. Machine learning is being applied within each of the four individual categories, but a solution to infer final vectors directly form hits in a single step (called end-to-end), that is, competitive in terms of precision and performance, is not considered feasible.

Pattern Recognition: Pattern recognition is the process of dividing all the space points from an event into track candidates, where each track candidate is just a group of space points. This process is very sensitive to the number and spacing of the layers in a detector and it is the greatest consumer of computing time in modern experiment track reconstruction. Space points in different layers are linked together by extrapolating "seeds," where a seed is any pair or triplet (different techniques use pairs or triplets) of space points that may or may not belong to the same track. Seeds are typically created combinatorially and can therefore reach astronomical numbers, but most of them will be spurious combinations that will be gradually rejected for having no additional points in their extrapolations. It is instructive to analyze the linking of hits of a hypothetical three-layer arrangement. In order to distinguish true tracks from random combinations, the probability of finding random combinations that align into tracks should be small. This "fake" probability is straightforward to calculate. Even when a particle track is known up to a certain layer, the extrapolation of that track to the next layer has an angular uncertainty α, with a lower bound given by multiple scattering (the multiple scattering angular error can be found in [1]). This uncertainty projects a circle of radius $x\alpha$ onto the next layer, where x is the distance to that next layer. The number of random coincidence hits within this circle is the area of the circle times the hit density in that layer, ρ. The hit density is given by the track density at that layer, so it is known regardless of detector details. Actually, to the extrapolation area, one must add a position resolution area, because the measured hit positions have their own uncertainty, given a resolution d (for simplicity, we assume equal resolution in two dimensions, but in practice, detectors can have different resolution in each dimension (notably strip detectors have excellent resolution in one dimension only). The probability of finding a random hit in the extrapolation of a track is therefore

$$P_{12} = \pi(\alpha^2 x^2 + d^2)\rho \tag{2.4}$$

where P_{12} denotes the probability of the extrapolation from layer 1 to layer 2 linking a fake hit, not related to the track in question. In the case of a three-layer tracker with total length L, two successive extrapolations (from layer 1 to layer 2 and from layer 2 to layer 3) must each find a random hit coincidence in order to result in a fake track. The fake probability from layer 1 to layer 2 was given in Eq. (2.4). The fake probability from layer 2 to 3 (P_{23}) is also given by the same equation but replacing the distance x (which was the distance from layer 1 to 2) by $L - x$. The probability of the two successive fakes is simply the product $P_{12}P_{23}$:

$$P_{\text{fake}} = P_{12}P_{23} = \pi^2 \rho_2 \rho_3 (\alpha_1^2 x^2 + d_2^2)(\alpha_2^2(L-x)^2 + d_3^2) \tag{2.5}$$

where the subscripts now refer to the layers since different layers can have different masses and resolutions. Note the multiple scattering in the extrapolation source layer and the position resolution in the destination layer determine the uncertainty circles. Figure 2.7 shows the relative probability for fake tracks vs. the position of layer 2, ranging from very close to layer 1 to very close to layer 3 (the separation between layers 1 and 3 is always the constant L), where we assumed for simplicity α and d are the same for all layers. Two cases are plotted: the extrapolation uncertainty at distance $L/2$ is dominated by the hit resolution or dominated by multiple scattering (dominated means twice as large). It can be appreciated that when multiple scattering is small (which is the case for high momentum particles) there is a slight pattern recognition advantage to evenly spaced layers, whereas if multiple scattering is important (the case for low momentum particles), there is a significant advantage to placing some layers close together (often called doublets, as in the self-triggered CMS tracker discussed earlier). This makes it clear that there is no one-size-fits-all tracker geometry, but the geometry must be optimized depending on the detailed requirements.

Track Fitting: Section 2.5 assumed particle trajectories are perfectly helical, the magnetic field is perfectly uniform, multiple scattering smears

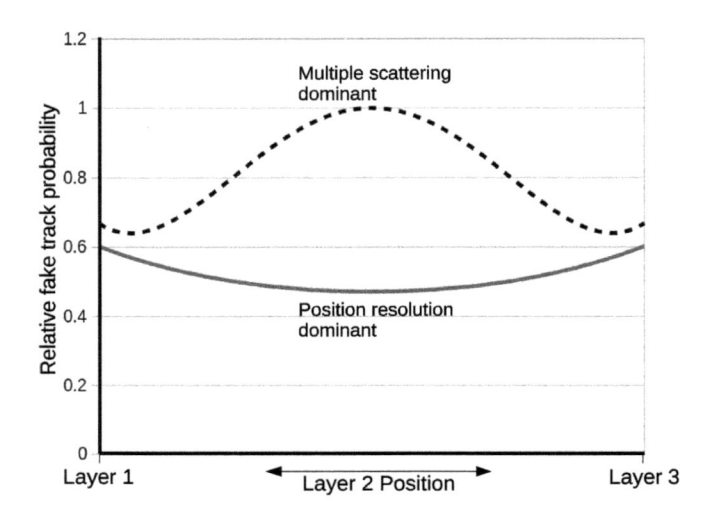

Figure 2.7. Relative fake track probability in three-layer tracker discussed in the text and given in Eq. (2.4). The positions of layers 1 and 3 are at the left and right of the plot, respectively, with constant separation L. The effect of varying the position of layer 2 between layers 1 and and 3 is plotted for the cases where the layer position resolution is the dominant uncertainty (solid) and multiple scattering is the dominant uncertainty (dashed).

all measurements equally, etc. But none of these conditions hold exactly. To obtain the best results, track fitting is not a simple least squares fit to an analytic formula. Starting from the initial momentum we wish to know, a particle's position in the first tracker layer is only smeared by the material entering the tracker (like the collider beam pipe), while at the last layer, it is smeared by a random walk through the entire tracker. The fitting algorithm that correctly unfolds these progressive errors, universally used in track reconstruction, is the Kalman filter [12]. An integral part of this method is comparing each hit position to the expectation of where it should be based on all prior hits. The track parameters are updated hit by hit. This requires a model of the detector and machinery to predict, from the track parameters, where the next hit should be. This is called track propagation. Track propagation performs the transport of track parameters and the associated covariance matrices through the detector geometry, taking into account interactions with the material (using the detector model) and the magnetic field (using a magnetic field map). Accurate simulation is an absolutely integral part of tracking.

2.7 Radiation Considerations

Sensors are fabricated from high purity single crystal silicon so that a significant depth can be depleted of charge carriers. Impurities and lattice defects degrade the sensor performance. Nuclear interactions from impinging hadronic particles can damage the crystal lattice, a mechanism called Displacement Damage (DD) or bulk damage. DD is quantified by Non-Ionizing Energy Loss (NIEL) in terms of the flux of 1 MeV neutrons that cause the same damage (abbreviated 1 MeV n eq.). DD leads to increased leakage current (by orders of magnitude), increase of reverse bias voltage needed to deplete the sensor, and reduction of free carrier lifetime (which reduces charge collection efficiency). Through fabrication techniques known as defect engineering, silicon sensors are currently being made to withstand NIEL doses of 2×10^{16} 1 MeV n eq.

Radiation damage in CMOS circuits is entirely due to charge carriers generated by ionization in the dielectric layers of the process and not displacement damage. Ionizing dose is delivered at hadron colliders by a combination of minimum ionizing particles (mainly pions) and background X-rays and termed Total Ionizing Dose (TID). The doping concentrations in CMOS transistors are so high (10^{15} cm^{-3} and higher) that compared to them, the defect density introduced by bulk radiation damage is negligible [13] (below 10^{14} cm^{-3} for HL-LHC inner layers after 3000 fb^{-1}). This means NIEL damage has no effect on CMOS electronics. However, there are many dielectric structures in a modern CMOS process and each one leads to its own radiation effect due to TID. It is not by accident that radiation tolerance requirements have kept pace with the logic density evolution in the ROIC generations. The reason is that both hit rate and radiation dose scale with particle flux. Required radiation tolerance went from 50 Mrad for the 1st generation, to 250 Mrad for the 2nd, to 1 Grad in the 3rd. 1 Grad corresponds to about 50 minimum ionizing particles crossing every Si lattice cell. Not all effects from charge generation in the dielectrics are equally important. As radiation dose increases, understanding and managing previously negligible effects become necessary.

In addition to long-term degradation due to accumulated dose, energy loss by ionizing particles leads to instantaneous soft errors called Single Event Upsets (SEUs). The most common SEU is the flipping of a stored bit in a memory. SEU also can produce voltage transients on signal or control lines that can result in accidental operations (for example, a single-level asynchronous line to reset logic or memory would be vulnerable to

SEU). Protection against SEU involves hardening of memory cells, avoiding designs with vulnerable control signals or hardening control signals where their use cannot be avoided, and circuit triplication. These techniques have been in use since 1st generation readout integrated circuits and they have not seen significant changes in the 2nd generation. However, as collider rate continues to increase and higher logic density translates into lower deposited charge needed for upset, these techniques will no longer be sufficient. An approach being introduced in 3rd generation readout integrated circuits is to design for reliable operation while a significant level of upsets is taking place. Fundamentally, this is abandoning the idea of circuit hardening as a solution to the SEU problem and instead designing all functions such that SEU is not a problem to begin with. In practice, a combination of hardening and SEU-friendly functionality will be used.

Extensive literature and experience exist on SEU of memory cells in the context of electronics used in space. This is not directly applicable to particle physics pixel detectors but is nevertheless a good starting point. SEU of a given circuit, like an SRAM cell, latch, or flip-flop, depends on the amount of energy deposited by an impinging ion, which is characterized by a Linear Energy Transfer (LET). It is important that this is meant to describe non-relativistic ions, which lose energy approximately uniformly along their path through electromagnetic interactions. The upset rate is characterized by the cross-section for causing a bit flip (SEU cross-section). Cross-section vs. LET is typically fit with a Weibull function, resulting in a threshold and saturation cross-section, as shown in Fig. 2.8. In submicron technologies, the typical LET threshold is of order $1 \text{ MeV} \cdot \text{cm}^2/\text{mg}$ pretty much regardless of

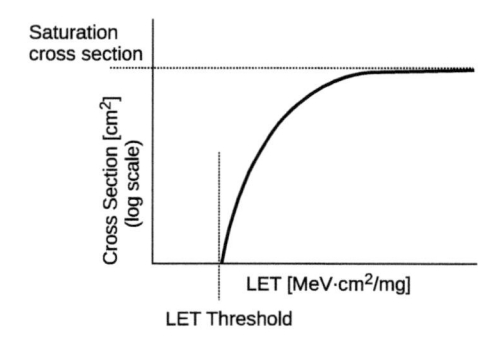

Figure 2.8. Conceptual plot of Single Event Upset (SEU) cross-section vs. Linear Energy Transfer (LET) for a typical memory cell.

memory cell type. Saturation cross-section varies with cell design but is of order 10^{-7}–10^{-8} cm² for common SRAM, latches, and flip-flops. However, an energetic proton (or pion) has an LET of order 0.01 MeV·cm²/mg, which immediately signals that it cannot upset memory cells by the same energy loss mechanism as ions (it is far below the LET threshold). Upsets in this case are due to nuclear interactions. This can be seen from the fact that SEU cross-sections are about the same for energetic neutrons and protons [14]. There is thus a kinematic threshold depending on the nuclei in the material rather than an LET threshold. Typical SEU cross-sections for protons are of order 10^{-13} cm² [14]. At relatively low energies, an adequate model has the proton imparting momentum to a nucleus which then becomes a traditional high LET heavy ion. But at the GeV energies of the LHC, inelastic collisions can produce showers of high LET particles, affecting a large area of silicon. This is important for hardening techniques.

2.8 System Aspects Overview

The mechanical and electrical system designs of a tracker are what ultimately limits its practical performance, not only due to cooling but also due to material that particles must pass through and to the alignment precision achieved. Achieving low mass, or more precisely low multiple scattering, is critical for tracker and heavily influences the system design. The multiple scattering detector mass is quantified in terms of electromagnetic radiation interaction length: the mean distance over which a high-energy electron loses all but $1/e$ of its energy by bremsstrahlung. Hybrid trackers achieve masses as low as 1–2% of a radiation length per layer, while MAPS trackers can be as low as 0.3–0.5%. One percent of a radiation length corresponds to a thickness of about 1 mm of either silicon or aluminum, or 0.16 mm of copper, or 3 mm of carbon fiber composites. This should convey the difficulty of producing detectors with 1% of a radiation length per layer or less, including micron-level mechanical precision mechanical structures, power cables, cooling, and readout. Simply in terms of energetics, the most power intensive hybrid systems can dissipate over 0.5 W/cm², which for a 10 m² tracker means 50 kW, yet this same tracker must operate at $-10°$C to control sensor leakage current and readout integrated circuits radiation damage. Given almost zero thermal mass, an exquisite balance must be maintained between 50 kW in through electrical wires and 50 kW out though circulating coolant. It is therefore not surprising that trackers make use of the highest performance to mass ratio materials and systems available.

All modern tracker structures are built of carbon composites. Composites allow choosing structural properties different from the intrinsic properties of individual bulk materials, such as thermal expansion coefficient or thermal conductivity. Furthermore, properties can be chosen to be different in different directions. Tracker structures have a near-zero thermal expansion coefficient so that trackers can be assembled at room temperature and maintain micron-level precision at their operating temperature 50°C colder.

Cooling, electrical power delivery, and data transmission are all system aspects that require custom solutions to meet tracker low mass requirements. Additionally, these elements must be radiation hard to different degrees, but this configuration is determined by the mass requirement. The technology of evaporative CO_2 cooling was developed for HEP trackers, as the first application that absolutely needed the extreme performance it offers. This is based on a typical two-phase refrigeration cycle, where the refrigerant is condensed and cooled outside of the active volume and evaporated though a pressure change inside the active volume. Heat is removed mainly through the latent heat of evaporation of the refrigerant. Liquid CO_2 has almost twice the density of water, which means a large mass flow can circulate in a small pipe, and more than twice the latent heat of evaporation as typical refrigerants. It has one significant challenge, however, which is that it only has a liquid phase at pressures above around 50 bar (depending on temperature). The implementation of systems that pumps cold, high-pressure liquid to the detector and extracts high pressure gas to be condensed outside, has allowed building of large low mass trackers with significant power consumption, not possible with any other cooling technology. The same cooling performance achieved with few mm diameter CO_2 pipes would require cm diameter conventional refrigerant pipes and several cm diameter pipes for simple water (single phase) cooling.

Just as important as removing the heat is supplying all that electrical power with low mass. Power supply to detector distance scales tend to be of order 50–100 m if nothing else because off-the-shelf electronics cannot be sited too close to a particle accelerator. State-of-the-art integrated circuit electronics operate at voltages below 2 V. Supplying 50 kW from 100 m away with cables that do not themselves produce more than 50 kW of heat due to their resistance (for a total wall power of 100 kW instead of 50 kW) would require 180 kg of copper per meter of cable length. Clearly, not good for making a low mass tracker. The well-known solution for delivering electrical power without massive cables, present everywhere around us in our power grid, is to do so at high voltage. The problem for trackers then becomes one of transforming "high" to low (under 2 V) voltage without significant mass, in a strong magnetic field, and with radiation tolerance. Two technologies have been developed: radiation hard, low mass DC-DC converters that operate in a high magnetic field and serial power. The former works with 10–12 V input and 1.5–2.5 V output, multiplying the input current by the voltage ratio. DC-DC conversion is adequate for relatively low power density detectors, such as hybrid strips, as it adds an amount of mass proportional to the power density (twice as much power in the same area needs twice as many DC-DC converters and therefore twice as much mass). Serial power can be implemented with readout integrated circuits designed to operate from a constant current rather than constant voltage power supply. Such devices can be chained in series and with a special type of power regulation [15] also placed in parallel. The total voltage drop of a serial chain is $N_d V_d$, where V_d is the voltage across a single device (of order 2 V). The mass of N_d devices is the same whether they are chained in series or powered in parallel, so serial power adds no mass to the tracker, making it suitable for high power density trackers, such as hybrid pixels. But serial power does add complexity in terms of operation, communication, and sensor bias. Values of N_d as high as 14 are being used in current trackers. Both DC-DC and serial technologies achieve an efficiency of 70–80%, which means an increase of the heat to be removed by the cooling system by 25%. This is acceptable thanks to the high capacity of CO_2 cooling.

Once power delivery and cooling mass have been minimized with the above techniques, the most significant mass contributions can be due to readout cables. While commercial high speed links can move vast data volumes with compact formats, a tracker presents a different readout problem, not well matched to the commercial solutions. Aside from radiation tolerance, a tracker contains a large number of data sources spread throughout a m^3 volume or larger. This is different from the commercial problem of a concentrated data source such as a processor. A 100 Gbps point-to-point link is not very useful for reading a tracker because that volume of data is produced by order 100 different devices over a 1 m length scale. There isn't a common solution to this problem: each detector finds a different optimization, depending of the data to be moved and geometrical constraints. Common components have been developed, notably a radiation hard, low mass optical data link capable of 10 Gbps. Custom electrical cables, including flexible circuit strip lines, aluminum conductors, and twin-axial transmission lines, are commonly produced for trackers.

References

[1] R. L. Workman et al. Review of particle physics. *PTEP*, 2022:083C01, 2022. Section 34: Passage of particles through matter. pdg.lbl.gov.

[2] M. Aleksa et al. Rate effects in high-resolution drift chambers. 446(3):435–443, 2000.

[3] W. Blum and L. Rolandi. *Particle Detection with Drift Chambers*. Springer-Verlag, 1993.

[4] W. Snoeys, J. Plummer, S. Parker, and C. Kenney. PIN detector arrays and integrated readout circuitry on high resistivity float zone silicon. 41:903–912, 1994.

[5] S. M. Sze, Y. Li, and K. K. Ng. *Physics of Semiconductor Devices*. John Wiley, 1969, 1981, 2007, 2021.

[6] P. A. M. Dirac. A theory of electrons and protons. *Proc. R. Soc. Lond. A*, 126(801):360–365, 1930.

[7] RD50 Collaboration. http://rd50.web.cern.ch/.

[8] M. Garcia-Sciveres and X. Wang. Data encoding efficiency in binary strip detector readout. 9(04):P04021, 2014.

[9] M. Garcia-Sciveres and X. Wang. Data encoding efficiency in pixel detector readout with charge information. 815:18–22, 2016.

[10] H. Spieler. *Semiconductor Detector Systems*, Vol. 12. Oxford University Press, Oxford, 2005.

[11] R. L. Gluckstern. Uncertainties in track momentum and direction, due to multiple scattering and measurement errors. 24:381, 1963.

[12] R. E. Kalman. A new approach to linear filtering and prediction problems. *Transactions of the ASME–Journal of Basic Engineering*, 82(Ser. D):35–45, 1960.

[13] R. Radu et al. Investigation of point and extended defects in electron irradiated silicon–dependence on the particle energy. *Journal of Applied Physics*, 117:164503, 2015.

[14] P. Roche et al. A commercial 65 nm CMOS technology for space Applications: Heavy ion, proton and gamma test results and modeling. 57(4):2079–2088, 2010.

[15] M. Karagounis et al. An integrated shunt-LDO regulator for serial powered systems. In *Proceedings of ESSCIRC '09*, 2009.

Chapter 3

Radiation Damage to Organic Scintillators

Sarah Eno

University of Maryland, Department of Physics,
College Park, MD 20782, USA
eno@umd.edu

Plastic scintillator is used in a wide variety of particle physics detectors due to its versatility and low cost. However, exposure to ionizing radiation decreases its light output. In this chapter, the chemistry and light emission mechanisms of plastic scintillators are reviewed. The causes and symptoms of its radiation damage are discussed.

3.1 Introduction

Organic scintillators are inexpensive, versatile materials that produce light when transversed by charged particles. In their liquid form, they are used extensively in neutrino experiments. As plastics, they are used in many collider experiment subdetectors [1], such as calorimeters, trackers, and beam luminosity monitors. Figure 3.1 shows some plastic scintillators, both in solid and fiber form, similar to those used in the calorimeter of the CMS detector [2–4] at the Large Hadron Collider.

When considering their use in detector elements in current and future hadron collider detectors, especially those close to the beamline, the intense radiation environment must be considered. Liquid scintillators generally continue to perform well even when subjected to large radiation doses [5–7], but flammability and containment issues limit their use. Plastic scintillators, on the other hand, are subject to significant radiation damage. For doses that are not too large (less than $\approx$100 kGy [8]), their light output

---

2024 © The Author(s). This is an Open Access chapter published by World Scientific Publishing Company, licensed under the terms of the Creative Commons Attribution 4.0 International License (CC BY 4.0). https://doi.org/10.1142/9789819801107_0003

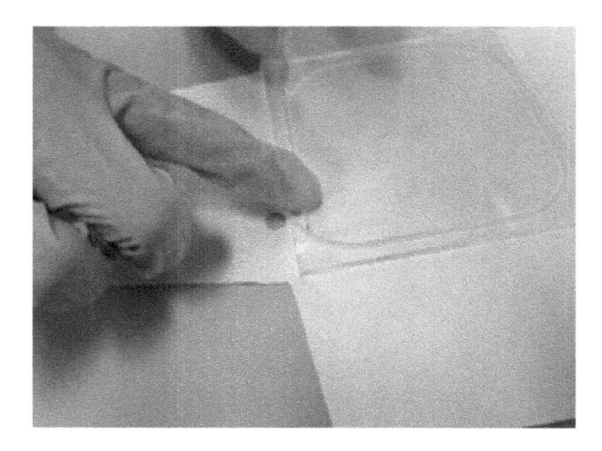

Figure 3.1. Plastic scintillating tiles with embedded wavelength shifting fiber similar to those used in the barrel, endcap, and outer hadron calorimeters of the CMS detector at the LHC. The photo shows the fiber being installed in a machined groove.

decreases approximately exponentially with the received dose

$$L(d) = L_0 \exp(-d/D) \qquad (3.1)$$

where $L(d)$ is the signal after irradiation to a dose d, L_0 is the signal before irradiation, and D is the "dose constant," a numeric parameter whose value depends on the scintillator geometry, the specific scintillator used, environmental factors, and on the dose rate $\frac{d}{dt}(d) \equiv \mathcal{R}$. Larger values of D correspond to greater radiation tolerance. Values of D for common commercial plastic scintillators based on polystyrene (PS) or polyvinyl toluene (PVT) at dose rates typical of current collider experiments at the LHC are on order of tens of kGy [8, 9]. Figure 3.2 shows a dose map for the proposed future collider FCC-hh [10], a proton–proton collider with a center-of-mass energy of 100 TeV. The inner regions of the calorimeters experience doses in significant excess of this value.

The purpose of this chapter is to describe the causes and results of radiation damage to plastic scintillators. In Section 3.2, I describe in greater detail the effects on radiation on plastic scintillators. In Section 3.3, I give an elementary review of the chemistry needed to understand the damaging interactions of ionizing particles with polymers. In Section 3.4, I give a more detailed chemical description of the polymers that are the basis of these materials. In Section 3.5, I discuss why these molecules scintillate.

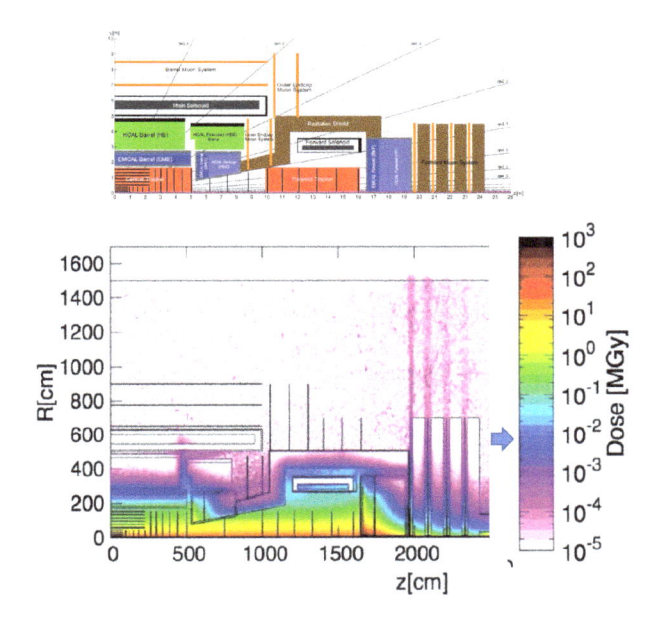

Figure 3.2. Schematic of a proposed detector for FCC-hh (top) and dose map (bottom) for 30 ab^{-1} of proton–proton collisions at a center-of-mass energy of 100 TeV from Ref. [10]. The blue arrow indicates the dose corresponding to significant damage to plastic scintillator.

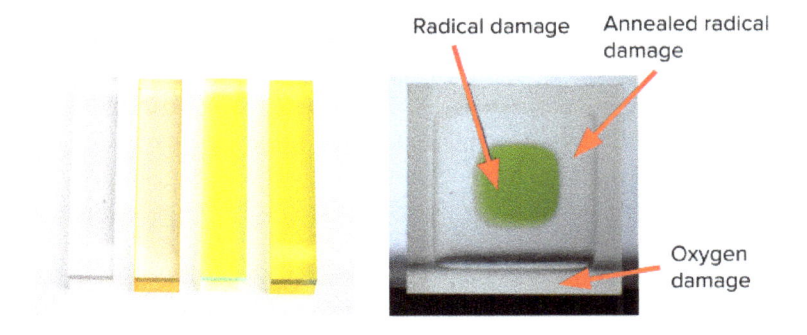

Figure 3.3. (left) From Ref. [8], plastic scintillators before and after irradiation. From left to right: Unirradiated EJ-200, EJ-200 irradiated to 500 kGy at 11 kGy/hr, unirradiated EJ-260, and EJ-260 irradiated to 500 kGy at 11 kGy/hr. (right) A piece of EJ200, a plastic scintillator from Eljen Corp. with a polystyrene matrix, that was irradiated to 70 kGy at 460 Gy/hr, 15 days after the end of the irradiation.

In Section 3.6, I discuss details of the chemistry of the interactions of charged particles with polymers. In Section 3.7, I discuss the special role oxygen plays in these interactions. In Section 3.8, I discuss some promising avenues for mitigating the damage, and finally in Section 3.9, I give some concluding remarks.

3.2 Radiation Damage Phenomenology

In this section, the basic symptoms of radiation damage are described qualitatively. In the following sections, the chemistry behind these behaviors and a more quantitative description are given.

Besides the exponential dependence of the light output on dose mentioned in Section 3.1, radiation damage in plastic scintillators can produce several other symptoms. First, as shown in Fig. 3.3 (left), while before irradiation the plastic scintillators shown (from Eljen Technology company[1]) are somewhat transparent, allowing light to easily pass to the photodetector. Afterwards, the plastic can become discolored or "yellowed." This is the same effect seen when common household plastics are left out in the sun. After the end of irradiation, the discoloration can lessen, a process called "annealing." During the annealing process, the plastic clears starting at the edges, moving toward the center. At room temperature in standard atmosphere for common commerical plastic scintillators based on PS and PVT that are a few mm thick, the annealing timescale is on the order of weeks.

When the decrease in light output is measured, two types of effects are noted. In general, both types of damage are seen. If a long piece of scintillator is irradiated, and a photodetector put at one end, the amount of detected light varies with the position of the transversing charged particle as shown in Fig. 3.4. There can be a part of the light loss that is independent of position, which will be called "initial light loss."

There is in general also a part that depends on the path length to the photodetector that will be called "color center formation." (The origin of this nomenclature is unclear, but perhaps the originator was thinking of small defects which absorb specific wavelengths or colors of light. The modern description is given in Section 3.6.) Note that some color centers exist before irradiation as well, as the material is not perfectly transparent.

---

[1] Eljen Technology, 1300 W. Broadway, Sweetwater, Texas 79556, United States.

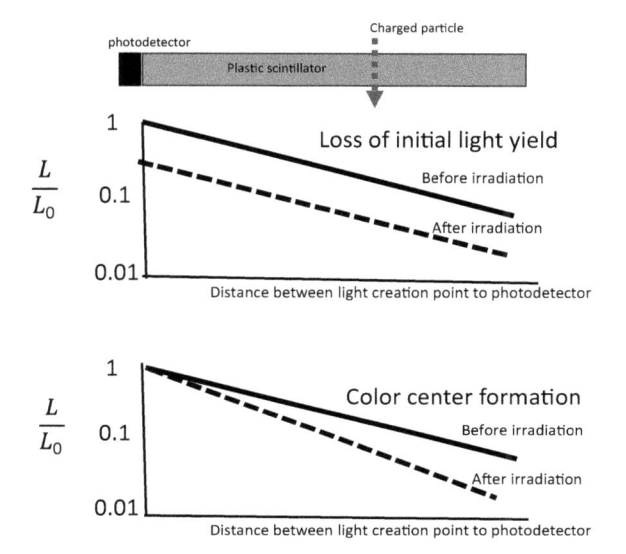

Figure 3.4. Illustration of the two fundamental types of light losses. The graphs show, as a function of the distance between the transversing charged particle, and the photodetector, the measured light L divided by the light output when the particle is very close to the photodetector, both before (solid) and after (dashed) irradiation. The top graph shows changes to the initial light loss, while the bottom shows color center formation. Typically, both kinds of damage occur.

For this second type, the detected light is an exponential function of the path length

$$L(x) = L_o e^{-x/\lambda} \qquad (3.2)$$

where $L(x)$ is the light output when the ionizing radiation is a distance x from the photodetector, L_o is the light measured for $x = 0$, and λ is the "attenuation length." The attenuation length is related to the number density of color centers μ that absorb light:

$$\lambda = \frac{1}{\mu \cdot \sigma_a} \qquad (3.3)$$

where σ_a is the light absorption cross-section of an individual color center. The cross-section σ_a has a strong wavelength dependence and generally is larger for shorter wavelengths. For ultraviolet light, the plastic is essentially opaque.

Radiation damage in plastic scintillators shows dose rate effects. Figure 3.5 (left) shows a plot from Ref. [9] of the light output from

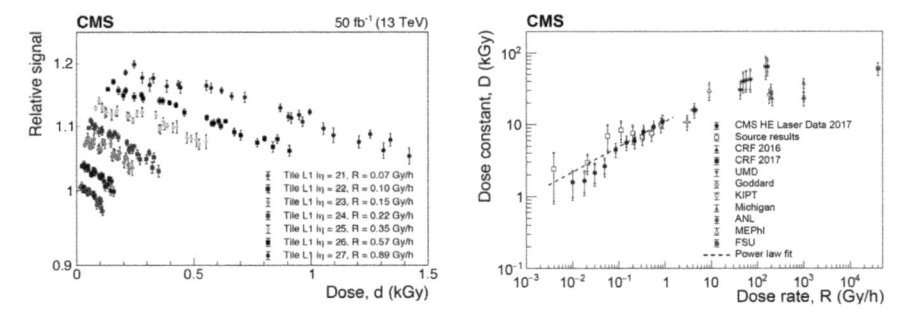

Figure 3.5. Figures from Ref. [9] on the light output from scintillating tiles with wavelength-shifting embedded fiber from the CMS endcap calorimeter. (left) Light output versus dose for tiles in different dose-rate (R) regions. (right) Fitted dose constant D versus dose rate from the CMS data, along with some measurements at high dose rate from ^{60}Co irradiations.

scintillating tiles with embedded wavelength shifting fiber in the CMS endcap calorimeter (similar to those shown in Fig. 3.1) versus dose for tiles at different distances from the beamline and thus in different dose rate environments. Tiles with lower dose rate show a larger decrease in light with dose than those at higher dose rate. This generally surprises those not familiar with plastics, as a naive guess might be that high dose rates would be more damaging. The reason low dose rates are more damaging is discussed in Section 3.7. A summary plot of the dose constant D versus dose rate is shown in Fig. 3.5 (right). The dose constant D exhibits a power-law behavior in dose rate, with an exponent close to 0.5. For both figures, the damage is to both the scintillator and the wavelength shifting fiber. Similar figures can also be seen in Ref. [8] for scintillator by itself.

Some of the dose rate effects are visible to the eye. Figure 3.3 (right) shows a piece of scintillator 15 days after the end of a low dose rate irradiation at room temperature. Three features can be seen: the edge is slightly cloudy and was like this all during the irradiation. Inside this is another clearer zone, which started to appear after the end of irradiation and which will continue to propagate toward the center of the tile until the entire inside of the cloudy region becomes clearish (annealing). The inside shows significant discoloration. The depth of the outer slightly cloudy region depends on the dose rate. Radiation damage also affects the material's index of refraction [11]. At low enough dose rate, the entire rod is slightly cloudy all though irradiation, and no annealing occurs at room temperature. Thus, barriers to oxygen should be avoided to limit discoloration. (Note that

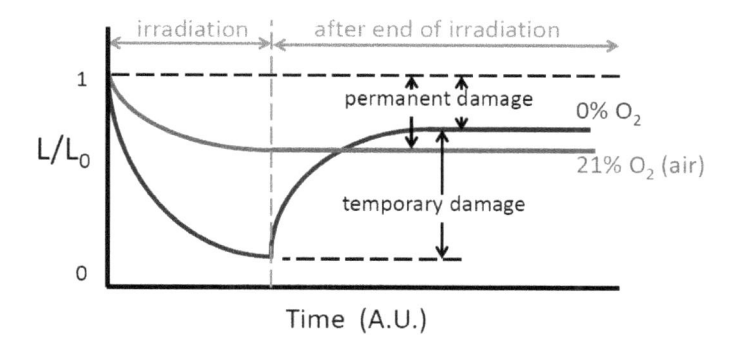

Figure 3.6. Sketch of light output versus time for a low dose-rate irradiation both during the irradiation and after the end of irradiation at room temperature.

the chemistry of this region has a strong temperature dependence, as is discussed in Section 3.7. Low temperatures should also be avoided when possible.)

Figure 3.6 shows a sketch of the light output as a function of time for a very low dose rate irradiation (where the definition of very low is given in Section 3.7) at room temperature both during irradiation and after its end for two different atmospheres: standard air and an oxygen-free atmosphere. The damage after annealing is less in the oxygen-free atmosphere, but severe damage occurs during the irradiation. The standard atmosphere at room temperature shows little annealing. The resulting damage after annealing is called "permanent damage," while the part that anneals is called "temporary damage." The exact behavior depends on the oxygen concentration and partial pressure of the atmosphere surrounding the scintillator during irradiation. It also depends strongly on the temperature [12].

Radiation damage, when severe, can also cause plastic to become brittle and a fine web of cracks ("crazing") to appear on its surface.

The causes of these behaviors have been understood by radiation chemists and are discussed in Sections 3.6 and 3.7.

3.3 Chemistry 101 for Particle Physicists

Understanding the effects of radiation on polymers requires a basic knowledge of chemistry. In this section, we remind the reader of the basic chemical notations and definitions needed to understand the subsequent sections.

There are three basic types of bonds between atoms in molecules. From strongest to weakest, they are covalent (where electrons are shared between atoms), ionic (where an electron is effectively transferred from one atom to another), and van der Waals. Covalent bonds are most relevant to plastics. A covalent bond between atoms A and B where one electron is shared is denoted A−B or A•B (for carbon, this is also called a "saturated" bond). A covalent bond where two electrons are shared is denoted A=B or A⦂B (for carbon unsaturated). A covalent bond where three electrons are shared is denoted A≡B (for carbon unsaturated). When a molecule interacts with photons during an irradiation, its electrons can be elevated to an excited state, and the resulting excited atom is denoted A^*.

The strength of a bond is an important factor in radiation chemistry, as weaker bonds are easier to break, producing "free radicals" (atoms, molecules, or molecular fragments which have one or more unpaired electrons available to form chemical bonds), and the resulting molecular fragment is denoted A. A radical created in a polymer can be a long or short, depending on the location of the break. Radicals and their chemical interactions are the main source of radiation damage to plastic scintillator. You can think of a radical as a molecule or molecular fragment that has an outer shell electron that would normally be in a covalent bond but is not because the bond was broken by the irradiation. This unbonded electron typically has strong interactions with nearby materials, especially other radicals. Note that the electron is a normal part of its molecule or molecular fragment and that a radical is electrically neutral: it is not an ion (although ions can also be produced during irradiation and are also very chemically active).

A key chemical structure in organic scintillators is the aromatic compound, which is a ring-shaped organic molecule. The simplest aromatic ring is benzene C_6H_6. The diagram for a benzene ring is shown in Fig. 3.7 (left). This diagram is also simplifed to Fig. 3.7 (right). Aromatic rings are quite stable (difficult to break their bonds), as the pi electrons (discussed in more detail in Section 3.5) distribute the energy over the whole system, so the pi electron's excitations do not concentrate energy at a particular location. They can also act as an energy sink for other parts of the molecule.

A polymer is an organic molecule built from long chains of a basic building block called a monomer. For the polymers of interest to us, each monomer will contain a benzene ring. The length can range from 1000 to 50000 carbon atoms. While crystal polymers exist, most affordable polymers are "glassy" and consist of long molecules with random bends and twists, often intertwining. As with other glasses, slow flows and other motions are

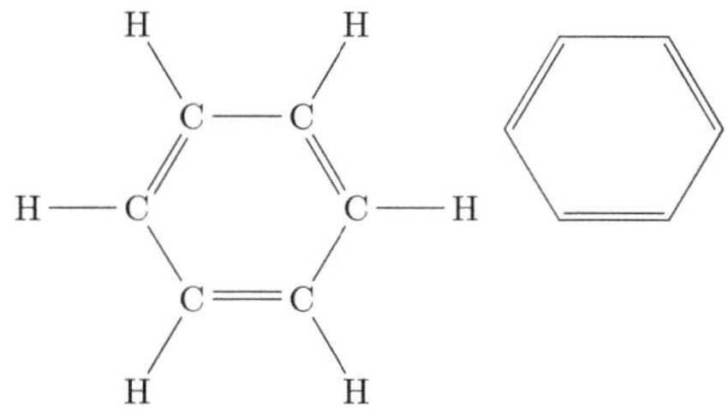

Figure 3.7. Benzene full diagram (left) and simplified (right).

possible. A key parameter in a polymer is its molecular weight, which is the sum of the atomic weights (a bit more than 1 for H, and a bit more than 12 for carbon) of its atoms and is a measure of the length of the molecule.

In articles in chemistry journals on radiation damage in plastic scintillator, the chemists will often talk about "g-values." A "g-value" is the number of molecules of reactant consumed or product formed per 100 eV of energy absorbed. Thus, it is a chemical constant needed to calculate the number of radicals produced, their types, and the rates at which chemicals such as carbon dioxide and water are produced when molecules are broken into radicals and interact, forming new chemicals ("products").

3.4 What Is a Plastic Scintillator?

Plastic scintillators are constructed from a "substrate" or "matrix" made of an aromatic polymer into which is dissolved a "primary" wavelength shifting fluorophore ("fluor" or "dopant") at a concentration of about 1–3% and a "secondary" wavelength shifting fluor with a concentration of about 0.01–0.2%. Inexpensive radiation-resistant aromatic polymers include PS and PVT. Their chemical diagrams can be seen in Figs. 3.8 and 3.9, respectively. While PS has extensive commercial uses, PVT is mostly used by nuclear and particle physicists. PVT produces about 30% more light than PS and is more resistant to radiation damage at high dose rates (in standard atmosphere and and standard (room) temperature). In addition, small amounts of other materials such as anti-oxidants can be added to improve mechanical and other properties.

Common fluors include 2,5-diphenyloxazole, p-terphenyl, 9,10-diphenylanthracene (9,10-DPA), 1,4-bis(2-methylstyryl)benzene (bis-MSB), and 1,4-bis(5-phenyl-2-oxazolyl)benzene (POPOP). The light output from PS and PVT-based scintillators typically is typically 1 photon per 100 eV of energy deposition [13], although the collected signal can be orders

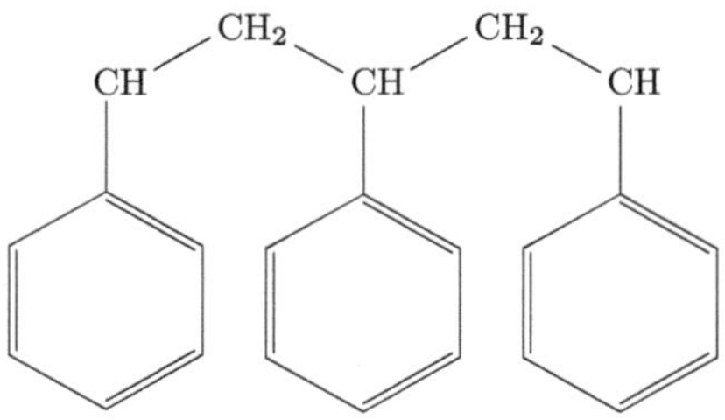

Figure 3.8. Chemical diagram of polystyrene.

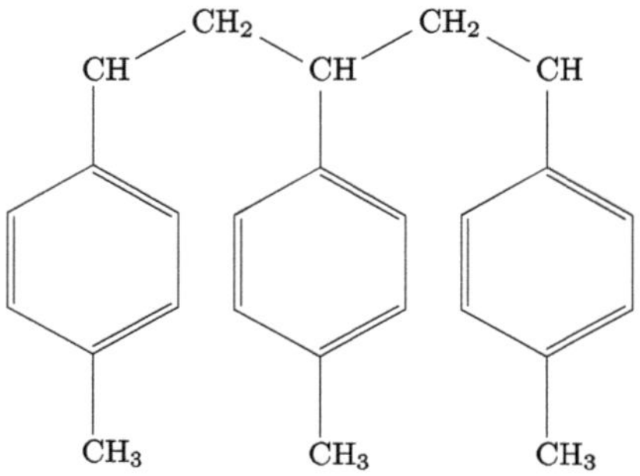

Figure 3.9. Chemical diagram of polyvinyl toluene.

of magnitude lower due to absorption in the materials, absorption upon reflection, and photodetector inefficiency. The emission wavelength of the secondary is chosen to minimize absorption in the matrix and to match the detection efficiency of the photodetector. Blue scintillators are the most popular, but green and red scintillators also exist. As the light absorption cross-section of color centers is larger at smaller wavelengths, redder scintillators have longer attenuation lengths.

An excellent detailed resource on plastic scintillators is Ref. [14]. Other useful reviews that focus on radiation damage include Refs. [15, 16].

3.5 Fundamentals of Scintillation

The primary source of both the scintillation and the radiation damage is the matrix. The processes whereby energy depositions by particles transversing bulk matter become emitted light were set out by Birks, and his 1964 book [17] remains a classic. A cartoon of the overall process is shown in Fig. 3.10.

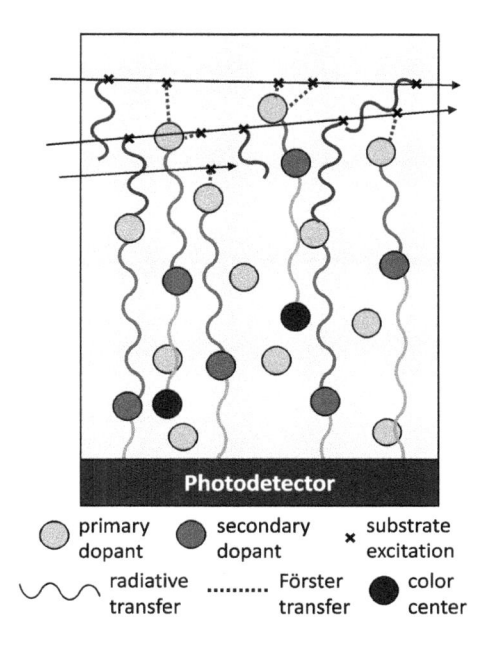

Figure 3.10. Cartoon summary of the production of light by scintillator when transversed by a charged particle. The solid lines represent the trajectory of charged particles, and the letter × indicates an interaction of a charged particle with the matrix, forming an excitation. The transfer of the excitation to the primary fluor (light grey circles) via the Förster mechanism is represented by the dashed lines, and the radiative transmission from the primary to the secondary fluor (medium grey circle) is represented by the waves. The photons emitted from the secondary (wave) travel through the matrix to the photodetector, sometimes being absorbed in their path by "color centers" (dark grey circle).

The process begins with the excitation of the matrix. The electron orbital structure of the benzene ring can be found in many standard chemistry texts, and is a key to its behavior. The p-orbital electrons in carbon form both "pi" and "sigma" bonds in the benzene molecule; the pi bonds are responsible for scintillation. Figure 3.11 (right) shows the pi energy levels. Scintillation involves the singlet (angular momentum $l = 0$) states, with decay times measured in ns. Typically, the excitation is not to the lowest energy excited state. The electron usually de-excites first to the lowest excited state and then to the ground state. Due to this, and because the energy levels shift when the molecule is excited, the emitted photon generally has longer wavelength than the absorbed. The wavelength

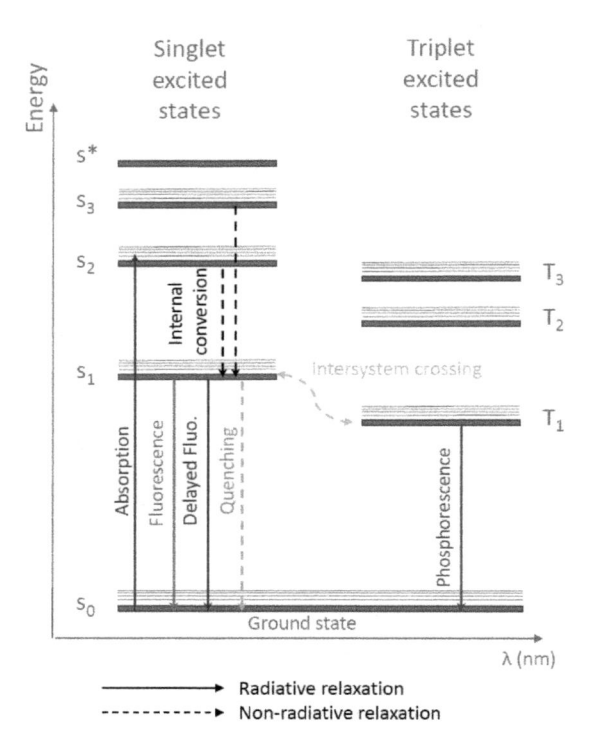

Figure 3.11. Pi orbital energy levels from Ref. [18].

difference is called the Stokes' shift. A large Stokes' shift is desirable because then the emitted photon is less likely to be re-absorbed by the matrix.

Thermal energies, collisions, and interactions between the excited molecules can transfer the excitation to the triplet (angular momentum $l = 1$) states that have significantly longer lifetimes, although these processes are rare. Photonic decays of triplet states are referred to as "phosphorescence" and produce lower-energy light emissions than the direct scintillation, as triplet states are the lowest energy excited state due to Hund's rule [19].

Photons emitted by the matrix are typically ultraviolet and are easily reabsorbed by the benzene ring. Fluors are thus introduced to shift the wavelength. At typical concentrations of the primary fluor, transfer from the matrix to the primary fluor is primarily via the Förster mechanism [20], a dipole–dipole interaction that decreases as the sixth power of the distance between molecules. To further decrease the absorption probability,

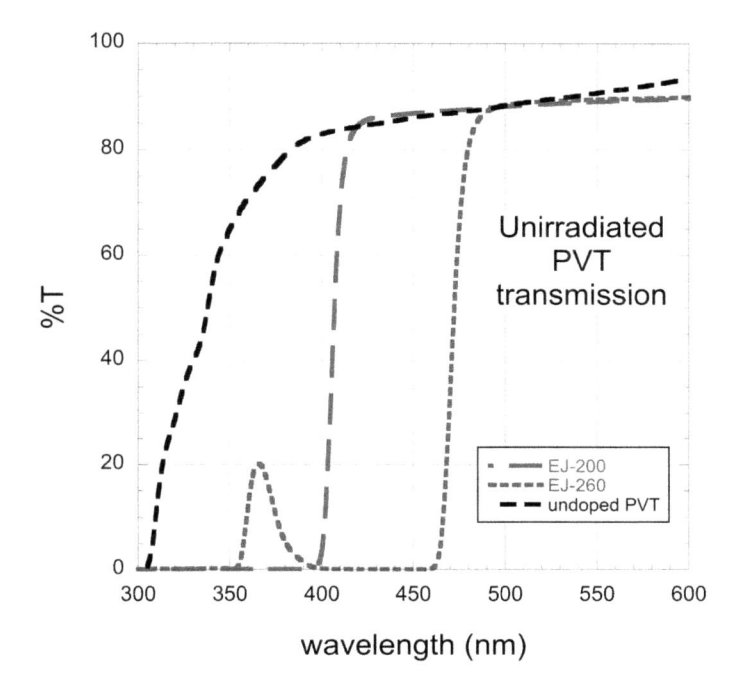

Figure 3.12. Typical transmission spectra for rods with a PVT matrix, with no fluors (black, medium dashes), EJ-260 fluors (green, short dashes), and EJ-200 fluors (blue, long dashes) from Ref. [8]. The sharp rise in transmission with wavelength moves to higher wavelength when the fluors are added because they strongly absorb at the lower wavelengths.

a secondary fluor is used to shift to even longer wavelengths. Figure 3.12, from Ref. [8], shows the transmission probability for a 1 cm thick piece of PVT versus wavelength for typical commercial scintillators with and without the fluors.

The light output of organic scintillators is not linear in the ionization density dE/dx. Tracks with large dE/dx (with slow speed and/or high charge) emit less light than expected, compared to minimum-ionizing particles. A widely used semi-empirical model by Birks posits that recombination and quenching effects between the excited molecules reduce the light yield [21]. Inter-molecular interactions can transfer excitations to the triplet states. These effects are more pronounced when the density of the excited molecules is larger. Birks' formula is

$$\frac{d\mathcal{L}(dE/dx)}{dx} = \mathcal{L}_0 \frac{dE/dx}{1 + kB\,dE/dx} \tag{3.4}$$

where dE/dx is the amount of energy deposited by a charged particle per unit path length, $\mathcal{L}$ is the resulting light output (luminescence), $\mathcal{L}_0$ is the luminescence at low specific ionization density, and the product kB is known as Birks' constant, which must be determined for each scintillator by measurement. The value of kB for polystyrene is 0.126 mm/MeV [22].

The light yield also depends on temperature. The temperature dependence of the refractive index and light yield was studied over a wide range of temperature in Ref. [23]. They find a 10/°C temperature rise would correspond to a 2% light yield loss.

3.6 Basics of Radiation Damage in Polymers

The effects of radiation on polymers are complex but were well studied in the 1950s. Excellent, detailed books [24, 25] exist that are well worth reading for anybody working with these materials. Parts of this section are an abbreviation of the material found in Spinks and Woods [24].

The start of the chain of processes that lead to radiation damage in plastic scintillator is the formation of free radicals. In general, radicals are formed when a bond in a molecule breaks, which can occur through thermal dissociation, photodissociation (via interaction with a photon, typical during irradiation), and oxidation–reduction processes. The photon energies required to break several relevant bonds can be found in Table A3.3 of the 1990 edition of Ref. [24]. When the energy of the electromagnetic interaction of the charged particle with the molecule is not enough to create a radical at the absorption bond, the excitation can move along the molecule or to another molecule until it finds a bond with low enough energy and create a radical there. Radicals can interact with the polymer, as is discussed. As the radicals are formed along the path of the charged particle, their density can be large, leading to radical–radical interactions in addition to radical–matrix interactions. Note that the triplet states shown in Fig. 3.11, with their parallel-spin electrons, act as di-radicals. Due to this and their long life time (10^{-4}–10^{-3} s), they can influence radiation chemistry as well.

While the chemistry behind the creation of radicals is fairly straightforward, the interactions of radicals with each other and with the polymer are complex. Here we give a sense of the kinds of interaction mechanisms that can influence how radiation-resistant a particular matrix will be and introduce some of the vocabulary needed to understand articles in chemistry journals. References [24, 25] give a more complete description. Once a

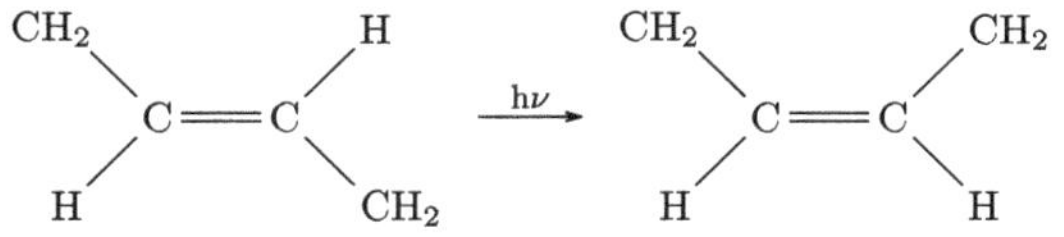

Figure 3.13. An example of a unimolecular reaction, involving an initial configuration with the hydrogens on opposite sides of the molecule changing to one where they are on the same side.

radical is created, it has a number of available interaction mechanisms: rearrangement, disassociation, addition reactions, abstraction reactions, radical combination, and disproportination. The first two of these are unimolecular reactions—where an excited polyatomic molecule reaches a stable stage through molecular rearrangement. An example is shown in Fig. 3.13. Unimolecular reactions include the breaking of a polymer into two molecules, with radicals at the breakage point. The rest of the interaction mechanisms are bimolecular. Bimolecular reactions involve a second molecule to form new chemical products and so depends on the radical density. Examples of bimolecular interactions are electron transfer, abstraction, addition, and Sterm–Volmer reactions. Addition and abstraction include a radical among the products, while combination, disproportionation, and electron transfer do not (they instead "terminate" the radical).

Radicals can move within the molecule, and they can propagate through the matrix, sometimes passing through a large number of molecules. The basic propagation mechanisms are electron transfer (producing two ions), abstraction (usually hydrogen abstraction) $A^* + B-C \longrightarrow A-B\cdot + C\cdot$, and addition $R\cdot + AB \longrightarrow R-AB\cdot$. An example of an addition diagram is shown in Fig. 3.14. As radicals propagate through a material, they also tend to become less reactive.

Due to their high reactivity, radicals typically are short-lived, forming new bonds quickly, via reactions like $R\cdot + S\cdot \longrightarrow RS$, sometimes reforming the bond that was originally broken. However, some long-lived radicals exist. An example of a long-lived radical is shown in Fig. 3.15. Due to the energy released via bond creation, small molecules typically immediately break back apart again, but larger molecules, like polymers, have other ways to distribute the energy and the bonds can hold.

The probability/timescale of rebonding ("annealing") depends on the radical's reactivity and selectivity, and strongly on temperature. The reactivity of a radical depends on several factors. The stability of organic

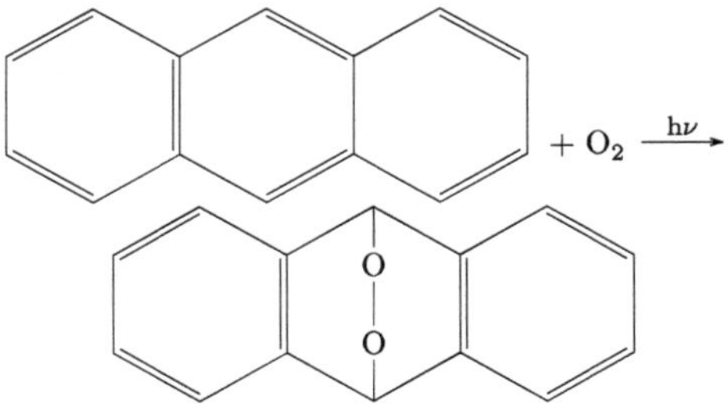

Figure 3.14. Addition to anthracene.

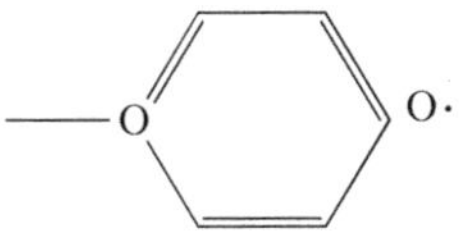

Figure 3.15. Semiquinone, a long-lived radical.

radicals is increased when a hydrogen attached to the carbon atom carrying the free electron is replaced by any other atom or group so that $CH_2\cdot$ is less stable then $=C\cdot$, which is less stable than $\equiv C\cdot$. Typical highly reactive radicals are $H\cdot$ and $\cdot OH$. Radical reactions tend to become less selective at higher temperatures because the increased motion increases the probability of encountering another radical or bonding site and interacting, forming bonds. Usually, chemical processes occur faster at higher temperature. Thus, annealing and radical migration have a strong temperature dependence.

Molecules and molecular fragments containing radicals have undesirable optical properties. They tend to strongly absorb visible light, especially at shorter (bluer) wavelengths. Even when they re-bond, the products can have undesirable optical properties: sometimes reformed bonds will absorb light of the relevant wavelengths; sometimes they will not. Ideally, the radical will reconnect to its original position. However, this is not always the case. Figure 3.16 shows some non-standard radical terminations with and without oxygen (the ones with oxygen are discussed in Section 3.7). Rebonding to the original position is more likely in liquids, where the molecules can be constrained from movement due to the "Franck-Rabinowtich" or "cage" effect [26].

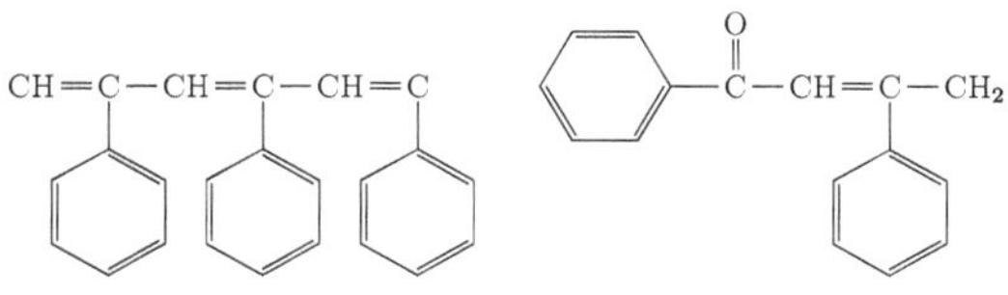

CH=C—CH=C—CH=C (with benzene rings)   —C(=O)—CH=C—CH$_2$ (with benzene rings)

Figure 3.16. Examples of changes to polystyrene undergoing irradiation. The change on the right can only occur in the presence of oxygen.

During irradiation, light can be absorbed by the radicals present in the material and by the new bad bonds. At the end of irradiation, the radicals continue to slowly rebond. Since many of the reformed bonds do not absorb the wavelengths of interest, the material becomes clearer. Thus, the observed strong initial discoloration after irradiation is due to unbonded radicals in the plastic, the annealing due to re-bonding. If the new bond interferes with the scintillation of the benzene ring, the transfer of its excitation to the primary fluor, or the transfer from the primary to the secondary, the initial light output decreases. If it absorbs in the emission wavelengths of the secondary fluor, it decreases the absorption length.

The reformed bonds in polymers contribute to two types of long-term physical damage. Irradiation changes the molecular weight via crosslinking and degradation. Crosslinking is when two separate long chain molecules become linked together into a single molecule. Crosslinking increases the materials viscosity and melting point. Degradation is the opposite: breaking a molecule into two molecules of lower weight. For PS and PVT, crosslinking dominates in oxygen-free environments, degradation when oxygen is present (more on the very important role of oxygen in all the chemistry of radiation damage in Section 3.7). Crosslinking can cause swelling. When the crosslinking is large enough that a molecule becomes crosslinked with itself via a chain of other crosslinked molecules, it is called a "gel" [25]. Although crosslinking is a dominant radiation damage mechanism in PS without oxygen, PS is one of the most radiation resistant polymers because the energy required to form a crosslink is about 100 times greater than for other molecules [25]. The benzene ring absorbs energy, reducing link formation.

Quantitatively, the rate of radical production during irradiation in an oxygen-free environment goes as [27]

$$\frac{d[Y]}{dt} = gQ\mathcal{R} - k[Y]^2 \qquad (3.5)$$

where $[Y]$ is the density of radicals, g is the radiation-chemical yield, Q is the scintillator density, and k is the reaction constant for the decay (which to a chemist means termination) of the radical. The first term represents the creation of radicals, while the second term represents two unterminated radicals recombining to neutralize ("second-order" termination). At short times, when the second term is small compared to the first, integration yields a radical density that is proportional to dose: $Y = gQd$. If the radical is not neutralized by, e.g., oxidation, then the second term grows with time, and eventually, a steady state is reached, when the two terms are equal. In this case, the radical density becomes constant with time and is no longer proportional to dose, and deviations from the postulated exponential behavior of Eq. (3.1) occur. When in this steady state, the number of radicals produced per unit energy absorbed is a nonlinear function of the dose rate. For a very simple propagation and termination chain, the yield goes as the inverse of the square root of the dose rate [24].

When the first term dominates, Eq. (3.2) becomes

$$L(d) = L_0 \exp(-gQd\sigma l) \tag{3.6}$$

where σ is the cross-section for absorption of light by the radicals and l is the light's path length through the scintillator to the photodetector, recovering Eq. (3.1).

3.7 Oxygen and Radiation Damage

The presence of oxygen inside a polymer completely changes the chemistry of its interactions with irradiation. Although plastic scintillators may look like impermeable solids to us, they actually contain gaps and voids into which gases can penetrate. For unirradiated plastic, the amount of gas present in a polymer is given by Henry's law

$$c = Sp \tag{3.7}$$

where c is the concentration of the gas in the polymer (usually quantified with pseudo-mass unit cm^3STP which corresponds to the mass of 1 cm^3 of gas at standard temperature and pressure STP), p is the partial pressure of the penetrant at the interface (usual unit is cmHg), and S is the solubility coefficient, also known as the Henry coefficient (usual unit $\frac{cm^3 STP}{cm^3 \cdot cmHg}$). The solubility has a strong temperature dependence, given by a Van't Hoff-type equation [28]:

$$S = S_0 e^{-\Delta H_s/(RT)} \tag{3.8}$$

where ΔH_s is the molar heat (enthalpy), R is the universal gas constant, and T is the temperature in degrees Kelvin. The author of this chapter only knows of one measurement of the solubility of PS [29] (8.9×10^{-6} cm^3 STP/cm^3/cmHg at 22° C) and none of PVT.

The reason oxygen is so important to radiation chemistry is that it readily adds to free radicals. Oxygen has a triplet ground state and therefore acts as a diradical. The chemical interactions of oxygen with polymers during irradiation have been carefully studied by chemists at Sandia National Laboratories, and papers with Clough or Gillen in the author list, such as Refs. [29–35], are well worth reading. (Papers by the Hamburg group are also very educational, for example Refs. [27, 36–40].) The Sandia group, though, was mainly interested in how oxidation affects the mechanical properties of scintillator. However, recent work [8] indicates the importance to its optical properties as well. A really excellent resource on studies of radiation damage in fibers at different oxygen concentrations is Ref. [27]. It is worth reading several times.

A typical reaction, producing the relatively stable peroxy radical, is R· + O$_2$ ⟶ R−O−O· Since this radical is relatively stable, it can linger for a long time after irradiation, acting as a color center. Another oxygen reaction involving hydrogen (a component of our plastics) is RO$_2$· + RH ⟶ RO$_2$H≡R··. These reactions lead to autooxidation of a wide variety of organic compounds. These reactions compete with crosslinking, explaining why degradation dominates in samples containing oxygen.

When an irradiation starts, the produced radicals will quickly interact with the oxygen currently in the sample, producing peroxides. As the oxygen in consumed, more oxygen will come in from the outside. If oxygen can enter fast enough to replenish that consumed, the plastic becomes cloudy due to the peroxides but does not develop the strong color typical of unterminated radicals. However, if the density of radicals is high enough, the oxygen will all be consumed before reaching the radicals in the center, and the cloudy region with peroxides is only on the edge, as seen in Fig. 3.3 (right), where the center is oxygen-free and unterminated radicals produce the green color. Note that this is a steady state condition: the position of the edge between the regions with and without oxygen becomes stable during the irradiation and does not change with time. The penetration depth depends on the radical density and thus on the dose rate, leading to dose rate effects. Quantitatively, the steady state penetration depth z_0 for oxygen into a rectangular slab of plastic is [41]

$$z_0^2 = \frac{2 M C_0}{\Upsilon \mathcal{R}} = \frac{2 M S P}{\Upsilon \mathcal{R}} = \frac{\gamma}{\mathcal{R}} \tag{3.9}$$

where M is the diffusion coefficient for oxygen, C_0 is the oxygen concentration at the matrix's surface on the matrix side, Υ $(= gQ)$ is the specific rate constant of active site formation, S is the oxygen solubility, and P is the external oxygen pressure. (In general, these parameters depend on strongly on temperature [42]. The diffusion constant for PS has been measured at room temperature in Ref. [29] and at several temperatures in Ref. [43].) The values in the literature for the constant γ at room temperature for PS vary considerably. Ref. [37] gives $99 \pm 10 \, \text{mm}^2 \cdot \text{Gy/hr}$ while Ref. [33] gives 400–$500 \, \text{mm}^2 \cdot \text{Gy/hr}$. Currently, this discrepancy is not understood. At the depth z_0, there is an abrupt transition between areas with and without oxygen. The oxygen concentration in the oxidized regions is almost uniform [33]. At low enough dose rates at room temperature, oxygen permeates the entire sample during irradiation. Since it binds quickly to any produced radicals, there will be no remaining radicals at the end of irradiation and no recovery. However, note that whether or not this occurs strongly depends on the temperature.

The diffusion of oxygen into the plastic after irradiation is the primary source of the observed annealing in normal air.[2] Thus, in Fig. 3.3 (right), the slightly cloudy part marks the oxygen penetration during irradiation, the clearer part the diffusion in the 15 days after irradiation, and the green part the portion oxygen has not yet reached.

During irradiation, the rate of polymer oxidation (oxygen consumption) is [29–31, 35, 44]

$$\frac{dC(x,t)}{dt} = -\frac{C_1\, C(x,t)}{1 + C_2\, C(x,t)} \tag{3.10}$$

where x is the depth relative to the surface of the material where the rate is being measured and $C(x,t)$ is the position-dependent concentration of oxygen within the matrix. The constants C_1 and C_2 depend on the kinematics of the chemical reactions. The constant C_1 is proportional to the square root of the dose rate for bimolecular reactions (leading to a dose-rate effect) and to the dose rate for unimolecular reactions (no dose-rate effect because integration yields a proportionality to dose).

---

[2]Note that the diffusion constant always has an exponential dependence on temperature.

3.8 Recovery of Damaged Scintillators

Recent research indicates that it may be possible to use visible light (from LEDs) to undo some of the permanent damage to scintillators. The first work is documented in Refs. [36, 38]. Recent work includes Refs. [45, 46]. Reference [46] studies the effect on soda lime glass and found up to 50% recovery. Ref. [45] studies EJ-260 after a high dose rate exposure to about 100 kGy but found negligible recovery. The effects could be related to breaking bad bonds, which then reform into better bonds. It could also be related to allowing long-lived radicals, ions, or electrons trapped in an energy well to escape. Much remains to be learned about the chemistry of this promising process.

3.9 Concluding Remarks

Plastic scintillators are useful, inexpensive, components of particle physics detectors. Care does need to be taken when using them in a high radiation environment, especially at low temperature. Damage is larger for a given dose at low dose rate, and the damage can have a characteristic depth profile, especially at high dose rates. In addition, the kinds and amounts of damage depend on the atmosphere, pressure, and scintillator geometry. While in principal the basic chemistry is known, currently many of the relevant constants have not been measured and can vary with the exact scintillator composition. In addition, the composition of scintillators is often information proprietary to the manufacturer. Thus, careful measurements at the relevant dose rate and atmosphere should be done to be sure the damage is not underestimated. There is ongoing research on how to reverse radiation damage that may extend the use of these materials to higher doses in the future.

Acknowledgments

The author wants to thank Tim Edberg and Christos Papageorgakis for providing figures.

References

[1] Y. Kharzheev. Scintillation detectors in modern high energy physics experiments and prospect of their use in future experiments. *Journal of Lasers, Optics & Photonics*, 4:1, 2017. DOI: 10.4172/2469-410X.1000148.

[2] S. Abdullin, et al. Design, performance, and calibration of CMS hadron-barrel calorimeter wedges. *European Physical Journal C*, 55:159, 2008. DOI: 10.1140/epjc/s10052-008-0573-y.

[3] CMS. The phase-2 upgrade of the CMS barrel calorimeters. Technical Report CERN-LHCC-2017-011, 2017. https://cds.cern.ch/record/2283187?ln=en.

[4] S. Chatrchyan, et al. The CMS experiment at the CERN LHC. *JINST*, 3:S08004, 2008. DOI: 10.1088/1748-0221/3/08/S08004.

[5] C. Zorn, S. Majewski, R. Wojcik, C. Hurlbut, and W. Moser. Preliminary study of radiation damage in liquid scintillators. *IEEE Transactions on Nuclear Science*, 37(2):487–491, 1990. DOI: 10.1109/23.106666.

[6] J. Klein, J. Gresset, F. Heisel, and G. Laustriat, Effets des rayonnements sur les caracteristiques des scintillateurs organiques. *The International Journal of Applied Radiation and Isotopes*, 18(6):399–406, 1967. http://dx.doi.org/10.1016/0020-708X(67)90143-3.

[7] I. B. Berlman. The effect of massive ^{60}Co doses on the light output of a scintillator solution. Radiological Physics Division Semiannual Report for July through December 1957 ANL-5829, Argonne National Laboratory, 1958. https://www.osti.gov/biblio/4213501.

[8] C. Papageorgakis, M. Al-Sheikhly, A. Belloni, T. Edberg, S. Eno, Y. Feng, G.-Y. Jeng, A. Kahn, Y. Lai, T. McDonnell, A. Mohammed, C. Palmer, R. Perez-Gokhale, F. Ricci-Tam, Z. Yang, and Y. Yao, Dose rate effects in radiation-induced changes to phenyl-based polymeric scintillators. *Nuclear Instruments and Methods in Physics Research Section A: Accelerators, Spectrometers, Detectors and Associated Equipment*, 1042:167445, 2022. https://doi.org/10.1016/j.nima.2022.167445.

[9] CMS Collaboration. Measurements with silicon photomultipliers of dose-rate effects in the radiation damage of plastic scintillator tiles in the CMS hadron endcap calorimeter. *JINST*, 15(06):P06009–P06009, 2020. DOI: 10.1088/1748-0221/15/06/p06009.

[10] A. Abada, et al. FCC-hh: The hadron collider: Future circular collider conceptual design report volume 3. *The European Physical Journal Special Topics*, 228:755, 2019. DOI: 10.1140/epjst/e2019-900087-0.

[11] C. Papageorgakis, M. Aamir, A. Belloni, T. Edberg, S. Eno, B. Kronheim, and C. Palmer, Effects of oxygen on the optical properties of phenyl-based scintillators during irradiation and recovery. *NIMA*, 168977, 2023. https://doi.org/10.1016/j.nima.2023.168977.

[12] B. Kronheim, A. Belloni, T. Edberg, S. Eno, C. Howe, C. Palmer, C. Papageorgakis, M. Paranjpe, and S. Sriram. Reduction of light output of plastic scintillator tiles during irradiation at cold temperatures and in low-oxygen environments. *NIMA*, 1059:168922, 2024. https://doi.org/10.1016/j.nima.2023.168922.

[13] P. Zyla, et al. Review of particle physics. *PTEP*, 2020(8):083C01, 2020, 2021. DOI: 10.1093/ptep/ptaa104.

[14] M. Hamel (Ed.). *Plastic Scintillators: Chemistry and Applications*. Springer, Cham, 2000. https://doi.org/10.1007/978-3-030-73488-6.

[15] C. Zorn. Plastic and liquid organic scintillators. In: F. Sauli (Ed.), *Instrumentation in High Energy Physics*. Advanced Series on Directions in High Energy Physics, Vol. 9. World Scientific, 1992, p. 218. DOI: 10.1142/9789814360333_0004.

[16] Y. Kharzheev. Radiation hardness of scintillation detectors based on organic plastic scintillators and optical fibers. *Physics of Particles and Nuclei*, 50(1):42–76, 2019. DOI: 10.1134/S1063779619010027.

[17] J. B. Birks. *The Theory and Practice of Scintillation Counting*. International Series of Monographs on Electronics and Instrumentation, Vol. 27. Pergamon Press, The Macmillan Company, New York, 1964. DOI: 10.1016/C2013-0-01791-4.

[18] R. L. Workman, *et al.* Review of particle physics. *PTEP*, 2022:083C01, 2022. DOI: 10.1093/ptep/ptac097.

[19] W. Kutzelnigg and J. Morgan. Hund's rules. *Zeitschrift für Physik D Atoms, Molecules and Clusters*, 36:197, 1996. https://doi.org/10.1007/BF01426405.

[20] T. Förster. Zwischenmolekulare energiewanderung und fluoreszenz. *Annalen der Physik* (Berlin), 437:55, 1947. DOI: 10.1002/andp.19484370105.

[21] J. B. Birks. Scintillations from organic crystals: Specific fluorescence and relative response to different radiations. *Proceedings of the Physical Society. Section A* 64:874–877, 1951. DOI: 10.1088/0370-1298/64/10/303.

[22] B. D. Leverington, M. Anelli, P. Campana, and R. Rosellini. A 1 mm scintillating fibre tracker readout by a multi-anode photomultiplier, 2011. DOI: 10.48550/ARXIV.1106.5649.

[23] C. C. Cowles, T. C. Kaspar, R. T. Kouzes, D. Li, Z. W. Bell, I. N. Ivanov, and E. D. Sword. Temperature-dependent properties of bc-412 polyvinyl toluene scintillator. *IEEE Transactions on Nuclear Science*, 69(4):942–951, 2022. DOI: 10.1109/TNS.2022.3154645.

[24] J. Spinks and R. Woods. *An Introduction to Radiation Chemistry*. Wiley-Interscience, 1990. https://doi.org/10.1002/bbpc.19910950346.

[25] A. Charlesby. *Atomic Radiation and Polymers*. Pergamon Press, 1960. https://doi.org/10.1016/C2013-0-07861-9.

[26] E. T. Denisov, *The Cage Effect*. Springer US, Boston, 1974, p. 109. DOI: 10.1007/978-1-4684-8300-0_3.

[27] W. Busjan, K. Wick, and T. Zoufal. Shortlived absorption centers in plastic scintillators and their influence on the fluorescence light yield. *Nuclear Instruments and Methods in Physics Research Section B: Beam Interactions with Materials and Atoms*, 152:89, 1999. DOI: 10.1016/S0168-583X(98)00974-4.

[28] J. E. Mark (Ed.). *Physical Properties of Polymers Handbook*, 2nd edn. Springer, Cincinnati, 2007.

[29] Wise, K. Gillen, and R. Clough, Quantitative model for the time development of diffusion-limited oxidation profiles. *Polymer*, 38:1929–1944, 1997.

[30] S. W. Shalaby and R. L. Clough. *Radiation Effects on Polymers*. American Chemical Society Symposium Series, Vol. 475, 1991, p. 457.

[31] K. Gillen and R. Clough. Rigorous experimental confirmation of a theoretical model for diffusion-limited oxidation. *Polymer*, 38:1929, 1992. DOI: 10.1016/0032-3861(92)90280-A.
[32] K. Gillen and R. Clough. Time-temperature-dose rate superposition: A methodology for extrapolating accelerated radiation aging data to low dose-rate conditions. *Polymer Degradation and Stability*, 24:137, 1989.
[33] K. Gillen, J. Wallace, and R. Clough. Dose-rate dependence of the radiation-induced discoloration of polystyrene. *Radiation Physics and Chemistry*, 41:101, 1993. DOI: 10.1016/0969-806X(93)90046-W.
[34] R. Clough, G. M. Malone, K. Gillen, J. L. Wallace, and M. B. Sinclair. Discoloration and subsequent recovery of optical polymers exposed to ionizing radiation. *Polymer Degradation and Stability*, 49:305, 1995. DOI: 10.1016/0141-3910(95)87013-X.
[35] K. T. Gillen, J. Wise, and R. L. Clough. General solution for the basic autoxidation scheme. *Polymer Degradation and Stability*, 47:149–161, 1995. https://doi.org/10.1016/0141-3910(94)00105-H.
[36] K. Wick, T. Gosau, R. Hornung, and A. Ziegler. Strong reduction of radiation damage in plastic scintillators by illumination with visible light. In: *IEEE Symposium Conference Record Nuclear Science*, Vol. 2, 2004, pp. 766–768. DOI: 10.1109/NSSMIC.2004.1462322.
[37] K. Wick, D. Paul, P. Schröder, V. Stieber, and B. Bicken. Recovery and dose-rate dependence of radiation damage in scintillators, wavelength shifters and light guides. *Nuclear Instruments and Methods in Physics Research Section B: Beam Interactions with Materials and Atoms*, 61:472, 1991. DOI: 10.1016/0168-583X(91)95325-8.
[38] K. Wick, R. Hornung, and T. Gosau. Reduction of the permanent radiation induced absorption by illumination of plastic scintillators during Î³-irradiation. *Nuclear Instruments and Methods in Physics Research Section A: Accelerators, Spectrometers, Detectors and Associated Equipment*, 538(1):668–671, 2005. https://doi.org/10.1016/j.nima.2004.09.023.
[39] B. Bicken, U. Holm, T. Marckmann, K. Wick, and M. Rohde. Recovery and permanent radiation damage of plastic scintillators at different dose rates. *IEEE Transactions on Nuclear Science*, 38:188, 1991. DOI: 10.1109/23.289295.
[40] B. Bicken, A. Dannemann, U. Holm, T. Neumann, and K. Wick. Influence of temperature treatment on radiation stability of plastic scintillator and wave-length shifter. *IEEE Transactions on Nuclear Science*, 39:1212, 1992. DOI: 10.1109/23.173180.
[41] A. V. Cunliffe and A. Davis. Photo-oxidation of thick polymer samples — Part II: The influence of oxygen diffusion on the natural and artificial weathering of polyolefins. *Polymer Degradation and Stability* 4:17, 1982. DOI: 10.1016/0141-3910(82)90003-9.
[42] B. Wang and P. R. Ogilby. Activation barriers for oxygen diffusion in polystyrene and polycarbonate glasses: Effects of codissolved argon, helium, and nitrogen. *Canadian Journal of Chemistry*, 73(11):1831–1840, 1995. DOI: 10.1139/v95-226.

[43] B. Wang and P. R. Ogilby. Activation barriers for oxygen diffusion in polystyrene and polycarbonate glasses: Effects of codissolved argon, helium, and nitrogen. *Canadian Journal of Chemistry*, 73(11):1831–1840, 1995. DOI: 10.1139/v95-226. https://doi.org/10.1139/v95-226.

[44] J. L. Bolland. Kinetic studies in the chemistry of rubber and related materials. I. The thermal oxidation of ethyl linoleate. *Proceedings of the Royal Society A* 186:218, 1946. DOI: 10.1098/rspa.1946.0040.

[45] J. Wetzel, E. Tiras, B. Bilki, Y. Onel, and D. Winn. Using leds to stimulate the recovery of radiation damage to plastic scintillators. *Nuclear Instruments and Methods in Physics Research Section B: Beam Interactions with Materials and Atoms*, 395:13, 2017. https://doi.org/10.1016/j.nimb.2017.01.081.

[46] K. Sahbaz, B. Bilki, H. Dapo, I. Karslioglu, C. Kaya, M. Kaya, and M. Tosun. Systematic study of LED stimulated recovery of radiation damage in optical materials. *Journal of Instrumentation*, 17:P05002, 2022. DOI: 10.1088/1748-0221/17/05/p05002.

Chapter 4
RICH Detectors

Sajan Easo

STFC-RAL, Particle Physics Department,
Rutherford-Appleton Laboratory, Harwell Campus,
Didcot, OX110QX, U.K.
Sajan.Easo@cern.ch

4.1 Particle Identification

Particle identification (PID) is an important aspect of many experiments in high energy physics and particle astrophysics. In general, the PID techniques aim to identify stable particles whose decay lengths are larger than 10^{-5} m. These include $p, n, K^{\pm}, \pi^{\pm}, e^{\pm}, \mu^{\pm}, \gamma$, etc. Each type of particle has a unique invariant mass and hence the PID involves measuring quantities which depend on the mass so that the particle type can be inferred from them.

4.1.1 *Identification of neutral particles*

Identification of neutral particles are based on the energy they deposit in calorimeters [15]. For example, in an experiment with a tracking system and an electromagnetic calorimeter, electrically neutral particles create signals only in the calorimeter and not in the tracking system. A charged particle, such as electron, would create signals in both. This feature is used to discriminate between the signals created by photons and electrons. Further information on neutral particle PID can be found in the chapter on electromagnetic and hadronic calorimeters.

2024 © The Author(s). This is an Open Access chapter published by World Scientific Publishing Company, licensed under the terms of the Creative Commons Attribution 4.0 International License (CC BY 4.0). https://doi.org/10.1142/9789819801107_0004

4.1.2 Identification of charged particles

Different techniques are used for identifying charged particles based on their electromagnetic interactions with matter via the exchange of photons. For identifying charged hadrons in a large momentum range, the detectors using the Cherenkov techniques described in the following sections are the most commonly used detectors. A Ring Imaging Cherenkov detector (RICH) is a type of Cherenkov detector that is used in many experiments.

Three other techniques for identifying charged particles are listed as follows, along with the corresponding references where further details can be found.

One method is to make use of the ionization in the a material as the particle traverses through it. The corresponding energy deposited by the particle in the material is related to the mass of the particle [15, 24]. For example, in a material such as silicon, one can deduce that the energy deposited per unit length (dE/dx) is proportional to $(m/p)^2$, where m and p are the mass and momentum of the particle, respectively. Therefore, measurements of the energy deposited and the momentum are used for inferring the particle type. Due to the fluctuations in the amount of energy deposited, the energy is measured from different detector layers to extract the most probable value. Typically, this method was used for particles whose momenta are below 5 GeV/c. The (dE/dx) technique has also been used for identifying particles traversing through a gas [29].

When a charged particle crosses the boundary between two materials with different dielectric properties, the electromagnetic field from the particle is reformed and this results in the emission of photons. Transition radiation detectors (TRD) make use of the fact that the intensity of this radiation is dependent on the (energy/mass) of the particle, in addition to the properties of the materials [11]. In order to produce sufficient number of photons, several interfaces are used, for example a stack of foils with gaps in between each of them. These detectors are used mainly for discriminating between pions and electrons in the momentum range 0.5–200 GeV/c.

The time-of-flight (TOF) detectors [19] measure the time of arrival of the particles on a detector plane. From this measurement and the distance travelled by the particle, one can determine the mass of the particle. Typically, these detectors provided a 3σ separation between pions and kaons up to 3 GeV/c. With the new photon detectors, which have excellent timing precision, new TOF detectors are being developed which can provide the similar separation up to 10 GeV/c.

4.1.3 Usefulness of Cherenkov detectors

These detectors are used mainly for discriminating between different charged hadrons. When trying to reconstruct the invariant mass of a particle which decays into hadrons, the PID provided by the Cherenkov detector can reduce the combinatorial backgrounds. This is illustrated in Fig. 4.1 where the invariant mass of the particle which may decay into K^+K^- is plotted from the data collected by the LHCb experiment at CERN. In this figure, one of the plots is made without using the data from the RICH system in LHCb, and in this case, the combinatorial background dominates over the signal. For the other plot, the kaons are selected based on the results of the PID provided by the RICH system, and in this case, the signal for the ϕ meson can be seen. In a similar way, the data from Cherenkov detectors are used for finding the signals from different mesons and baryons.

Some RICH detectors provide very good capability to identify muons and electrons in addition to hadrons. The RICH detector in NA62 experiment at CERN is an example of this [20].

The Cherenkov detectors are a crucial part of the physics programme of many experiments.

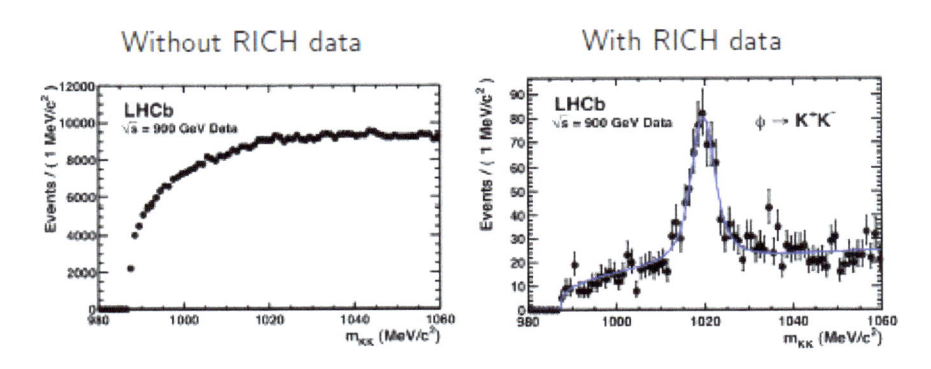

Figure 4.1. Plots of K^+K^- invariant mass spectrum with (right) and without (left) particle identification made possible by RICH data. The signal for the ϕ meson can be seen in the plot which uses the PID provided from the RICH data.
Source: Ref. [14].

4.2 Introduction to Cherenkov Detectors

4.2.1 *Historical perspective*

Although Cherenkov radiation was predicted in 1889 by Heaviside from Maxwell's equations and a blueish luminescence was observed in radioactive solutions by Marie Curie 100 years ago, it was Pavel Cherenkov who investigated this effect using a simple apparatus in 1934. His observations were fully explained by Frank and Tamm using classical electromagnetic theory and this resulted in these three scientists being awarded the Nobel Prize in 1958. The apparatus used for the discovery of antiproton in 1955 [10] made use of the Cherenkov radiation. This was followed by the development of a few Cherenkov detectors named "differential counters" in the 1970's.

At present, there are two major categories of Cherenkov detectors that are used for particle identification. The first category is the "imaging counters" that are used in accelerator-based experiments and in Cherenkov telescopes. An example of such counters is the RICH detector whose development was pioneered by Ypsilantis and others [30, 31] in the 1980's. The second category is the "Cherenkov calorimeters" that are used in astroparticle and long baseline experiments.

4.2.2 *Fundamentals of Cherenkov radiation*

The phase velocity of light (c_M) in a transparent dielectric material is lower than the speed of light (c) in vacuum. Passage of a charged particle with velocity (v) through such a material results in local polarization of the molecules and as they return to their original states, electromagnetic radiation is emitted. When $v < c_M$, this radiation is evanescent. When $v > c_M$, this radiation forms a coherent electromagnetic wave and it is called the Cherenkov radiation. Here, the photons are emitted uniformly in azimuth (ϕ) around direction of the particle. The polar angle (θ) between the particle and the photons is called the Cherenkov angle where

$$\cos(\theta) = 1/(n\beta) \qquad (4.1)$$

Here, $\beta = v/c = 1/\sqrt{(1+(mc/p)^2)}$, $n(\lambda) = c/c_M$ = phase index of refraction, λ = photon wavelength, m = mass of the particle, and p = momentum of the particle. This is illustrated in Fig. 4.2 where the trajectories of a charged particle and a photon can be seen. The photons produced along the path of the particle form a wavefront, as indicated by the dashed line in this figure. If n is a constant, the photon path is normal to the wavefront. In this case, the lines in this figure form a right triangle

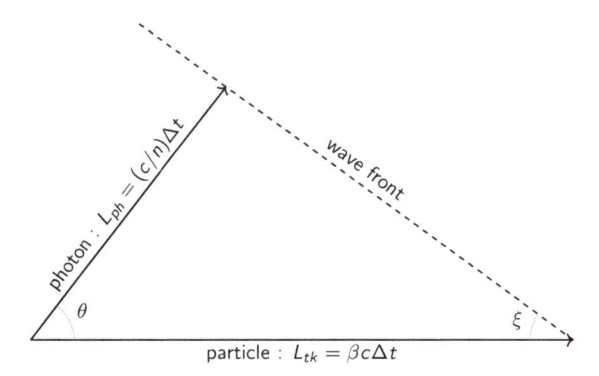

Figure 4.2. Illustration of Cherenkov photon production. Here L_{tk} and L_{ph} refer to the distances traveled during a time interval Δt by the charged particle and photon, respectively.

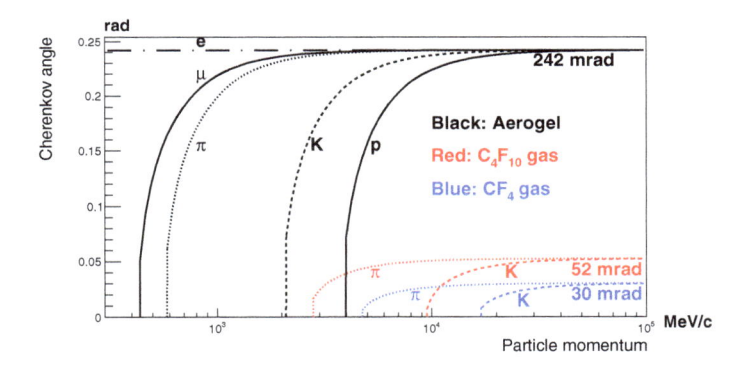

Figure 4.3. Plots of Cherenkov angle versus the particle momentum for different particles in three different radiators named Aerogel, C_4F_{10} gas, and CF_4 gas.

and therefore one can derive Eq. (4.1) by dividing the photon path length (L_{ph}) with the particle path length (L_{tk}) in this figure. In general, n varies with λ and this is discussed in Section 4.3.2.

The Cherenkov photons are not emitted below a velocity threshold corresponding to $\beta_t = 1/n$ and the corresponding threshold in momentum is $p_t = m/\sqrt{(n^2 - 1)}$. This threshold is specific for each combination of material and particle type, as illustrated in Fig. 4.3. As β approaches 1.0, the θ tends to become $\cos^{-1}(1/n)$, regardless of the particle type, as shown in this figure, and the corresponding charged track is called a saturated track. For particle identification, the momentum range of the particles needs to be well below saturation in the radiator material and this is taken into account when designing Cherenkov detectors. The amount of energy radiated by a particle per unit length as Cherenkov radiation is very small and is given by

$$dE/dx = (q/c)^2 \int_{v>c/n(\omega)} \omega(1 - (1/(\beta n(\omega))^2))d\omega \qquad (4.2)$$

where ω = photon frequency = c/λ and q is the charge of the particle. To first order, the energy loss is proportional to the photon frequency and hence there is more intense radiation at low wavelengths compared to high wavelengths. Indeed, most of the photons are emitted in the UV part of the spectrum and the photons in the visible part of the spectrum are predominantly blue. In a typical Cherenkov detector in accelerator-based experiments, particles whose momenta are a few GeV/c emit photons which have a few eV of energy.

The number of photons produced (N_{prod}) by a particle along a path length L in a material for a photon energy range from E_1 to E_2 is predicted by the Frank–Tamm theory to be

$$N_{\text{prod}} = (\alpha/\hbar c)q^2 L \int_{E_1}^{E_2} \sin^2(\theta(E))dE \qquad (4.3)$$

where $(\alpha/(\hbar c))$ = 370 eV^{-1}cm^{-1} and $E = hc/\lambda$ = photon energy. In a Cherenkov detector, only a fraction of these photons are detected. For example, if they are reflected by a mirror with reflectivity $R(E)$ and go through a quartz window with transmission T(E) and detected by a photon detector with efficiency Q(E), the number of photons detected is

$$N_{\text{det}} = (\alpha/\hbar c)q^2 L \int_{E_1}^{E_2} \sin^2(\theta(E))R(E)T(E)Q(E)dE \qquad (4.4)$$

For the case where the variation of n with photon energy is small and $q = 1$, $N_{\text{det}} \approx N_0 L \sin^2(\theta)$, where N_0 is called the "figure of merit" of a Cherenkov detector and is indicative of the fraction of photons detected compared to what was produced. Typical values of N_0 are in the range 30–200 cm^{-1}.

The effective resolution of a Cherenkov detector is expressed in terms of $\Delta\beta/\beta$.

4.2.3 Overview of PID using Cherenkov detectors

This overview uses RICH detectors as an example. Many of the items described here are valid for the other types of counters also. Further information regarding PID from Cherenkov calorimeters can be found in a later section and in the chapter on "Large scale Cerenkov neutrino detectors".

Typically, in an accelerator-based experiment, the direction and momentum (p) of each charged track are obtained from the tracking system. The RICH detector measures the positions (X_{ph}) of Cherenkov photons on a detector surface. Combining the X_{ph} and the information from the tracking system on the track direction, one can calculate the Cherenkov emission angle (θ). In principle, from the θ, the knowledge of the index of refraction (n) and the momentum (p), one can estimate the mass of the particle using Eq. (4.1) and thus identify the particle type.

However, when there are many charged tracks that give rise to many overlapping Cherenkov rings, one does not know which photon signals (recorded hits) are associated with a given track. Hence, one starts with many combinations of tracks and hits. In this case, one needs the algorithms described in Section 4.6 for PID.

Some detectors also measure the time of arrival (TOA_{ph}) of the photons. In the absence of dispersion, the n is a constant and the photons created in a radiator by a track that are expected to arrive in the same region of the detector plane would have the same TOA_{ph}, regardless of their emission points. The dispersion typically results in a spread up to 100 ps on TOA_{ph}. Therefore, using the TOA_{ph}, one can reduce the backgrounds from spurious combinations of tracks and photon hits. One option for this is to select only those combinations which have the TOA_{ph} within a few hundred picoseconds to a few nanoseconds of the expected time of arrival of the corresponding photons. Various options are adopted for combining the information from X_{ph} and TOA_{ph} measurements in different detectors.

The presence of dispersion also implies that one does not know the exact index of refraction (n) corresponding to any specific photon and this also contributes to the uncertainty in the PID results as described later.

Some of the techniques described in Section 4.6 can combine the information from X_{ph}, momentum, track direction, and TOA_{ph} measurements. In the Cherenkov calorimeters described in Section 4.5, the tracking system does not exist and hence track direction also needs to be estimated from Cherenkov hit data.

4.3 Main Components of Cherenkov Detectors

A Cherenkov detector has a radiator for creating photons, optical elements such as mirrors and quartz plates for facilitating the photon transport, and a set of photon detectors with their readout system.

4.3.1 Cherenkov radiators

A radiator can be any transparent dielectric medium. The PID in a certain momentum range depends on the refractive index of the radiator. Choosing a material as a radiator is primarily based on refractive index and radiation length. In practice, before choosing a material, one also considers many other factors. A flammable gas can cause safety concerns. If the material absorbs other gases, such as water vapor, its refractive index could change with time. The probability for Rayleigh scattering and Mie scattering processes by the photons needs to be as small as possible in the material since the scattering loses the photon emission angle information. The material must have excellent optical transparency in the wavelength ranges considered.

Examples of radiators used in accelerator-based experiments are helium gas, C_4F_{10} gas, C_6F_{14} liquid, silica aerogel, and quartz. In some astroparticle experiments, the atmosphere, the ocean, and the ice in Antarctica are used as radiators, and the photon transport is contained in these radiators. The Cherenkov emission angles from all these radiators are acute angles. There is R&D proposed for designing radiators with desired refractive index using periodic structures, and in case this materializes, the emission angle can be an acute or obtuse angle [22].

4.3.2 Cherenkov radiation in dispersive materials

When the radiator medium is dispersive, n is not a constant as seen in Fig. 4.4 and the value of θ depends on the wavelength of the photon emitted. The corresponding variation in the measured Cherenkov angles for each track is called the chromatic error. The strategies to reduce this error include the following:

(1) Filtering off low wavelength photons since the variation of refractive index is worse at low wavelengths compared to that at high wavelengths. However, this option reduces the number of detected photons. The feasibility of this option also depends on the sensitivity of the photon detectors in different wavelengths, as indicated in the following sections.
(2) Choosing a material that has only a small variation in refractive index.

Another consequence of the dispersion is that the conical wavefront in Fig. 4.2 is not quite normal to the photon direction and the mach cone angle ξ is not a complement to θ in that figure. Using the notations in Fig. 4.5, the time taken by a photon to travel from its emission point to a

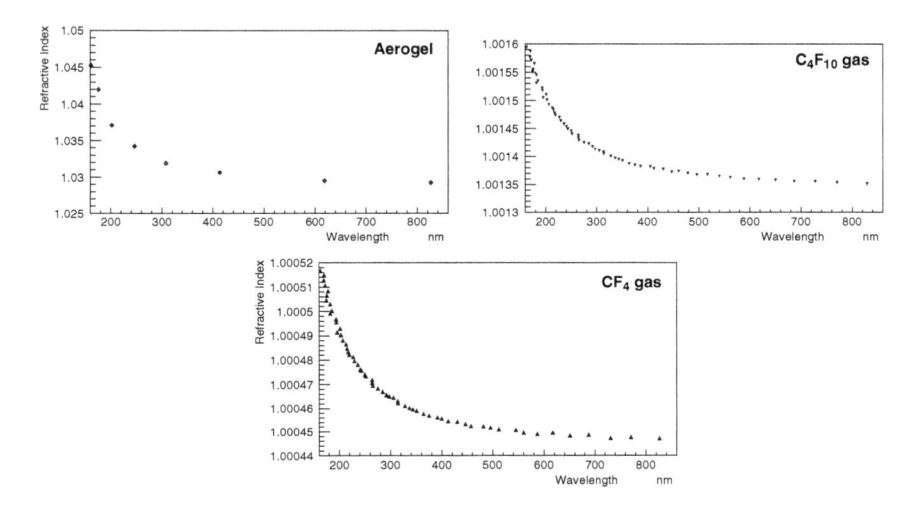

Figure 4.4. Refractive index of three different radiator materials as a function of photon wavelength.

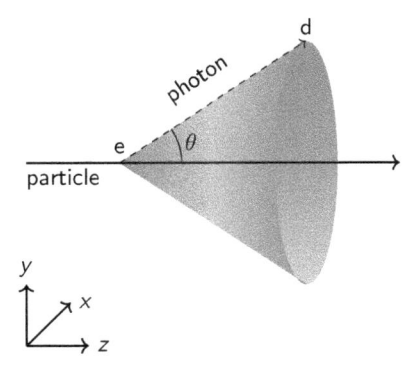

Figure 4.5. Schematic picture of a Cherenkov cone. Here a photon is emitted at e and detected at d so that it travels a distance $L_{\rm ph} = (d_z - e_z)/\cos(\theta)$.

plane where it is detected is given by

$$t_{\rm ph} = (L_{\rm ph})n_g/c = (d_z - e_z)(n_g/c)(1/\cos(\theta)) \quad (4.5)$$

Here one uses the fact that the photon energy travels with the group velocity and the group index of refraction (n_g) is related to the phase index of refraction (n) using the equation

$$n_g(\lambda) = n(\lambda) - \lambda dn(\lambda)/d\lambda \quad (4.6)$$

As a result, different photons from the same track arrive at the detector plane at different times, depending on their wavelengths. This can have an observable effect on the time resolution of the Cherenkov detectors which use fast timing if, in these detectors, the time resolution for the measurement is expected to be already in the range of a few tens of picoseconds. In the past, it was difficult to observe this effect since the time resolution was dominated by that of the photon detectors and readout system. However, some of the modern photon detectors can have time resolutions in this range.

There have been attempts in recent years to calibrate the variation in the emission angle with the time of arrival (TOA_{ph}) of photons at the detector plane and thus correct for the chromatic error contribution to the Cherenkov angle resolution [7]. For this, the photon detectors need to have a timing resolution typically better than 100 ps.

4.3.3 Mirrors and focusing

In some RICH detectors, Cherenkov photons are focused with spherical mirrors. The Cherenkov photons produced at the same azimuthal angle (ϕ_c) around the particle direction, are parallel to one another if we neglect the effects of scattering and dispersion. In this case, they are focused by the mirror to a point at the focal plane and thus the photons produced at different ϕ_c create a ring at the focal plane as can be inferred from Fig. 4.6. Ideally, the focal plane of the mirror with radius of curvature R also has curved shape and the reflected photons travel a distance $R/2$. The radius (r) of this Cherenkov ring is $r = (R/2)\tan(\theta)$. In practice, the photon detectors are placed in a flat plane near the ideal focal plane and, in some cases, the spherical mirrors are tilted to keep those detectors outside the acceptance of the charged particles. These give rise to a variation of the measured Cherenkov angle from different photon hits at different azimuthal angles (ϕ_c) around the same track and this is called the emission point error.

Special coatings are applied on the mirror surface to get an average reflectivity above 90% over the wavelength range of interest. In general, the mirrors are made from quartz or light materials, such as carbon fibre [23], which are radiation hard.

When the mirrors are installed in the RICH detector, there can be residual misalignments between the tracking system and RICH detector system. As a result, the projection of a track on the detector plane would be at a point away from the center of the Cherenkov ring and therefore

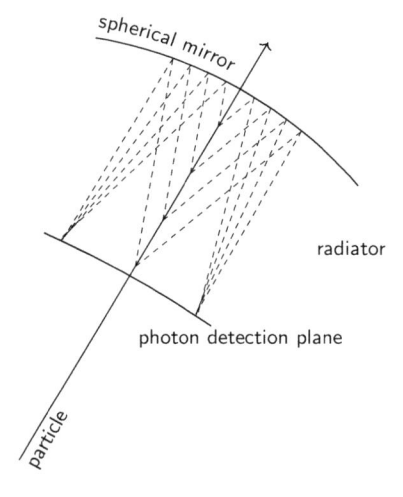

Figure 4.6. Schematic picture of the optical configuration of a RICH detector. The dotted lines are photon paths. This shows how a spherical mirror is used for focusing the Cherenkov photons.

the reconstructed Cherenkov angle as a function of ϕ_c will be a sinusoidal distribution. An alignment procedure would need to correct for this so that it does not degrade the Cherenkov angle resolution.

In some experiments, the Cherenkov photons travel only a small distance in the radiator before reaching the photon detectors. This avoids the need to have the focusing mirrors, and this arrangement is called "proximity focusing". In these cases, the radiator is normally a solid or liquid, which produces sufficient number of photons even when the distance traversed by the particle in the radiator is small. Figure 4.7 [25] shows an example of this. This configuration was used by the ALICE experiment at CERN.

4.3.4 *Photon detectors*

These detectors are required to be sensitive to single photons with high efficiency and low noise. The photons are converted into electrons using the photoelectric process or photoconductivity. The quantum efficiency (QE) is the probability for a photon to convert into a photoelectron from any of these processes.

Different photocathodes are sensitive to different wavelength ranges, from UV to near IR. Many of the commercially available solid photocathodes are made from mixtures of alkali metals since they have a

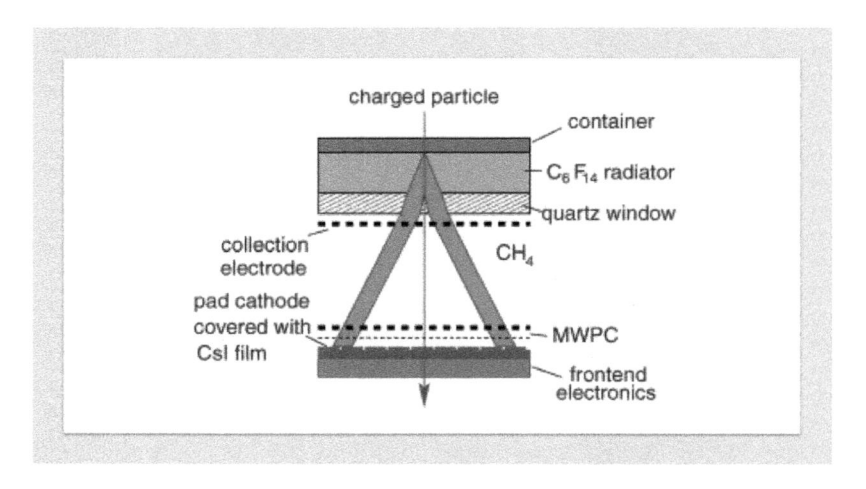

Figure 4.7. Example of proximity focusing. Here, the Cherenkov photons are created in the C_6F_{14} liquid radiator and they are then incident on a photocathode made of CsI. *Source*: Reproduced with permission of IOP Publishing from [1]; permission conveyed through Copyright Clearance Center, Inc.

relatively low "work function" to liberate the electrons. In some detectors, photosensitive gases are used for creating the photoelectrons. The silicon photomultipliers (SiPMs) can be sensitive to photons up to about 1100 nm using photoconductivity.

The signal from each electron is amplified so that this becomes a detectable signal ready to be sent to the readout system. This is done using different methods in different photon detectors. The probability for initiating this amplification process is labeled here as the triggering efficiency (P_{Trig}).

Some of the applications also require excellent timing resolution, low sensitivity to magnetic fields, radiation hardness, capability for high readout rate, and good pixel segmentation. These requirements can drive up the costs and hence they need to be reviewed when designing the Cherenkov detectors. Although several types of photon detectors are available these days, developing the best performing photon detector is an active area of R&D, done in collaboration with industry.

Arrays of photon detectors, such as photomultiplier tubes (PMTs), may have gaps between adjacent pixels and between adjacent PMTs. The active area fraction (P_A) is the ratio of the sensitive area to the total area on the

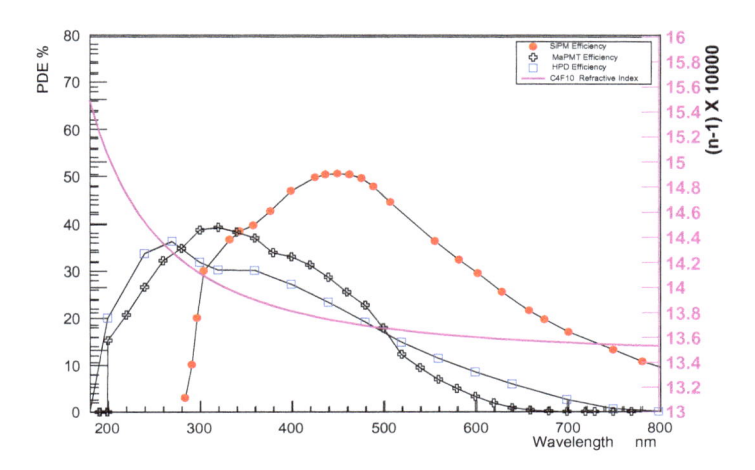

Figure 4.8. Typical photon detection efficiencies of SiPM, MaPMT, and HPD. These efficiencies are plotted for different photon wavelengths. A plot of (refractive index-1) × 10000 for C_4F_{10} gas is also shown as a function of photon wavelength.

detector plane. Some of the recent versions of PMTs can reach a P_A near 80%. The readout also can have an efficiency (E_r) for signal detection.

The detection efficiency (PDE) of a photon detector is defined as the product of QE, P_{Trig} and P_A, as appropriate. Figure 4.8 shows the typical detection efficiencies of three types of photon detectors as a function of the wavelength. The refractive index of a gas radiator is also superposed in this figure. From this it can be inferred that, one of the ways to mitigate the chromatic error is to choose a photon detector whose efficiency distribution is shifted to large wavelengths.

The photon detectors can be broadly categorized into (a) vacuum-based detectors, (b) gaseous detectors, and (c) solid state detectors.

Examples of vacuum-based detectors are PMTs with dynodes, hybrid photodiodes (HPD), and microchannel plate PMTs (MCP-PMT). These devices have a thin layer of photocathode in a vacuum tube. The signal created by the photoelectron is amplified to be well above the noise level in the readout system. In a multi-anode photomultiplier tube (MaPMT) [16], this is achieved through a set of dynodes. In an HPD, the photoelectron is accelerated within the tube so that they are focused on a silicon detector anode, as shown in Fig. 4.9. This figure also has a photograph of a typical HPD that was used in the LHCb experiment until 2018. In an

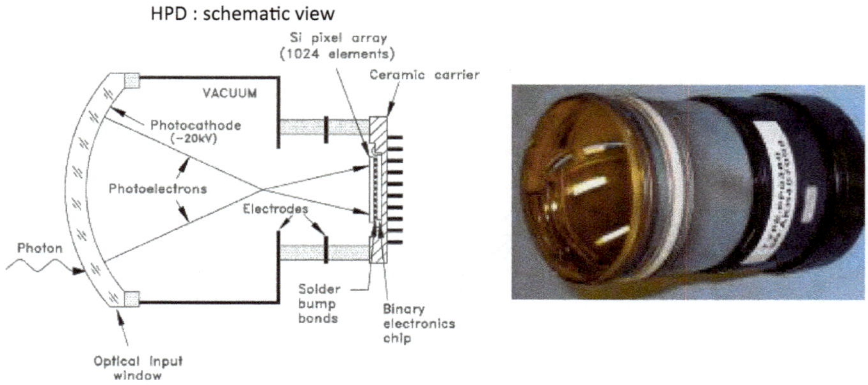

Figure 4.9. HPD: Schematic picture and a photograph.
Source: Reproduced with permission of IOP Publishing from [4]; permission conveyed through Copyright Clearance Center, Inc.

MCP-PMT [16], the dynodes are replaced by couple of thin glass plates named microchannel plates, which have arrays of thin holes where the amplification takes place.

The PMTs and HPDs have been used in large Cherenkov detector systems, although there are opportunities for further improvements. Many versions of them can work at high rates such as 40 MHz and some type of PMTs can be operated in remote locations such as below the surface of the ocean. The MCP-based detectors can offer better spatial and time resolutions and they are more resistant to magnetic fields compared to HPDs and PMTs. For example, some MCPs have achieved a time resolution of 30 ps with a low dark count rate; however, they suffer from restricted lifetime and R&D is in progress to mitigate this. One problem of the MCP is that residual gas in the detector's channels gets ionized. The ions and photons produced from this gas gradually destroy the photocathode. Measures that are implemented to alleviate this problem include improving the level of vacuum in the tube and using atomic layer deposition (ALD) for making the microchannel plate [15]. Recently, MCP-PMTs also started to be used in Cherenkov detectors in accelerator-based experiments.

Gaseous detectors [28] offer a relatively inexpensive and traditional option to cover large areas with photon detectors. Many of the first generation of RICH detectors were based on gas technology. In these devices, the photoelectrons are detected by a wire chamber or TPC (Time Projection

Chamber). For the photon conversion, one option is to mix the TPC gas with approximately 0.05% TMAE (tetrakis(dimethylamino)ethylene), as was done in the RICH in the DELPHI experiment at CERN. The CLEO-III RICH at CESR mixed the wire chamber gas with TEA (triethylamine). The ALICE experiment used a solid photocathode composed of CsI for the photon conversion and the photoelectrons were detected by an MWPC (multi-wire proportional chamber) in the configuration shown in Fig. 4.7. In these devices, the secondary photons and ions created in the gas can reduce efficiency and create backgrounds. These ions can also accelerate the aging of solid photocathodes. Recently, gas electron multipliers (GEM and THGEM) have been developed to solve these problems by directing the secondary electrons and ions created in the gas to another electrode [27, 28]. The PHENIX HPD [5] is one of the first RICH detectors to use CsI photocathode with GEM. Since the photon conversion in the photocathodes in these detectors are optimized for the UV region, there can be significant chromatic errors and the photon yields can be reduced, compared to those from the photocathodes operating in the visible region.

These detectors can work in a magnetic field. The detectors which use TEA and CsI are moderately faster than those which TMAE. However, the readout rate for these detectors are normally too low for Cherenkov detectors which require fast timing at the level of a few nanoseconds or better.

Silicon photomultipliers (SiPMs) [15] have already been used in Cherenkov telescopes and there is R&D underway for using them in other detectors that are in high luminosity environments. A silicon detector is essentially a reverse-biased p-n junction and it is widely used for the tracking of charged particles. Photons with energy greater than the bandgap energy of silicon (1.12 eV at 300 K) can create electron–hole pairs in such a detector. In order to provide an amplification above 10^5 to this signal, a bias voltage is applied which is above the breakdown voltage. This results in the creation of an avalanche of electron–hole pairs and this mode of operation is called the "limited Geiger mode". A quenching resistor is installed in series with the diode so that after the single-photon detection, the Geiger avalanche is quenched and the state of the diode is restored to be able to detect another photon. A SiPM contains arrays of such single-photon avalanche diodes (SPADs). SiPMs are compact, resistant to magnetic fields, and have excellent spatial and time resolutions. The design of the SiPMs can be optimized such that the PDE can peak at high wavelengths, as seen in Fig. 4.8, and this can help reduce the chromatic error. However, the typical dark count rate (DCR) is about 500 kHz/mm² at room temperature, which is too high to use in many Cherenkov detectors. If the SiPMs were to be installed in some of the experiments at the Large Hadron Collider (LHC), they could be subjected to fluences of approximately $10^{14}/\text{cm}^2$ 1-MeV-equivalent neutron fluence. This can also increase the DCR. Operating SiPMs at very low temperatures, for example at $-100°\text{C}$, is considered as an option to mitigate this problem.

4.4 Types of Cherenkov Detector Designs

Cherenkov detectors can be classified into threshold counters and imaging counters.

4.4.1 *Threshold counters*

Threshold counters do not measure the Cherenkov angle θ, but they can give a decision on the particle type using of the fact that only particles whose momenta are above the Cherenkov threshold create signals in the detector. For example, in the case of a mixture of two particle types "a" and "b" with masses m_a and m_b, where $m_a > m_b$, their momenta would need to be such that the "a" is below threshold and "b" is above threshold. As an improvement, one may use the number of photoelectrons (or calibrated pulse heights) observed to discriminate between particle types and this can result in a modest increase in the momentum range covered. The typical figure of merit (N_0) is 90 cm^{-1}. If the momentum threshold for "a" is p and the radiator length is L, the number of photons created by "b" at this momentum per unit length is $N_b/L = N_0 \times (m_a^2 - m_b^2)/(p^2 + m_a^2)$. The efficiency of such a counter is influenced by Poisson fluctuations, which can be challenging when there is one dominant particle type in the mixture. Since the rejection of a particle type is based on not seeing a signal, various sources of noise and the probability that a particle which does not create signal may decay into another particle which is "above threshold" would need to be considered. One of the recent experiments that used a threshold counter is the BELLE experiment in Japan [17].

4.4.2 *Imaging counters*

Imaging counters measure the positions (X_{ph}) of the photons as they arrive at the photon detector, and from this, the Cherenkov emission angles (θ) are calculated. Some of these also measure the time of arrival (TOA$_{\text{ph}}$) of the photons and the number (N_p) of photons detected. There are two main types of such counters, namely, the RICH detector and Cherenkov tracking calorimeters.

4.4.2.1 Features of RICH detectors

All the RICH detectors in accelerator-based experiments measure X_{ph} and N_p and some of them measure also measure TOA_{ph}. The momentum and direction of the charged particles are provided by the tracking systems in those experiments. In a typical RICH detector, the photons are focused onto the image plane, as shown schematically in Fig. 4.6. The spatial resolution obtained from RICH per charged track is given by

$$\delta\beta/\beta = \tan(\theta) \times \sigma_{\theta_t} \qquad (4.7)$$

where $\sigma_{\theta_t} = (<\sigma_{\theta_{\text{ph}}}>/N_p) \oplus C$. Here $\sigma_{\theta_{\text{ph}}}$ is the Cherenkov angle resolution for single photon, N_p is the number of photons that created signals from a charged track, and σ_{θ_t} is the Cherenkov angle resolution per charged track. The $\sigma_{\theta_{\text{ph}}}$ can have contributions from chromatic error, detector pixel granularity, optical arrangement, and other sources. In this equation, the term "C" includes the resolutions associated with the measurement of the directions of charged particles, multiple scattering, and the curvature of the corresponding tracks in the magnetic field in the radiator.

In general, the resolution obtained from a RICH detector is better than that from a threshold counter since the RICH measures both X_{ph} and N_p, whereas the threshold counter measures only N_p. An estimate of the number of sigma separation between two particle types with masses m_a and m_b at momentum p is given by

$$N_\sigma = (|m_a^2 - m_b^2|)/(2p^2 \sigma_{\theta_t} \sqrt{n^2 - 1}) \qquad (4.8)$$

This equation is useful for the initial design of a RICH detector. In practice, various other factors, such as the presence of multiple tracks giving overlapping rings and various types of backgrounds, can influence the actual discrimination capability between particle types.

4.4.2.2 Features of Cherenkov tracking calorimeters

Cherenkov tracking calorimeters [18] use large radiator volumes, such as the large tanks of water or the ice in Antarctica. They search for rare processes, such as the conversion of neutrinos into electrons, muons, and tau in these radiators. The Cherenkov signals created by these charged particles are detected by photon detectors, which are typically PMTs.

These photon detectors have good timing resolution, which allows them to measure TOA in addition to X_{ph} and N_p. These are used for determining direction of the photons and the charged tracks. Considering there can be many charged tracks in these events, the pattern recognition can become a complex procedure. In this case, one makes use of the fact that non-showering particles such as muons and protons create sharp rings, whereas particles such as electrons create diffused rings. The number of photons detected is a measure of the particle energy, and in this context, a careful energy calibration is needed.

4.5 Examples of Cherenkov Detector Systems

There are several accelerator-based experiments which have a Cherenkov detector for PID. These include the RICH detectors in DELPHI at LEP at CERN, HERMES at HERA at DESY, NA62 at CERN, BELLE2 in Japan, and many others [20]. The salient features of three RICH detectors described in the following are indicative of the properties of many other similar detectors.

4.5.1 *Threshold detector at BELLE*

The BELLE detector at the asymmetric KEKB e^+e^- collider studied CP violation in the decays of particles produced at and around the $\Upsilon(4S)$ resonance. The Cherenkov detector had to discriminate between pions and kaons in the hadronic decays of B mesons. To accomplish this, it used a detector named Aerogel Cherenkov Counter (ACC) [17], which used silica aerogel as the radiator. Fine mesh PMTs were used for photon detection in a 1.5 T magnetic field from a solenoid. In this detector, most pions produced Cherenkov radiation above threshold, while most kaons were below Cherenkov threshold. This detector collected data successfully for more than 10 years and achieved a kaon identification efficiency up to 90% with a pion misidentification probability of 6%, for a momentum range up to 3.5 GeV/c.

4.5.2 *DIRC at BABAR*

The BABAR experiment at the asymmetric PEP II e^+e^- collider also studied CP violation in $\Upsilon(4S)$ decays. For this detector, pion/kaon separation for particle momenta up to 4 GeV/c was provided by a Cherenkov detector called DIRC (Detection of Internally Reflected Cherenkov light) [2]. In this detector, quartz bars were used as both a radiator and a light pipe. These

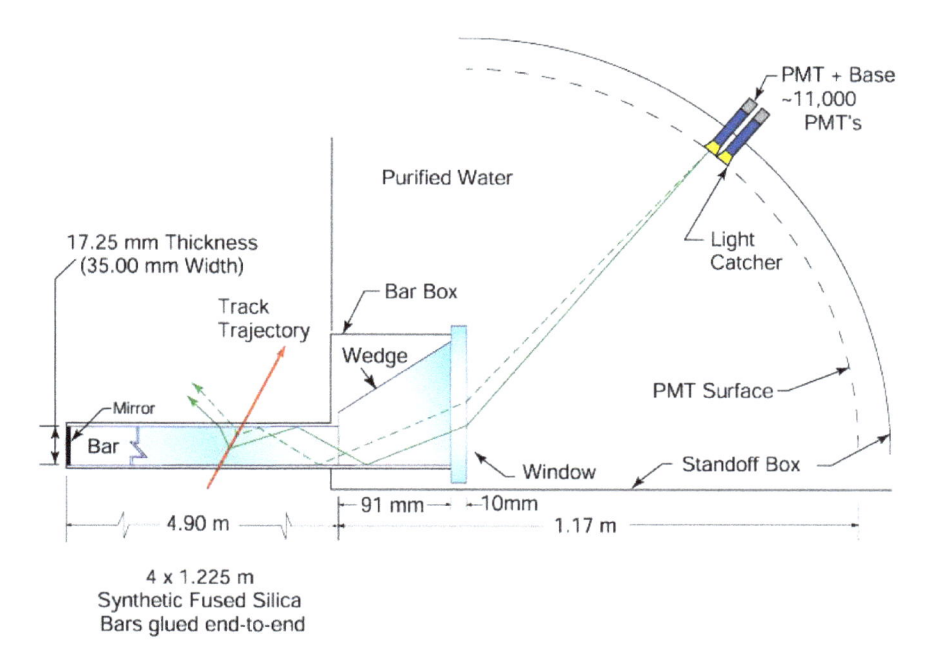

Figure 4.10. Schematic picture of the DIRC at BaBAR experiment.
Source: Reproduced with permission of IOP Publishing from [21]; permission conveyed through Copyright Clearance Center, Inc.

bars were rectangular in shape and hence the photons were transported toward the edge of the bar via total internal reflection, ensuring that the photon detectors could be kept outside the acceptance of the charged particles. The photons that reached the edge of the bar were imaged via a "pin hole" through an expansion region filled with purified water and then onto an array of photomultiplier tubes located 1.2 m from the end of the bar. These tubes recorded the location and time of arrival (TOA$_{ph}$) of the photons. Figure 4.10 shows a schematic picture of this configuration. This detector collected data for more than 8 years until 2008. For a pion identification probability around 85%, the kaon misidentification probability was well below 1%, for momenta up to 3 GeV/c. This corresponds to a π/K separation of at least 2.5 σ for momenta up to 4 GeV/c.

New versions of DIRC detectors are expected to use MaPMTs or MCP-PMTs, which provide better pixel granularity and improved time resolution compared to the photon detectors used in BaBAR. This can improve the precision in the measurement of the Cherenkov angle. In some cases, they would be able to measure the particle time of flight and correct for the chromatic dispersion in the radiator as indicated in Section 4.3.2. For example, the Barrel DIRC for the PANDA detector at FAIR [20] will use MCP-PMTS with a timing precision of 100 ps. In this case, the photons created in silica radiator bars will be focused with a lens and this detector is expected to provide a π/K separation at the level of 3σ, up to a momentum of 3 GeV/c.

4.5.3 LHCb RICH detectors

The LHCb experiment installed at the LHC has two RICH detectors, covering the momentum range 2–100 GeV/c [4]. The original version had three radiators, whereas since 2015, only two gas radiators are used. Their refractive indices are shown in Fig. 4.4 and these values determined the ranges of momenta covered by each of these radiators for particle identification. Table 4.1 shows some of the features of these radiators, including their momentum coverage and the RICH detector in which they are installed. The detector named RICH1 is installed upstream of the LHCb magnet and it is designed for the PID of the low momentum particles. The detector named RICH2 is installed downstream of the LHCb magnet and it is designed for the PID of the high momentum particles. A large fraction of the photons produced in aerogel were subject to Rayleigh scattering within the aerogel, thereby losing the Cherenkov angle information. The fraction of the photons transmitted without getting scattered in aerogel is given by:

$$T = Ae^{(-Ct/\lambda^4)} \qquad (4.9)$$

Table 4.1. Features of LHCb RICH radiators.

Radiator	Units	Aerogel[a]	C_4F_{10} gas	CF_4 gas
Length	cm	5	86	196
θ_c^{max}	mrad	242	52	30
π_{th}	GeV/c	0.6	2.6	4.4
K_{th}	GeV/c	2.0	9.3	15.6
Momentum coverage	GeV/c	<10	<70	<100
Detector		RICH1	RICH1	RICH2

Note: [a]Aerogel not used since 2015.

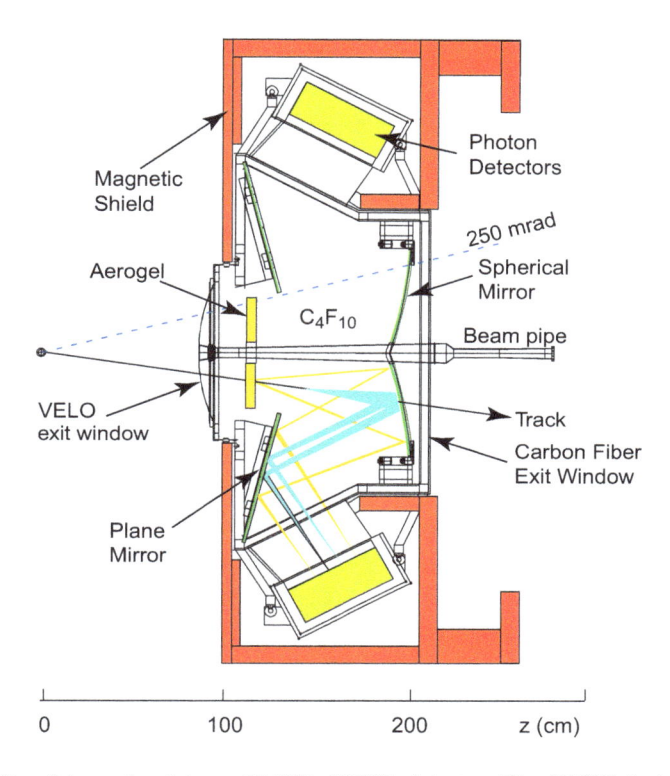

Figure 4.11. Schematic picture of LHCb RICH1 detector. The RICH1 is downstream of the LHCb vertex detector named VELO. A magnetic shield surrounds the arrays of photon detectors.
Source: Reproduced with permission of IOP Publishing from [4]; permission conveyed through Copyright Clearance Center, Inc.

where t = thickness of the aerogel tile, λ = photon wavelength, typically $A = 0.94$, and $C = 0.0059$ mm^4/cm.

The schematic picture of the RICH1 detector is shown in Fig. 4.11. In this figure, one can see examples of Cherenkov cones created in the radiators and how the photons are focused onto the detector plane. The optical system consists of spherical and flat mirrors. The spherical mirrors are tilted to keep the photon detectors outside the acceptance of the charged particles. The flat mirror helps reduce the length of the detector along the beam axis. Until 2018, HPDs were used as photon detectors. The RICH system was later upgraded and, beginning in 2022, MaPMTs have been used. Considering that the residual magnetic field in the region of the photon detectors could affect the path of the photoelectrons, the arrays of photon detectors are surrounded by a magnetic shielding structure as shown in Fig. 4.11. In addition to this, local shielding is provided to each photon detector.

The detector has a binary readout and hence the data collected are a set of channel numbers. From these, the locations (X_{ph}) where the photons arrive at the detector plane are reconstructed. Until 2018, the readout had a time window of 25 ns, which corresponds to the LHC collision frequency. However, beginning in 2022, the readout is expected to have a time window in the range of 3–6 ns. Further upgrades to the readout are planned in the future, which would enable the detector to read out the $\mathrm{TOA_{ph}}$ of the photons so that one can apply tighter time windows around the signal during data reconstruction. These time windows are expected to be approximately 600 ps and will be limited mainly by the transit time spread of the MaPMTs. With the introduction of new photon detectors such as the SiPM, these windows could be further tightened to a level below 100 ps in the future.

The two gas radiators were kept at ambient pressure (P) and temperature (T). Since their refractive index varies with (P/T), these parameters were monitored continuously. The photons produced by scintillation in CF_4 in RICH2 are quenched by adding a small amount of CO_2 to the CF_4 so that the resulting mixture had about 5% of CO_2 [8]. In this case, the scintillation was a cascade-free emission in the wavelength range used by RICH2, and this quenching resulted in a radiation-less transition from one molecule to the other, with a subsequent emission in the infrared.

Since 2015, the RICH PID software has been running in LHCb online as part of the high-level trigger in the data acquisition system, in order to select events that are useful for different physics channels. This required the RICH calibration and alignment also to run as part of the data acquisition system.

In previous runs, the single photon Cherenkov angle resolution achieved from the C_4F_{10} radiator in RICH1 was 1.62 mrad and that from the CF_4 radiator in RICH2 was 0.68 mrad [3]. The number of photon signals collected per saturated track was approximately 20 from C_4F_{10} in RICH1 and 16 from CF_4 in RICH2. This performance is expected to be further improved in the runs beginning in 2022 [12]. As described in earlier sections, the contributions to the single photon resolution come from the pixel granularity of the photon detectors, chromatic error, and emission point error.

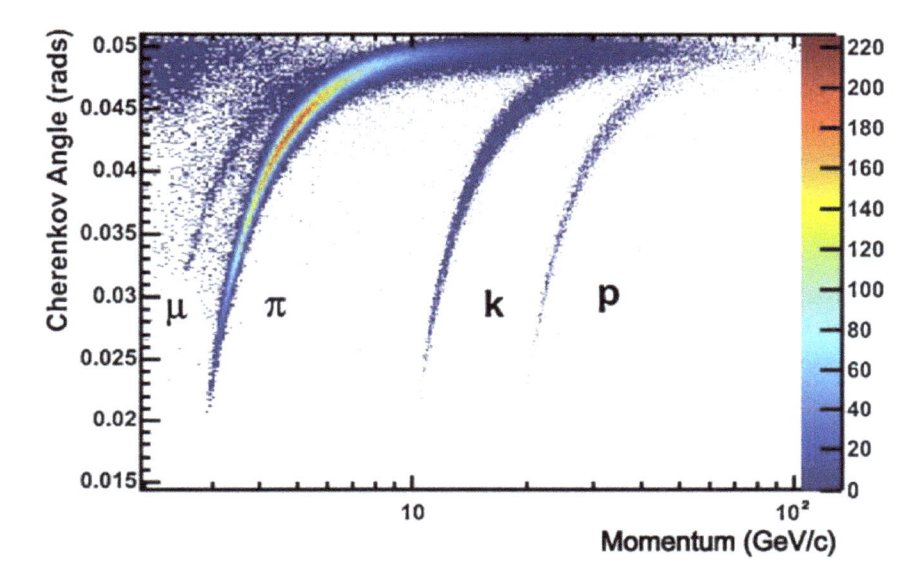

Figure 4.12. Plots of Cherenkov angle versus particle momentum from the real data collected using C_4F_{10} gas radiator in LHCb-RICH1.
Source: Ref. [3].

In Fig. 4.12, for isolated tracks in the RICH1 detector, the Cherenkov angle from the C_4F_{10} gas radiator measured in real data is plotted against the corresponding particle momentum measured using the LHCb tracking system. From this, one can see the signals for the different particle types listed in this figure. The shapes of these signals are similar to those expected from Fig. 4.3.

4.6 Algorithms for PID in RICH Detectors

Events with a large number of charged tracks give rise to many overlapping rings at the photon detector surface. In this environment, it is difficult for the software algorithms to find the photon hits associated with each track. Depending on the optics, the image of the hits at the detector plane from each track may or may not be circular in shape. Even if the images are expected to be circular rings, they may not have perfect circular shapes, due to the tilts of the mirrors.

The Hough transform method [6, 13] projects each track to the detector plane and accumulates the distance of each hit from the projection point in case the images are expected to be circular rings. The peaks in the distribution of these distances are used to associate the hits with tracks. From the Cherenkov angle reconstructed using these hits and the tracks, one can use Eq. (4.1) to determine the mass, thus identify the particle. This method is used for the Cherenkov detector in the ALICE experiment at CERN.

In the global log-likelihood method [9, 26], the reconstructed tracks are projected to the detector plane to determine the set of plausible combinations of tracks and hits corresponding to each particle type hypothesis. These combinations are called "photon candidates", and the Cherenkov angle is reconstructed for each of them. Photons are generated for each track and for each "particle type hypothesis". Using a fast simulation procedure, they are projected to the photon detectors to determine the "expected signal" at the pixels and predict a corresponding uncertainty. The average level of background expected at each pixel is also added to this. This "expected signal" is compared with the signal from the corresponding "photon candidate", in order to create a "likelihood". This comparison takes into account the uncertainties in the measurement and it is done in the Cherenkov angle space in order to mitigate the problems associated with the imperfect signal shapes on the detector plane. Using this procedure, a "global likelihood" is determined for each combination of track and "particle type hypothesis" in an event. This is used to find the combination with the best "global likelihood". The results of PID for each track are quoted in terms of the logarithm of the likelihood ratios which is defined as:

Delta Log Likelihood (DLL) = log (likelihood for a "particle type hypothesis"/likelihood for pion hypothesis).

This procedure does not involve reconstruction of Cherenkov rings. The global algorithm takes into account the signals from all tracks in an event, in order to find the optimal likelihood.

A version of this algorithm, called the "local likelihood method", uses the signals from each track separately. This is used for the data from the detectors where the track multiplicities are modest.

Figure 4.13 shows an example of PID performance from LHCb, in terms of the kaon identification efficiency and pion misidentification probability over a large momentum range, obtained using the global likelihood algorithm. In order to evaluate the PID performance from real data, samples of events from a calibration channel are selected using cuts based on decay kinematics and not using the information provided by the RICH. For Fig. 4.13, the typical calibration channel used is $D^* \to D^-\pi^+$, where

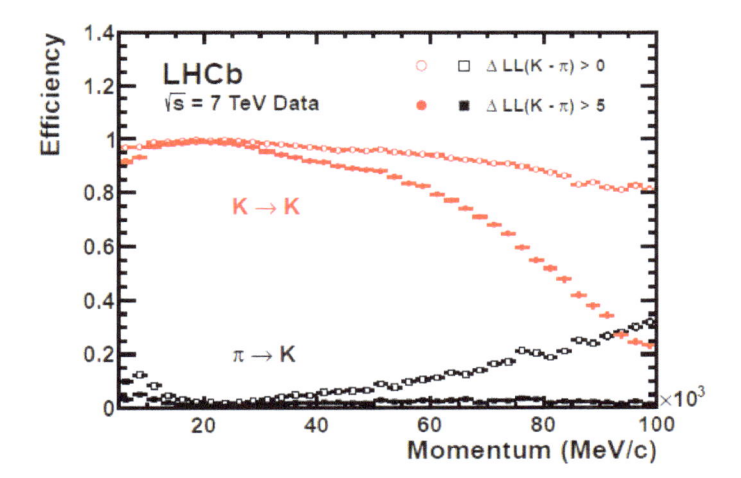

Figure 4.13. Example of particle identification performance from LHCb as function of particle momentum. The red plots represent the efficiency of identifying a kaon as a kaon when the selection cuts listed in the figure are applied on the DLL. The black plots represent the probability of misidentifying a pion as a kaon when the selection cuts listed in the figure are applied on the DLL. For illustration, these plots are shown with two different cuts on DLL: one for the plots with solid squares and another for the plots with empty squares. The DLL parameter is defined in the text.

Source: Ref. [3].

$D^- \to K^- \pi^- \pi^+$. Here, the narrow mass difference between D^* and D^- helps apply a tight cut to select a clean sample of events. The kaons and pions in the final states of such events are compared with the results of the PID from the RICH.

4.7 Summary

The Cherenkov radiation in bulk materials is described by the classical theory of electromagnetism. Cherenkov detectors make use of the different properties of this radiation for particle identification. These include the existence of the Cherenkov threshold, the relationship between the Cherenkov angle and the mass of the particle, and the dependence of the number of photons produced on the radiator length and the Cherenkov angle.

Cherenkov detectors have benefited from various advances in technology in recent years. These include the development of photon detectors that have single-photon detection capability and fast timing.

The RICH detectors offer better particle identification performance compared to those from the earlier versions of Cherenkov detectors. The design of these detectors continue to be improved. It is expected that these improvements will be made possible by future advances in the vacuum-based photon detectors and solid state photon detectors.

Cherenkov detectors are used for particle identification in many accelerator-based experiments and astrophysics experiments. They have contributed to important discoveries in particle physics in the last 70 years.

References

[1] K. Aamodt, et al. The ALICE experiment at the CERN LHC. *Journal of Instrumentation*, 3(8):S08002, 2008. DOI: 10.1088/1748-0221/3/08/S08002.

[2] I. Adam, et al. The DIRC particle identification system for the BaBar experiment. *Nuclear Instruments and Methods in Physics Research Section A: Accelerators, Spectrometers, Detectors and Associated Equipment*, 538(1):281–357. https://doi.org/10.1016/j.nima.2004.08.129.

[3] M. Adinolfi, et al. Performance of the LHCb RICH detector at LHC. *The European Physical Journal C*, 73:2431, 2013.

[4] A. A. Alves Jr. et al. The LHCb detector at the LHC. *JINST*, 3:S08005, 2008. DOI: 10.1088/1748-0221/3/08/S08005.

[5] W. Anderson, et al. Design, construction, operation and performance of a hadron blind detector for the PHENIX experiment. *Nuclear Instruments and Methods in Physics Research Section A: Accelerators, Spectrometers, Detectors and Associated Equipment*, 646(1):35–58, 2011. https://doi.org/10.1016/j.nima.2011.04.015.

[6] D. D. Bari. Pattern recognition techniques of charged hadron identification with HMPID at LHC. *Nuclear Instruments and Methods in Physics Research Section A: Accelerators, Spectrometers, Detectors and Associated Equipment*, 595(1):241–244, 2008. https://doi.org/10.1016/j.nima.2008.07.017, RICH 2007.

[7] J. Benitez, D. W. G. S. Leith, G. Mazaheri, B. N. Ratcliff, J. Schwiening, J. Vavra, L. L. Ruckman, and G. S. Varner. Status of the fast focusing DIRC (fDIRC). *Nuclear Instruments and Methods in Physics Research Section A: Accelerators, Spectrometers, Detectors and Associated Equipment*, 595: 104–107, 2008. DOI: 10.1016/j.nima.2008.07.042.

[8] T. Blake, C. D'Ambrosio, S. Easo, O. Ullaland, et al. Quenching the scintillation in CF_4 Cherenkov gas radiator. *Nuclear Instruments and Methods in Physics Research Section A: Accelerators, Spectrometers, Detectors and Associated Equipment*, 791:27–31, 2015. https://doi.org/10.1016/j.nima.2015.04.020.

[9] C. P. Buszello. LHCb RICH pattern recognition and particle identification performance. *Nuclear Instruments and Methods in Physics Research Section A: Accelerators, Spectrometers, Detectors and Associated Equipment*, 595:245–247, 2008. DOI: 10.1016/j.nima.2008.07.101.

[10] O. Chamberlain, E. Segrè, C. Wiegand, and T. Ypsilantis. Observation of antiprotons. *Physical Review*, 100:947–950, 1955. DOI: 10.1103/PhysRev.100.947.

[11] B. Dolgoshein. Transition radiation detectors. *Nuclear Instruments and Methods in Physics Research Section A: Accelerators, Spectrometers, Detectors and Associated Equipment*, 326:434–469, 1993.

[12] S. Easo, Overview of LHCb-RICH upgrade. *Nuclear Instruments and Methods in Physics Research Section A: Accelerators, Spectrometers, Detectors and Associated Equipment*, 876:160–163, 2017. https://doi.org/10.1016/j.nima.2017.02.061.

[13] D. Elia, N. Colonna, *et al.* A pattern recognition method for the RICH-based HMPID detector in ALICE. *Nuclear Instruments and Methods in Physics Research Section A: Accelerators, Spectrometers, Detectors and Associated Equipment*, 433(1):262–267, 1999. https://doi.org/10.1016/S0168-9002(99)00307-1.

[14] S. Erhan, R. Fleischer, and N. Harnew (Eds.). *Proceedings, 13th International Conference on B-Physics at Hadron Machines (Beauty 2011)*: Amsterdam, Netherlands, 4–8 April 2011, 24 p. https://pos.sissa.it/129/024/pdf.

[15] I. Fleck, M. Titov, C. Grupen, and I. Buvat. *Handbook of Particle Detection and Imaging*, 2nd edn. Springer, Switzerland, 2021. https://doi.org/10/1007/978-3-319-93785-4.

[16] K. K. Hamamatsu Photonics. Photomultiplier Tubes, 2007. https://www.hamamatsu.com/content/dam/hamamatsu-photonics/sites/documents/99_SALES_LIBRARY/etd/PMT_handbook_v3aE.pdf.

[17] T. Iijima, I. Adachi, *et al.* Aerogel Cherenkov counter for the BELLE detector. *Nuclear Instruments and Methods in Physics Research Section A: Accelerators, Spectrometers, Detectors and Associated Equipment*, 453(1):321–325, 2000. *Proceedings of 7th International Conference on Instrumentation for Colliding Beam Physics*.

[18] U. Katz. Cherenkov light imaging in astroparticle physics. *Nuclear Instruments and Methods in Physics Research Section A: Accelerators, Spectrometers, Detectors and Associated Equipment*, 952:161654, 2020. https://doi.org/10.1016/j.nima.2018.11.113. *10th International Workshop on Ring Imaging Cherenkov Detectors* (RICH 2018).

[19] W. Klempt. Review of particle identification by time-of-flight-techniques. *Nuclear Instruments and Methods in Physics Research Section A: Accelerators, Spectrometers, Detectors and Associated Equipment*, 433:542–553, 1999.

[20] P. Krizan, E. Nappi, *et al. International Workshop on Ring Imaging Cherenkov Detectors. Nuclear Instruments and Methods in Physics Research Section A: Accelerators, Spectrometers, Detectors and Associated Equipment*, 876:1–300, 2017. Proceedings of the 9th Workshop listed here. The Proceedings of Other Workshops in the same series can also be found in Nuclear Instruments and Methods A.

[21] P. Križan. Advances in particle-identification concepts. *Journal of Instrumentation*, 4(11):P11017, 2009, DOI: 10.1088/1748-0221/4/11/P11017.

[22] X. Lin, S. Easo, et al. Controlling Cherenkov angles with resonance transition radiation. *Nature Physics*, 14:816–821, 2018.
[23] F. C. D. Metlica. Development of light-weight spherical mirrors for RICH detectors. *Nuclear Instruments and Methods in Physics Research Section A: Accelerators, Spectrometers, Detectors and Associated Equipment*, 595: 197–199, 2008. DOI: 10.1016/j.nima.2008.07.026.
[24] S. Paganis and J.-L. Tang. Momentum resolution improvement technique for silicon tracking detectors using dE/dx. *Nuclear Instruments and Methods in Physics Research Section A: Accelerators, Spectrometers, Detectors and Associated Equipment*, 469:311–315, 2001.
[25] C. Pastore, I. Sgura, G. De Cataldo, A. Franco, and U. Fratino. The Cherenkov radiator system of the high momentum particle identification detector of the ALICE experiment at CERN-LHC. *Nuclear Instruments and Methods in Physics Research Section A: Accelerators, Spectrometers, Detectors and Associated Equipment*, 639:231–233, 2011. DOI: 10.1016/j.nima.2010.10.069.
[26] R. Forty. RICH pattern recognition for LHCb. *Nuclear Instruments and Methods in Physics Research Section A: Accelerators, Spectrometers, Detectors and Associated Equipment*, 433:257–261, 1999. https://doi.org/10.1016/S0168-9002(99)00310-1.
[27] F. Sauli. The gas electron multiplier (GEM): Operating principles and applications. *Nuclear Instruments and Methods in Physics Research Section A: Accelerators, Spectrometers, Detectors and Associated Equipment*, 805: 2–24, 2016. https://doi.org/10.1016/j.nima.2015.07.060.
[28] F. Tessarotto. Evolution and recent developments of the gaseous photon detectors technologies. *Nuclear Instruments and Methods in Physics Research Section A: Accelerators, Spectrometers, Detectors and Associated Equipment*, 912:278–286, 2018. https://doi.org/10.1016/j.nima.2017.11.081, New Developments in Photodetection 2017.
[29] J. Va'vra. Particle identification methods in high-energy physics. *Nuclear Instruments and Methods in Physics Research Section A: Accelerators, Spectrometers, Detectors and Associated Equipment*, 453:262–278, 2000.
[30] T. Ypsilantis. Theory of ring imaging Cherenkov counters. *Nuclear Instruments and Methods in Physics Research Section A: Accelerators, Spectrometers, Detectors and Associated Equipment*, 343:30–51, 1994.
[31] T. Ypsilantis and J. Seguinot. *Development of Ring Imaging Cherenkov Counters for Particle Identification*. Springer US, Boston, 1996, pp. 551–592. DOI: 10.1007/978-1-4613-1147-8_29.

Chapter 5

Machine Learning for Analysis and Instrumentation in High Energy Physics

Javier M. Duarte* and Dylan S. Rankin[†]

*University of California San Diego, La Jolla, CA 92093, USA
[†]University of Pennsylvania, Philadelphia, PA 19104, USA

5.1 Introduction

In high energy physics (HEP), the study and use of machine learning (ML)—the practice of solving problems by allowing machines to "discover" algorithms using data or experience without explicit programming— have been exploded in recent years. According to the INSPIRE HEP database, the number of articles in HEP and related fields that refer to ML and related topics has grown twenty times compared to ten years ago.[1] Notwithstanding this recent surge of interest, ML has deep ties to HEP, especially instrumentation, with early work dating back to the late 1980s and early 1990s [31, 33–35, 87–89]. In these early days, the most popular techniques, including cellular automata and multi-layer perceptrons, helped shape experimental particle physics. As deep neural networks have achieved human-level performance for various tasks, such as image classification [59, 79] in the early 2010s, they were adopted more regularly in particle physics [14, 56, 61, 106]. Unlike traditional approaches, deep learning techniques operate on lower-level information to extract higher-level patterns directly from the data.

2024 © The Author(s). This is an Open Access chapter published by World Scientific Publishing Company, licensed under the terms of the Creative Commons Attribution 4.0 International License (CC BY 4.0). https://doi.org/10.1142/9789819801107_0005

[1]https://inspirehep.net/literature?q=%28deep%20learning%29%20OR%20%28neural%20network%29%20OR%20%28machine%20learning%29%20OR%20%28artificial%20intelligence%29.

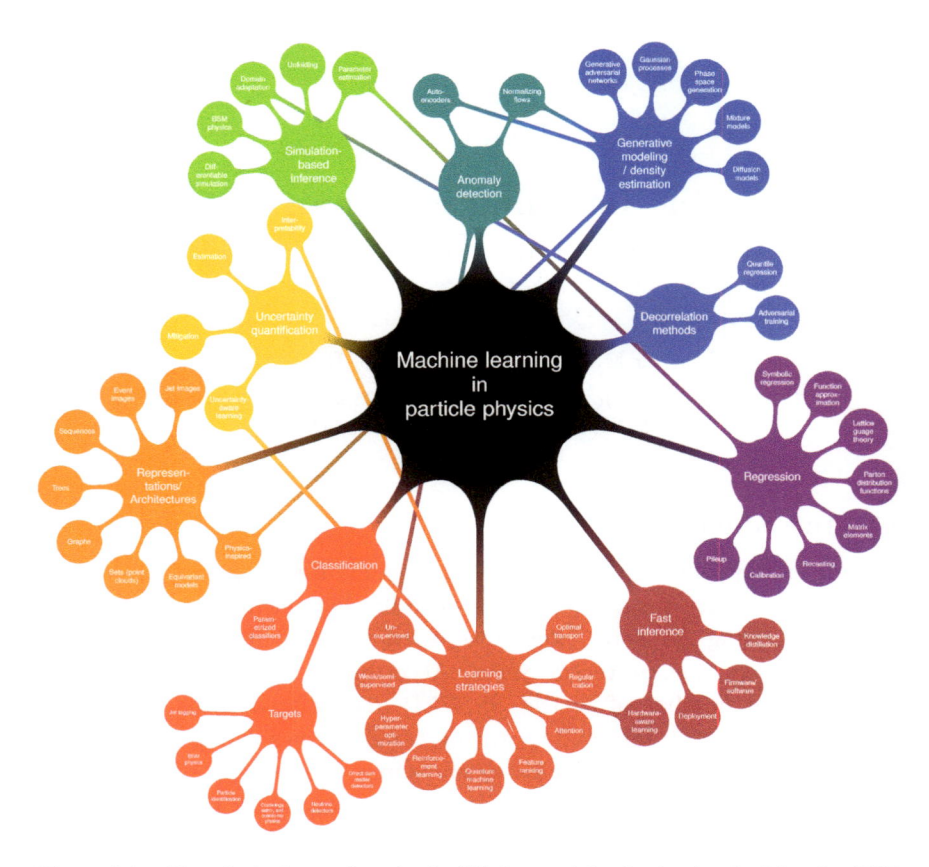

Figure 5.1. Nomological net of topics in ML in particle physics inspired by the HEP ML Living Review [61].

ML in particle physics has become more than a tool and has emerged as a subfield worthy of intense academic study in its own right. This can be seen through the HEP ML Living Review [61], which as of January 2024[2] categorizes 1,252 articles, proceedings, reviews, book chapters, and other contributions in this subfield. Inspired by this classification, we can visualize the different topics of ML in particle physics as a *nomological net* in Fig. 5.1. Use cases range from standard classification and regression to simulation, uncertainty quantification, and real-time inference.

---

[2]https://github.com/iml-wg/HEPML-LivingReview/blob/2c7cd26/HEPML.bib.

This chapter is meant to introduce the reader to the basic concepts of ML that are widely used in HEP. After reviewing these concepts, we survey popular applications in HEP.

5.2 Machine Learning Basics

5.2.1 *Types of learning*

The basic premise of ML is to use a set of observations to uncover an underlying process corresponding to an unknown target function mapping the inputs to the correct outputs. Within this framework, there are several different types of learning paradigms, which differ in the information contained in the dataset and how that information is used. When observations are coupled with correct outputs, known as *labels*, based on reliable information from simulation or empirical observation (ground truth), and the learning process that uses them, this is known as *supervised learning*. This is the most prevalent and well-studied form of learning in HEP and beyond, but other types are increasingly being applied. For example, in *unsupervised learning*, the training data do not contain any desired output or label information at all. For the remainder of this chapter, we focus primarily on supervised learning, but we discuss some applications of unsupervised learning.

5.2.2 *Supervised learning*

Within supervised learning, different tasks require different types of outputs. Tasks that require producing continuous, real-valued predictions, for example for quantities like mass, temperature, or energy, are known as *regression*. On the contrary, the main goal of *classification* is to assign, among a set of fixed options, the category to which a data sample belongs. Typically, the output of the model is a set of values $p_i \in [0, 1]$, one for each class, that represent the probabilities that the data sample belongs to a particular class i.

Given a training dataset $S = \{(x_1, y_1), \ldots, (x_N, y_N)\}$ consisting of data samples in an input domain $x_i \in \mathcal{X}$ and labels in a target domain $y_i \in \mathcal{Y}$, where i indexes the sample in the dataset, the goal is to learn a function from the input to the output domain $f\colon \mathcal{X} \to \mathcal{Y}$, parameterized by a vector of *parameters* θ, that best approximates the labels. We denote the output of the function for a given input x as $f(x|\theta)$. The space of functions under consideration is known as the *model or hypothesis class*.

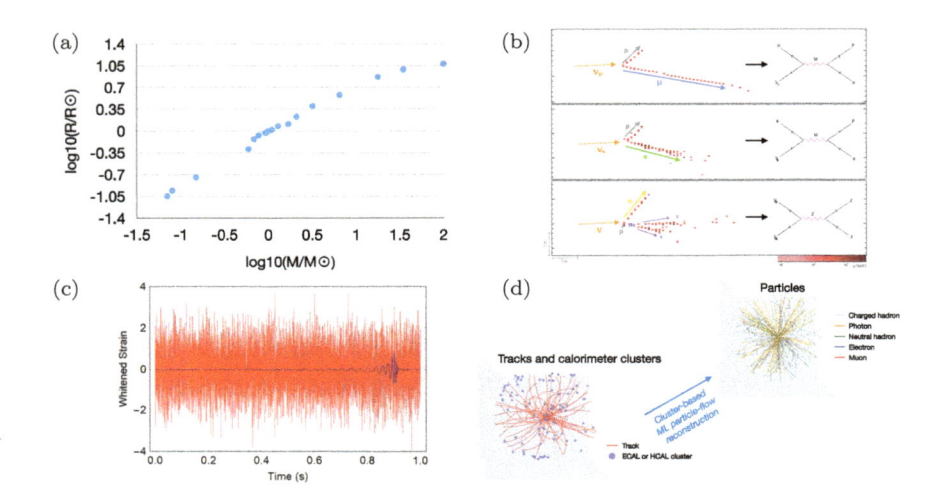

Figure 5.2. Examples of data representations and supervised learning tasks in physics, including (a) predicting the mass of a star given a measurement of its radius, (b) classifying image data from the NOvA experiment as one of the four types of neutrino interactions [6], (c) reducing noise in time series data to better identify gravitational wave signals [101], and (d) reconstructing particles based on detector measurements in a collider experiment [103, 104].

Examples of supervised learning are illustrated in Fig. 5.2:

(a) Predicting the mass of a star given a measurement of its radius. In this case, the input domain corresponds to the set of real numbers $\mathcal{X} = \mathbb{R}$ and $\mathcal{Y} = \mathbb{R}$.
(b) Classifying image data from the NOvA experiment as one of four types of neutrino interactions [6]. In this case, $\mathcal{X} = \mathbb{R}^{100 \times 80 \times 2}$ because there are two detector views ($x - z$ and $y - z$) with each image featuring 100 by 80 pixels of information. The target domain is a set of labels $\mathcal{Y} = \{\nu_\mu \text{ CC}, \nu_e \text{ CC}, \nu_\tau \text{ CC}, \nu \text{ NC}\}$, where each element is a different type of neutrino interaction.
(c) Reducing noise in time series data to better identify gravitational wave signals [101]. For this task, $\mathcal{X} = \mathbb{R}^{8192}$ and $\mathcal{Y} = \mathbb{R}^{8192}$, corresponding to 8 s of the data sampled at a rate of 1024 Hz, before and after noise reduction.
(d) Reconstructing particles based on detector measurements in an LHC experiment [103]. Here, $\mathcal{X} = \prod_{i=1}^{6,400} \mathbb{R}^7 \times \{\text{track}, \text{cluster}\} \times \{-1, 0, +1\}$, where there are seven continuous features and two discrete features

(whether the measurement is a calorimeter cluster or a track and the measured charge of the track) for up to 6,400 measurements per event. The target domain $\mathcal{Y} = \prod_{i=1}^{6,400} \mathbb{R}^4 \times \{\text{charged hadron, neutral hadron,} \gamma, e^{\pm}, \mu^{\pm}\} \times \{-1, 0, +1\}$ because there are four continuous features (four momentum of the particle) and two discrete features (particle type and charge) for up to 6,400 particles per event.

5.2.3 Objective function

The *objective function*, often called the *loss* or *cost function*, $L(y_i, f(x_i|\theta))$ measures the quality of predictions made by an ML algorithm. For example, a simple choice for regression problems is the squared loss $L(y, y') = (y' - y)^2$. The farther away the predicted value y' is from the true value y, the larger the value of the loss function. The more accurate an ML algorithm is, the smaller the loss value should be, on average, for a given set of data. Therefore, our goal is to minimize the loss function.

The *learning objective* is to find the parameters that minimize the loss function averaged over the entire training dataset, which we denote $l(\theta)$. These optimal parameters, denoted θ^*, can be expressed using the arg min operator, which returns the value where a given function attains its minimum:

$$\theta^* = \arg\min_{\theta} l(\theta) \equiv \arg\min_{\theta} \frac{1}{N} \sum_{i=1}^{N} L(y_i, f(x_i|\theta)) \tag{5.1}$$

Roughly speaking, θ^* is the set of parameters that minimizes the difference of the output of the algorithm and ground truth label.

Depending on the type of optimization process, there are additional requirements for the loss function. For example, the *gradient descent* algorithm discussed in Section 5.2.6 requires calculating the gradient of the loss function with respect to the model parameters to determine how to modify the parameters to reduce the loss function. Thus, the loss function must be *differentiable* in the model parameters.

Training an ML algorithm is closely related to statistical inference via the method of maximum likelihood [49]. In the maximum likelihood method, observed data are modeled by a probability distribution function with some free parameters. To estimate those parameters, we find their values such that the observed data are the most probable under this statistical model.

There is a correspondence between commonly used loss functions and likelihood functions. For example, minimizing the squared loss corresponds to maximizing a Gaussian likelihood. A Gaussian likelihood with observed value y' and expected mean y and standard deviation σ_y is given by

$$G(y'|y, \sigma_y) = \frac{1}{\sigma_y \sqrt{2\pi}} \exp\left(\frac{-(y'-y)^2}{2\sigma_y^2}\right) \quad (5.2)$$

If we take the negative logarithm of this likelihood,

$$-\ln G(y'|y, \sigma_y) = (y'-y)^2/(2\sigma_y^2) + \ln(\sigma_y\sqrt{2\pi}) \quad (5.3)$$
$$= c(y'-y)^2 + b \quad (5.4)$$

we see that up to a multiplicative constant c and an additive constant b, this is equivalent to the squared loss.

Another common loss function appropriate for binary classification tasks is the binary cross-entropy (BCE), which can be derived from the Bernoulli likelihood. Given two true classes, $y = 0$ or $y = 1$, and a model output y' defined between 0 and 1, which represents the probability that the data sample belongs to the $y = 1$ class, the Bernoulli distribution defines the likelihood

$$B(y'|y) = (y')^{\delta[y=1]}(1-y')^{\delta[y=0]} \quad (5.5)$$

where the δ operator evaluates to 1 or 0 if the argument is true or false, respectively. Note that only one of these two terms appears, depending on the true value of y. Taking the negative logarithm of the likelihood yields the BCE loss function:

$$L_{\text{BCE}}(y, y') = -\ln B(y'|y) = -\delta[y=1]\ln y' - \delta[y=0]\ln(1-y') \quad (5.6)$$

This can also be generalized to the categorical cross-entropy (CCE) for classification tasks with more than two target classes.

Figure 5.3 compares the squared loss and BCE for a given example whose true label is $y = 0$. Although both losses increase the farther y' is from y, BCE is more appropriate for classification problems because it takes into account that $y' = 1$ is an extremely incorrect prediction and the loss grows without bound as $y' \to 1$.

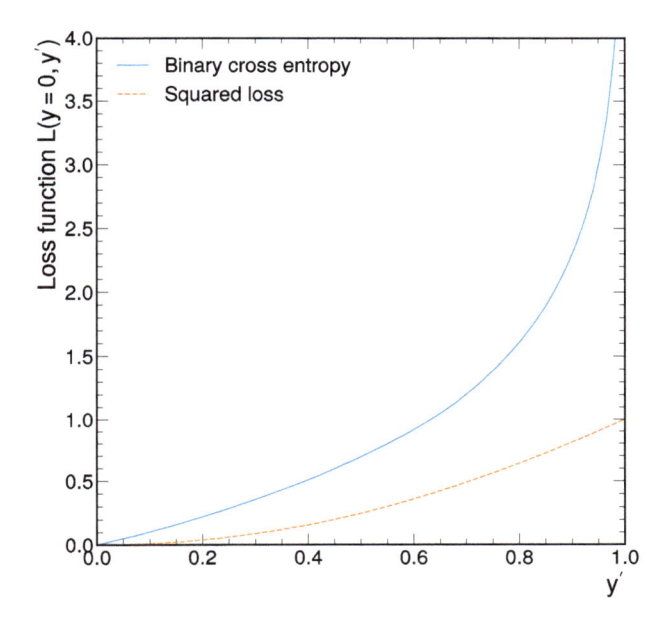

Figure 5.3. Comparison of the squared and binary cross-entropy loss functions for a true value of $y = 0$. The BCE loss grows without bound as the prediction y' approaches 1.

5.2.4 Linear models

Despite their simplicity, linear models are the workhorse of machine learning. Given a set of D features, each data point is a vector in D-dimensional space $x \in \mathbb{R}^D$, and a linear model can be expressed as

$$f(x|\theta, b) = \theta^\mathsf{T} x + b \tag{5.7}$$

where the *weight* $\theta \in \mathbb{R}^D$ and *bias* $b \in \mathbb{R}$ are unconstrained parameters of the model. These parameters are chosen to minimize the loss function on the training data. For notational convenience, we can absorb the bias into the weight vector by extending the input vector with a constant feature $x^{(0)} = 1$ and setting the corresponding entry of the weight vector equal to the bias $\theta^{(0)} = b$. This allows us to express linear models directly as

$$f(x|\theta) = \theta^\mathsf{T} x \tag{5.8}$$

To expand on the foundational ML concepts, we introduce an explicit example of regressing the logarithm of the radius of stars in the so-called main sequence as a function of the logarithm of the star mass. This means

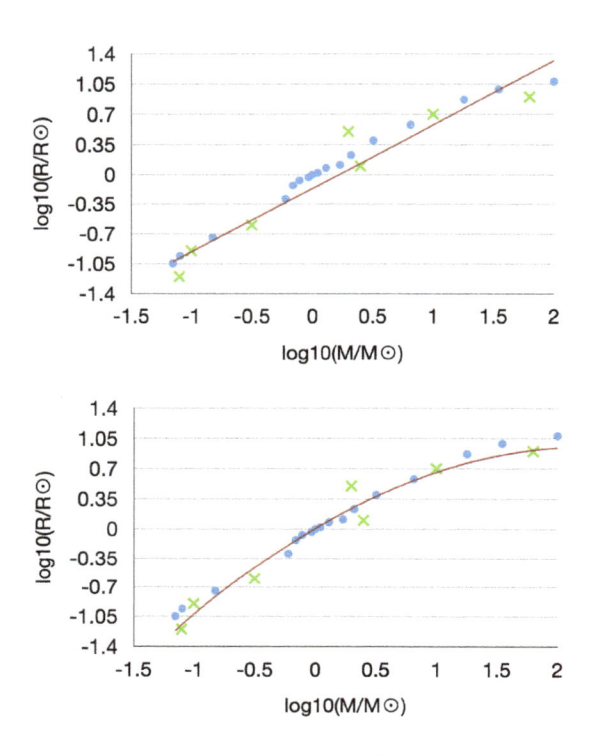

Figure 5.4. Training data points (crosses), testing data points (dots), and linear models (lines) fit to the data. A linear model using the original features x (upper) and a linear model after using a polynomial embedding $\phi(x) = (1, x, x^2)$ (lower) are shown.

we will train a model to predict $\log_{10}(R/R_\odot)$ given $\log_{10}(M/M_\odot)$. Sample data, split into training data (crosses) and testing data (dots), and a trained linear model (line) are shown in Fig. 5.4 (upper).

Linear models can perform more challenging tasks by replacing our input x with a transformation or *embedding* of x called $\phi(x)$. To illustrate this, consider a classification task in which we want to separate the two classes of data points in the (x_1, x_2) plane, represented by + and ○ symbols, respectively, as shown in Fig. 5.5 (left). The two classes could be separated by a circular boundary. Unfortunately, linear models can only create boundaries that are straight lines. Thus, no linear model can perfectly separate these two classes of data in the original input space of (x_1, x_2). However, we can apply a simple transformation squaring both components of the input $\phi(x_1, x_2) = (x_1^2, x_2^2)$, as shown in Fig. 5.5 (lower). Now, the two

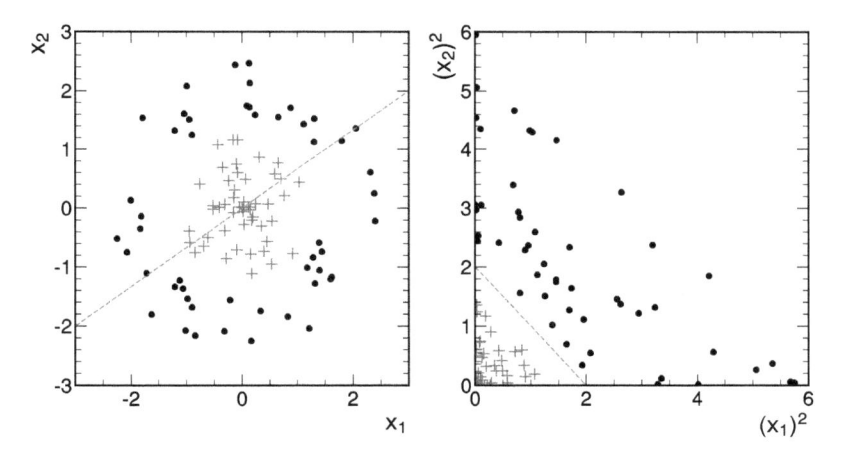

Figure 5.5. Example of embedding for a classification task. Two classes of data points in the (x_1, x_2) plane, represented by + and ∘ symbols, respectively, cannot be separated by a straight line (left). After transforming the data $\phi(x_1, x_2) = (x_1^2, x_2^2)$, the two classes can be separated by a straight line (right).

classes are separable by the straight line shown, which we can implement with a linear model.

More quantitatively, we can return to our regression task. If we use a polynomial embedding $\phi(x) = (1, x, x^2)$, then our model becomes

$$f(\phi(x)|\theta) = \theta^\mathsf{T} \phi(x) = \theta_0 + \theta_1 x + \theta_2 x^2 \qquad (5.9)$$

This model achieves a smaller training error than a linear model with the original feature x, as shown in Fig. 5.4 (lower). We say that this model is *more expressive* because it can represent a wider variety of functions. Although this is equivalent to polynomial regression in the original feature x, it is still a linear model in the new embedded features $\phi(x)$. For certain models, it is even possible to use the discriminating power of the embedded features without explicitly calculating them through the so-called "kernel trick." Further discussion of kernel methods can be found in Hofmann et al. [67], Scholkopf and Smola [114].

Although we have not yet defined neural networks, we can already try to build some intuition for how they work based on the concepts discussed already. As shown in Fig. 5.6, neural networks have linear models as their basic building block. A neural network can be thought of as a linear model after inputs are mapped to features through a nonlinear transformation. The initial layers of a neural network act as "automatic featurizers," where

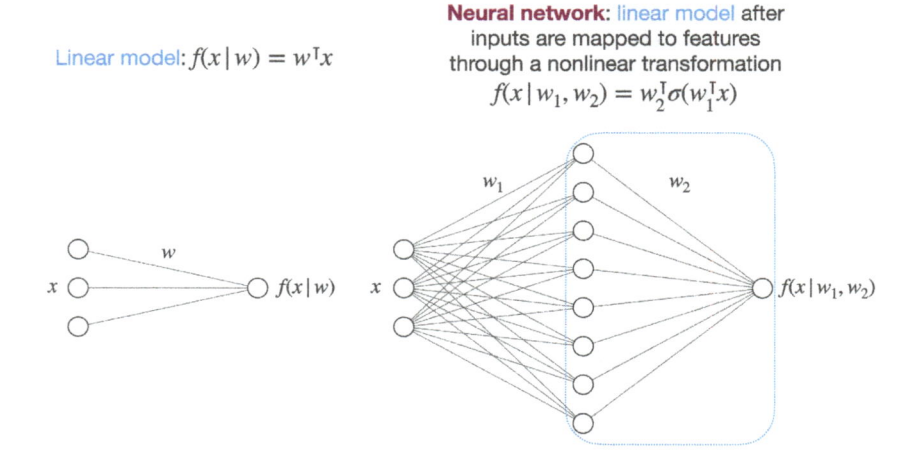

Figure 5.6. More complex model classes like neural networks have linear models as their basic building block. A neural network can be thought of as a linear model after inputs are mapped to features through a nonlinear transformation. Neural networks are "automatic featurizers."

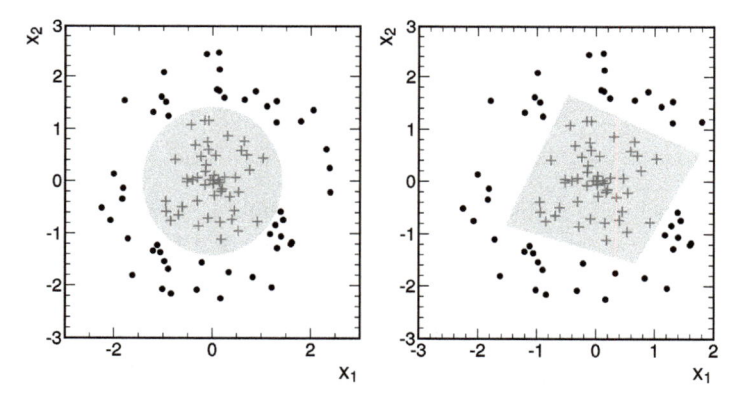

Figure 5.7. Decision boundary of a linear model after an embedding $\phi(x_1, x_2) = (x_1^2, x_2^2)$, corresponding to $x_1^2 + x_2^2 = 2$ (left). Decision boundary for a simple two-layer neural network with three hidden features (right).

instead of us guessing a well-suited embedding of our input features, the model learns one directly.

Revisiting the classification task of Fig. 5.5, a simple two-layer neural network can map the two input features to three "hidden features" where the two classes are separable. Figure 5.7 displays the decision boundaries for the linear model after the embedding described previously and a simple neural network. Since the embedding is hand-tuned for this dataset, its decision boundary can be thought of as ideal. The neural network's decision boundary is an imperfect approximation with jagged corners, but it has the advantage that no feature engineering was necessary—the features were learned by the neural network automatically. To gain intuition for neural networks before we describe them fully in Section 5.3.2, readers are encouraged to explore a visualization tool called TensorFlow Playground at https://playground.tensorflow.org.

5.2.5 *Generalization and bias-variance decomposition*

One of the central goals of ML is to train models that *generalize*, meaning that they perform well on test data outside the training set. But what exactly does that mean? Generally, it means the *expected test error* is small. As we see, two main sources of error prevent ML algorithms from generalizing beyond their training set. One is *bias* arising from erroneous assumptions in the ML algorithm and the other is *variance* arising from sensitivity to statistical fluctuations in the training set. A graphical visualization of bias and variance is shown in Fig. 5.8.

These ideas are connected to *underfitting*, when a model is unable to capture the relationship between the inputs and labels accurately, resulting in a large error rate in both training and test data, and *overfitting*, when a model fits exactly (or nearly so) in training data but does not perform accurately on test data. Explicit examples of both underfitting and overfitting are shown in Fig. 5.9. In this case, either a zeroth-order (upper) or fifth-order polynomial (lower) is used to fit the training data. The zeroth-order polynomial underfits the training data, resulting in a large

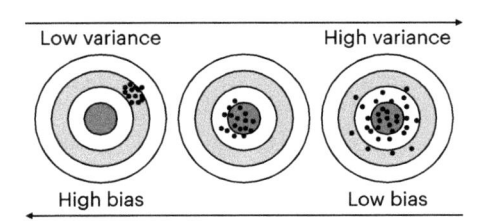

Figure 5.8. Graphical visualization of bias and variance using a bulls-eye diagram. Each hit represents a different, individual training of an ML model. The proximity to the center of the bulls-eye target indicates how low the test error is. Three different cases representing different combinations of high and low bias and variance are shown.

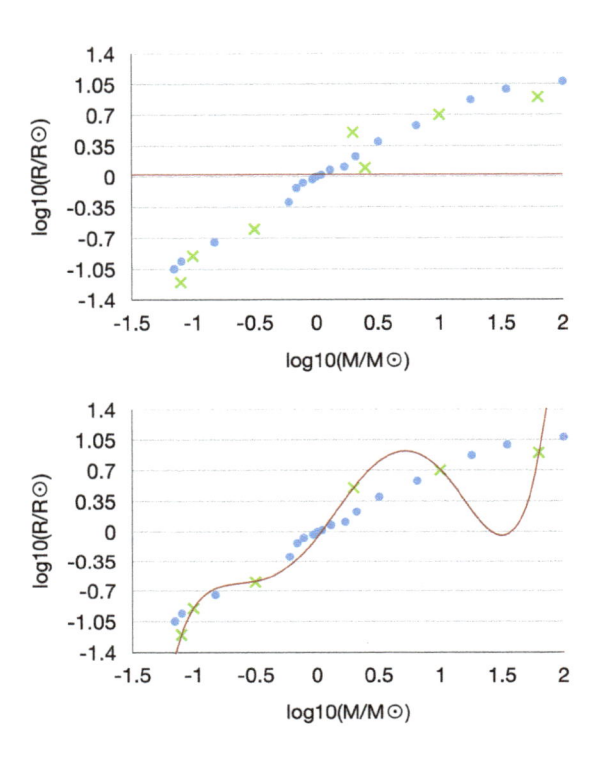

Figure 5.9. Examples of underfitting with a zeroth-order polynomial (upper) and overfitting with a fifth-order polynomial (lower). Training data points (crosses), testing data points (dots), and models (solid lines) fit to the data.

test error due to its high bias. Correspondingly, the fifth-order polynomial overfits the training data, also resulting in a large test error due to its high variance.

The *bias-variance decomposition* is a way of analyzing an ML algorithm's expected test error as a sum of bias and variance terms. To formalize this concept, we must introduce some statistical concepts and notation. For a random variable x sampled from a probability density function (PDF) $P(x)$, which we denote $x \sim P(x)$, we can define its expected value as

$$\mathbb{E}_{x \sim P(x)}[x] = \int_{-\infty}^{\infty} x' P(x') dx'. \qquad (5.10)$$

The expectation operator $\mathbb{E}$ is a generalization of the weighted average, where a subscript usually denotes the random variable(s) being sampled. Informally, the expected value is the arithmetic mean of a large number of independently selected outcomes of a random variable. For a continuous random variable, we effectively weight the integral by the PDF. For an integrable function $f(x)$ of the random variable, we can obtain its expected value in an analogous way:

$$\mathbb{E}_{x \sim P(x)}[f(x)] = \int_{-\infty}^{\infty} f(x')P(x')dx' \tag{5.11}$$

Returning to the question of the generalizability of our models, we examine the test error. Assuming each training data point (x_i, y_i) is sampled independently from $P(x, y)$ the "true" unknown probability distribution, then a trained model $f(x|\theta)$ has a true test error

$$L_P(f) = \mathbb{E}_{(x,y) \sim P(x,y)}[L(y, f(x|\theta))] \tag{5.12}$$

In general, we cannot compute this quantity, but we can estimate it using a test set of independent samples from $P(x, y)$. The training error is generally smaller than the test error. Overfitting occurs when the test error is much larger than the training error, while underfitting corresponds to the case when the training and test error are similar, but both are high.

The optimal set of model parameters θ_S^* is a function of the training dataset S. We can rewrite Eq. (5.1) to make this dependence explicit,

$$\theta_S^* = \arg\min_\theta \frac{1}{|S|} \sum_{(x,y) \in S} L(y, f(x|\theta)) \tag{5.13}$$

that is, if we change the training dataset S, the optimal set of parameters may change as well. The optimal parameters θ_S^* are themselves random variables because the training dataset S is randomly sampled.

We can write the expected test error over all possible training datasets as

$$\mathbb{E}_S[L_P(f(x|\theta_S))] = \mathbb{E}_S[\mathbb{E}_{(x,y) \sim P(x,y)}[L(y, f(x|\theta_S))]] \tag{5.14}$$

If L is the squared loss, we leave it as an exercise to the reader to show that we can decompose this expected test error into two terms:

$$\mathbb{E}_S[L_P(f(x|\theta_S))] = \mathbb{E}_{(x,y) \sim P(x,y)}\left[\underbrace{\mathbb{E}_S[(f(x|\theta_S) - F(x))^2]}_{\text{variance}} + \underbrace{(F(x) - y)^2}_{\text{bias}}\right] \tag{5.15}$$

where $F(x) \equiv \mathbb{E}_S[f(x|\theta_S)]$ can be thought of as the "average" prediction of our model over different possible training datasets.

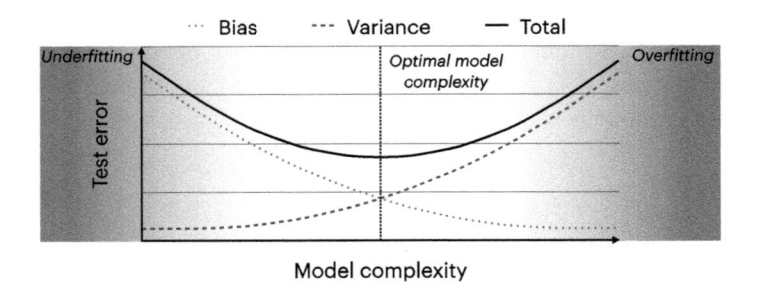

Figure 5.10. Bias-variance decomposition of test error as a function of model complexity.

How can we interpret Eq. (5.15)? The first term inside the expectation operator quantifies the *variance*: the difference in predictions when training on different datasets. The second term quantifies the *bias*: the difference of the average prediction from the ground truth. Thus, there is naturally a tradeoff: models with high variance tend to have low bias and vice versa.

We can relate overfitting and underfitting to the concepts of bias and variance. Overfitting implies high variance: the model class is too complex and retraining yields vastly different models. Variance tends to increase with model complexity and decrease with more training data. Underfitting implies high bias: the model class is too simple and has a large error rate. This relationship is shown schematically in Fig. 5.10.

5.2.6 *Optimization*

Gradient descent is a first-order iterative optimization algorithm for finding a local minimum of a differentiable function. It is the basis for many of the optimization algorithms commonly used in modern ML. "First order" means it only requires first derivatives of the function. The idea is to start with some (possibly random) initial values for all the parameters and then compute the gradient of the function with respect to all the parameters. The gradient represents the direction of the steepest ascent of the function in parameter space. Since we want to minimize the function, we take a small step in the opposite direction of the gradient by updating the parameter values. Then, we repeat this process until we reach a minimum.

More precisely, the gradient descent algorithm proceeds as follows. Each *iteration* of the algorithm is indexed by an integer t, starting with $t = 0$, and the current values of the parameters are θ_t. We set the parameters to some initial values, for example $\theta_0 = 0$ or randomly sampled from a Gaussian distribution $\theta_0 \sim \mathcal{N}(\mu = 0, \sigma = 1)$ or some other distribution specific to a particular type of learning algorithm. At iteration t, the parameters are updated using the negative of the loss function gradient:

$$\theta_{t+1} = \theta_t - \eta \nabla_\theta l(\theta_t) \tag{5.16}$$

$$= \theta_t - \frac{\eta}{N} \nabla_\theta \sum_{i=1}^{N} L(y_i, f(x_i|\theta_t)) \tag{5.17}$$

where η is a hyperparameter known as the step size or *learning rate*. The learning rate controls how large a step the algorithm takes during each update.

Unfortunately, we cannot determine *a priori* the optimal learning rate for a given model on a given dataset. Instead, a good (or good enough) learning rate must be discovered through trial and error. Typical values to consider are in the range of $\eta \in [10^{-6}, 1]$, while a good starting point is generally 10^{-3} or 10^{-2}. If you set the learning rate too high, your training may not converge because the weight updates "overshoot" the minimum of the loss function. If you set the learning rate too high, your model may also not converge (or converge too slowly) because the weight updates are tiny. Hyperparameter optimization procedures, like grid search, Bayesian optimization, or the asynchronous successive halving algorithm [84], can help find a good learning rates.

Note that in Eq. (5.17), the entire "batch" of training data is used to determine the gradient. In principle, this can give a more accurate estimate of the test loss that is less susceptible to statistical fluctuations, at the cost of more computation, that is, iterating over the full training dataset for each update. We repeat these updates until we reach some predefined convergence criteria.

A popular variant of this algorithm is *stochastic gradient descent (SGD)*. In this case, the true gradient over the entire dataset is approximated by that for a single data point. In other words, the update rule is modified to consider only one, usually shuffled, data point (x_i, y_i) at a time:

$$\theta_{t+1} = \theta_t - \eta \nabla_\theta L(y_i, f(x_i|\theta_t)) \tag{5.18}$$

Although this is much more computationally efficient, it can be subject to large statistical fluctuations.

At this point, it may be helpful to work through an end-to-end example of SGD for a regression problem, as shown in Fig. 5.11. Consider a training dataset consisting of two labeled data points $(x_1 = (1,1), y_1 = 1)$ and

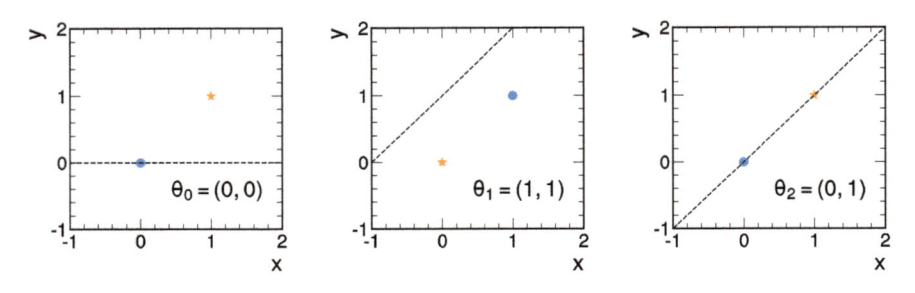

Figure 5.11. Example of stochastic gradient descent with two data points. Each frame from left to right represents an SGD iteration. The dotted line represents the current model with the current parameters listed on the canvas. The starred data point represents the one being used to compute the next parameter update. SGD converges after the second iteration.

$(x_2 = (1,0), y_2 = 0)$, where we have augmented the input with the "dummy" feature of 1 to simplify notation as described earlier. We use the squared loss function, a learning rate of $\eta = 0.5$, and an initial set of parameters $\theta_0 = (0,0)$, which includes the bias as the first component.

First, we can calculate the gradient of the loss with respect to the parameters:

$$\nabla_\theta L(y, f(x|\theta)) = \nabla_\theta (y - \theta^\mathsf{T} x)^2 = -2(y - \theta^\mathsf{T} x)x \tag{5.19}$$

Now we can write the SGD update rule of Eq. (5.18) as

$$\theta_{t+1} = \theta_t + 2\eta(y - \theta_t^\mathsf{T} x)x \tag{5.20}$$

$$= \theta_t + (y - \theta_t^\mathsf{T} x)x \tag{5.21}$$

where in the second line we use the fact that $\eta = 0.5$. Performing the first update with the data point (x_1, y_1) yields

$$\theta_1 = \theta_0 + (y_1 - \theta_0^\mathsf{T} x_1) = \theta_0 + x_1 \tag{5.22}$$

$$= (0,0) + (1,1) = (1,1) \tag{5.23}$$

Similarly, the second update with the data point (x_2, y_2) gives

$$\theta_2 = \theta_1 + (y_2 - \theta_1^\mathsf{T} x_2) = \theta_1 - x_2 \tag{5.24}$$

$$= (1,1) - (1,0) = (0,1) \tag{5.25}$$

which is exactly the optimal set of parameters. In this example, SGD converges after two iterations and will not give any further updates to the parameters because the loss is now zero for all data points, i.e., the data are fit perfectly. We note that the example here was carefully chosen, and, in general, many more updates are required.

A compromise between batch and stochastic gradient descent is *mini-batch stochastic gradient descent*, where the gradient is approximated by the average over a mini-batch of N_b samples:

$$\theta_{t+1} = \theta_t - \frac{\eta}{N_b} \nabla_\theta \sum_{i=1}^{N_b} L(y_i, f(x_i|\theta_t)) \qquad (5.26)$$

This is more computationally efficient and may result in smoother convergence, as the gradient computed at each step is averaged over more training samples. The hyperparameter N_b is known as the *mini-batch size*, which is typically taken to be a power of 2. It has been observed that choosing a large mini-batch size to train deep neural networks appears to deteriorate generalization [82]. One explanation for this phenomenon is that large mini-batch SGD produces "sharp" minima that generalize worse [64, 74]. Specialized training procedures to achieve good performance with large mini-batch sizes have also been proposed [55, 66, 126].

Many alternatives to SGD have been developed to improve training dynamics and avoid common pitfalls, such as slow progress along shallow parameter dimensions, "jitter" or oscillations along steep parameter dimensions, sensitivity to parameter initialization, excessively noisy gradient estimates, and getting stuck in local or sharp minima. SGD with momentum, named by analogy with physical momentum, remembers previous updates in an attempt to accelerate training, reduce the impact of statistical fluctuations, and prevent getting stuck in local minima [100, 110, 118].

Adaptive momentum estimation (Adam) [76] is an extremely popular SGD variant that combines many improvements from its predecessors [44, 62, 127] to make it more robust. In particular, it uses an adaptive learning rate specialized for each parameter. Figure 5.12 illustrates a comparison of SGD-based methods. Momentum can be seen as a ball running down a slope while Adam behaves like a heavy ball with friction that prefers flat minima in the error surface.

5.2.7 *Regularization*

Regularization refers to the practice of applying constraints, either implicitly or explicitly, to a model in order to guide optimization toward a simpler solution to prevent overfitting and improve generalization. As the complexity, capacity, and sheer number of parameters of ML models have

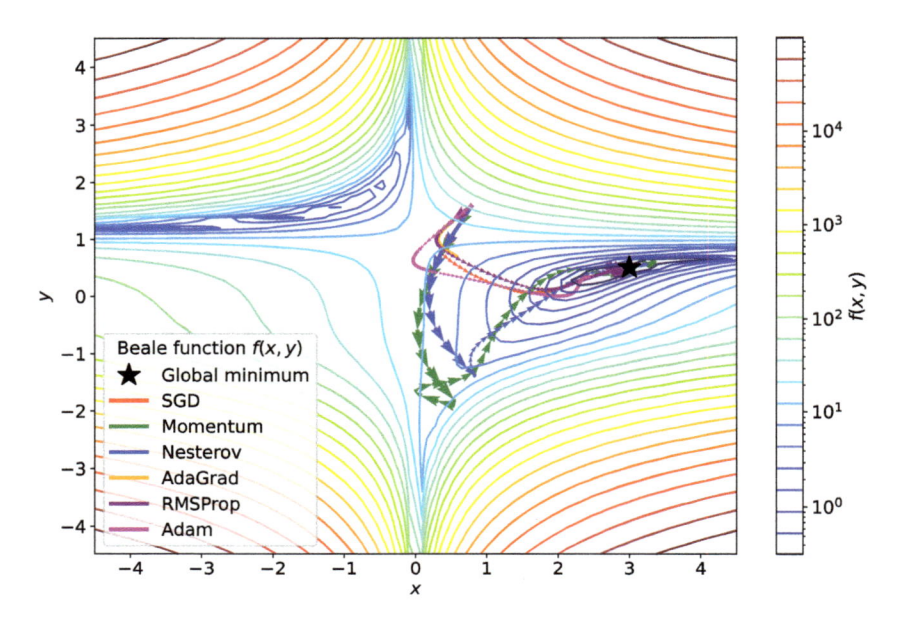

Figure 5.12. Comparison of different SGD methods optimizing the Beale function $f(x, y) = (1.5 - x + xy)^2 + (2.25 - x + xy^2)^2$ with global minimum $f(3, 0.5) = 0$.

grown in recent years, the likelihood of overfitting becomes greater, making regularization a critical component of modern ML. Explicit regularization refers to when an explicit term is added to the loss function, while implicit regularization includes other forms of regularization, for example, early stopping, using a robust loss function, and discarding outliers. Implicit regularization is ubiquitous in modern ML approaches, including stochastic gradient descent for training deep neural networks, and ensemble methods (such as random forests and gradient boosted trees).

The most common type of explicit regularization is L_n regularization, in which a term is added to the loss that penalizes large weights and biases:

$$L_n = -\lambda_1 \sum_{i=1}^{N_\theta} |\theta_i|^n \qquad (5.27)$$

where θ_i is parameter of the model. Usually, $n = 1$ (called L_1 regularization or lasso regression) or $n = 2$ (called L_2 regularization or ridge regression) is chosen. L_1 regularization naturally induces sparsity, whereas L_2 regularization tends to keep all parameters with lower magnitudes. The reason for this is illustrated in Fig. 5.13. In these two parameters, the constraint region for L_1 regularization is diamond-shaped, while for L_2, it is elliptical. Since

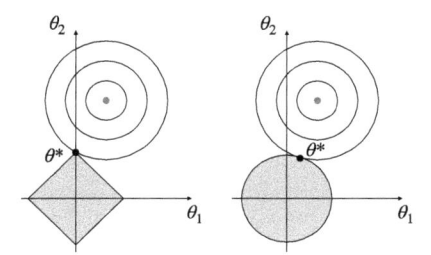

Figure 5.13. Depiction of L_1 (left) and L_2 (right) regularization constraint regions. The contours of an unregularized loss function are shown. The intersection with the constraint region from L_1 regularization gives an optimum value θ^* that is sparse, i.e., $\theta_1 = 0$. On the other hand, L_2 regularization yields an optimum value θ^* where both θ_1 and θ_2 are small but non-zero.

L_1 regularization sets certain weights to zero, it is often used as part of feature selection and model compression techniques. On the other hand, L_2 regularization reduces the contribution of high outlier nodes and distributes the weight given to correlated features, potentially leading to a more robust model.

A popular implicit regularization method is known as dropout [116], in which certain units are randomly dropped (along with their connections) from a neural network during training. This prevents units from co-adapting too much. During training, dropout samples from an exponential number of different "thinned" networks. At test time, a single "unthinned" network is used that effectively averages the predictions of all these thinned networks. Dropout introduces a new hyperparameter p (typically between 0.1 and 0.5) that specifies the probability of dropping units in a given layer.

To illustrate the effectiveness of regularization, we use a highly overparameterized neural network (three hidden layers of 100 nodes each) to classify data generated according to spiral patterns, both with and without dropout ($p = 0.15$). The results are shown in Fig. 5.14. The unregularized network (left) overfits the data as the decision boundary encircles single data points. The regularized network (right) learns a decision boundary that is much more faithful to the underlying spiral pattern.

5.2.8 Compression

In recent years, ML models have grown dramatically in their computational complexity, from thousands of parameters and operations to millions or even billions. However, many real-world and HEP applications require real-time on-device processing capabilities. The main challenge is that the devices

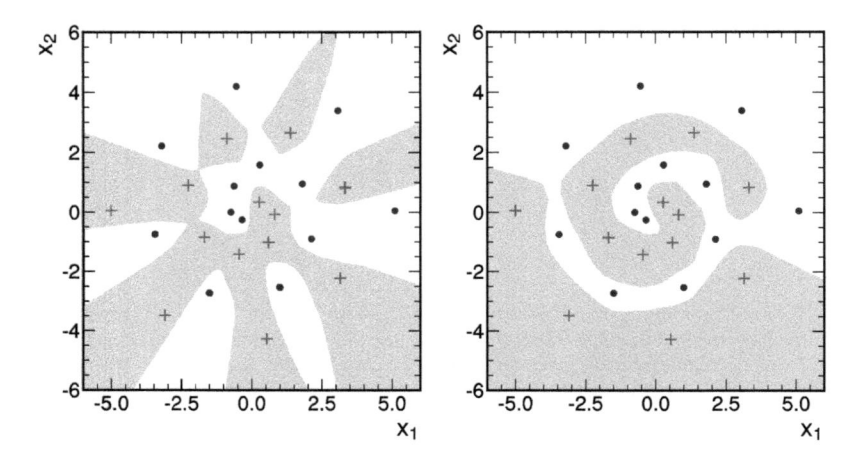

Figure 5.14. Decision boundary for a highly overparameterized network fitting spiral data with (right) and without (left) dropout ($p = 0.15$). The unregularized network (left) overfits the data as the decision boundary encircles single data points. The regularized network (right) learns a decision boundary that is much more faithful to the underlying spiral pattern.

used in these scenarios are resource-constrained, with limited memory, processing capabilities, and usually a strict latency budget. Reducing the size of ML models with *compression* can enable their use.

Compression techniques aim to improve the computational efficiency of models while keeping the performance as close as possible to the original. The two most ubiquitous methods are *quantization* [7, 30, 36, 41–43, 57, 70, 83, 92, 93, 95, 98, 107, 123, 128, 129, 131, 132], which modifies the number of bits used to calculate and store results in the model, and *pruning* [8, 46, 57, 81, 108, 130], which removes connections in a neural network.

In CPU- and GPU-based ML inference, it is common to use 32-bit floating-point precision. This allows the network to capture a very large range of values; the largest magnitude number that can be stored in 32-bit floating point format is $3.402823466 \times 10^{38}$ and the smallest is $1.175494351 \times 10^{-38}$. However, for many applications, the full floating-point precision range may not be required. Reduced-precision formats, such as integer or fixed-point precision, are commonly used instead, as shown in Fig. 5.15.

One disadvantage of reduced-precision formats with respect to floating point is a reduced dynamic range. Thus, care must be taken to ensure that weights or outputs of the ML model do not underflow or overflow in the

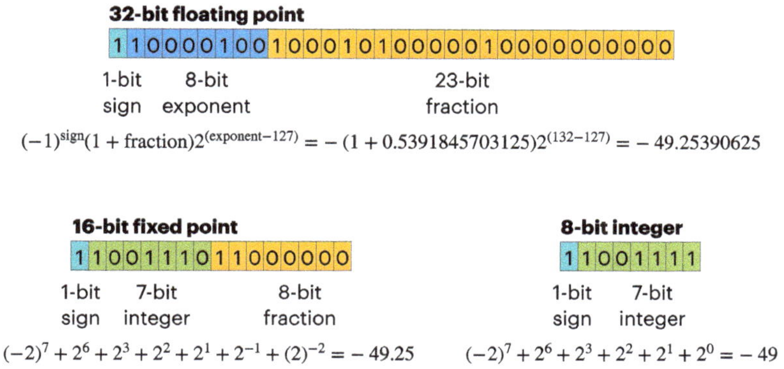

32-bit floating point

| 1 | 1 0 0 0 0 1 0 0 | 1 0 0 0 1 0 1 0 0 0 0 0 1 0 0 0 0 0 0 0 0 0 0 0 |

1-bit sign 8-bit exponent 23-bit fraction

$(-1)^{\text{sign}}(1 + \text{fraction})2^{(\text{exponent}-127)} = -(1 + 0.5391845703125)2^{(132-127)} = -49.25390625$

16-bit fixed point

| 1 | 1 0 0 1 1 1 0 | 1 1 0 0 0 0 0 0 |

1-bit sign 7-bit integer 8-bit fraction

$(-2)^7 + 2^6 + 2^3 + 2^2 + 2^1 + 2^{-1} + (2)^{-2} = -49.25$

8-bit integer

| 1 | 1 0 0 1 1 1 1 |

1-bit sign 7-bit integer

$(-2)^7 + 2^6 + 2^3 + 2^2 + 2^1 + 2^0 = -49$

Figure 5.15. Comparison between 32-bit floating-point (upper), 16-bit fixed-point (lower left), and 8-bit integer (lower right) representations.

reduced-precision format. However, reduced-precision representations are much more amenable to computations on specialized hardware, such as field-programmable gate arrays (FPGAs).

We can distinguish *post-training quantization* (PTQ), in which model parameters are quantized after a traditional training is performed with 32-bit floating-point precision, and *quantization-aware training* (QAT), in which training is performed with a modified procedure designed to emulate reduced precision formats.

Pruning is the removal of unimportant weights, quantified in some way, from a neural network. The two main categories are *unstructured pruning*, where weights are removed without considering their location within a network, and *structured pruning*, where weights connected to a particular node, channel, or layer are removed. These are depicted in Fig. 5.16. Pruning reduces the number of computations that must be performed to produce an inference result, thus reducing the hardware resources or algorithm latency. There are many different ways to decide which connections can be removed in a network, and the development of pruning algorithms and understanding their behavior are active areas of research.

One relatively simple method of pruning weights is called iterative, magnitude-based pruning, illustrated in Fig. 5.17. In this process, an L_1 regularization term is added to the loss that penalizes large weights. Training with this loss term typically produces two populations in the weights for a given layer. The weights that are deemed unnecessary by

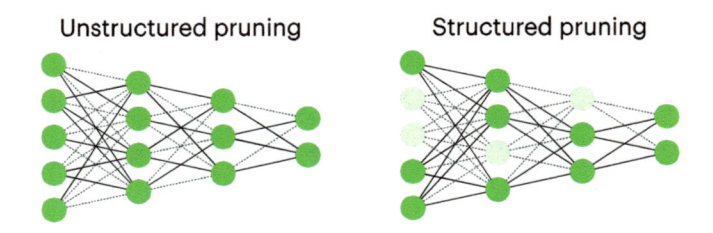

Figure 5.16. Pruning removes "unimportant" parameters and operations from a neural network. Removed connections are illustrated as gray dotted lines, while the remaining connections are solid black lines. Unstructured pruning (left) removes weights without considering their location within a network. Structured pruning (right) removes weights connected to a particular node, channel, or layer.

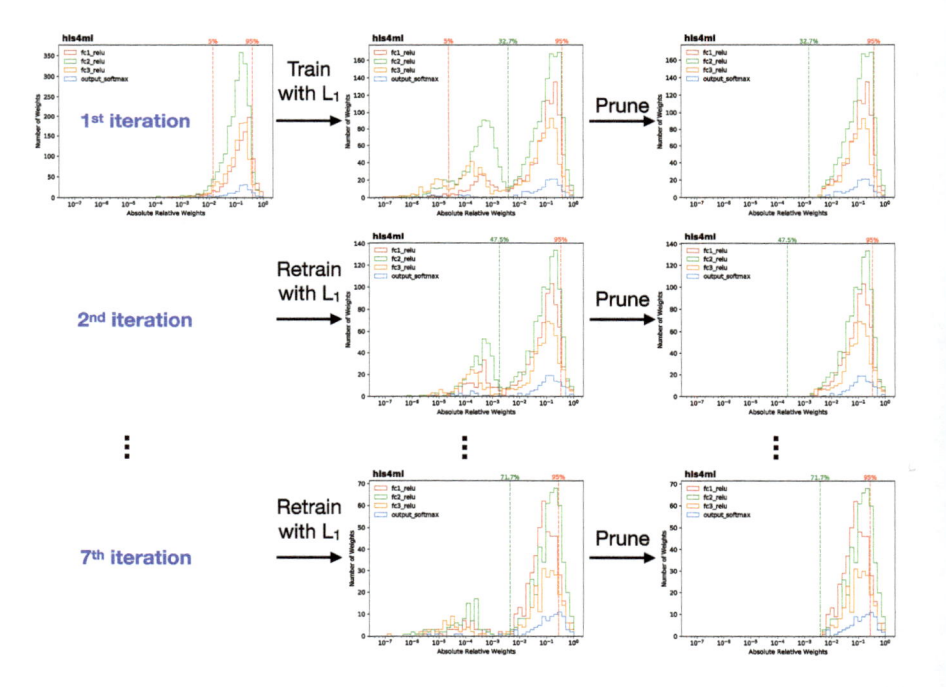

Figure 5.17. Illustration of the iterative magnitude-based parameter pruning and retraining with L_1 regularization procedure [43]. The distribution of the absolute value of the weights relative to the maximum absolute value of the weights is shown after each step of the pruning and retraining procedure. In the top left, the distribution before compression is shown, while in the bottom right, the distribution after compression is displayed.

the training will have very small values, while the weights that are deemed necessary will have larger values. Then, those weights with small values can be fixed to 0 (thereby removing that connection from the network), and training can be repeated. In many cases, successive training will identify additional weights that can be made small and thus removed. Repetition of this procedure can remove more weights until the desired reduction in connections, or *sparsity*, is achieved. This process usually results in networks that have slightly reduced performance compared to the full network, although the performance loss can be negligible depending on the target sparsity.

Both quantization and pruning can be applied together or individually depending on the problem at hand and implementation requirements, and the exact tradeoff between performance and sparsity or quantization is model-specific and depends on the model size, complexity, and task.

5.3 Models

In this section, we explore some of the most frequently used models in HEP.

5.3.1 *Decision trees*

Decision trees are among the simplest and most robust nonlinear models first invented in the context of data mining and pattern recognition as classification and regression trees (CART) [16]. Roughly speaking, they ask a series of yes-or-no questions based on individual features in order to categorize data. An example of a simple decision tree is shown in Fig. 5.18 to differentiate electron neutrino signal interactions ($\nu_e n \to pe^-$) from muon neutrino background interactions ($\nu_\mu n \to p\mu^-$) in the MiniBooNE detector [109]. In this case, the features used are relevant for this classification task, including the number of photomultiplier tube (PMT) hits, the total deposited energy, and the radius of the Cherenkov radiation ring. Distinguishing these two classes is essential to measure the quantum mechanical phenomenon of *neutrino oscillation*, in which a neutrino of one flavor (electron, muon, or tau) can later be measured to have a different flavor [53].

Formally, decision trees consist of a set of *internal*, or *branch*, *nodes*, that lead to two further nodes, and *terminal*, or *leaf*, *nodes* with no further branching. Every branch node i has a binary query function $q_i(x)$ that maps the input x to 0 or 1 and determines the subsequent node. The basic form

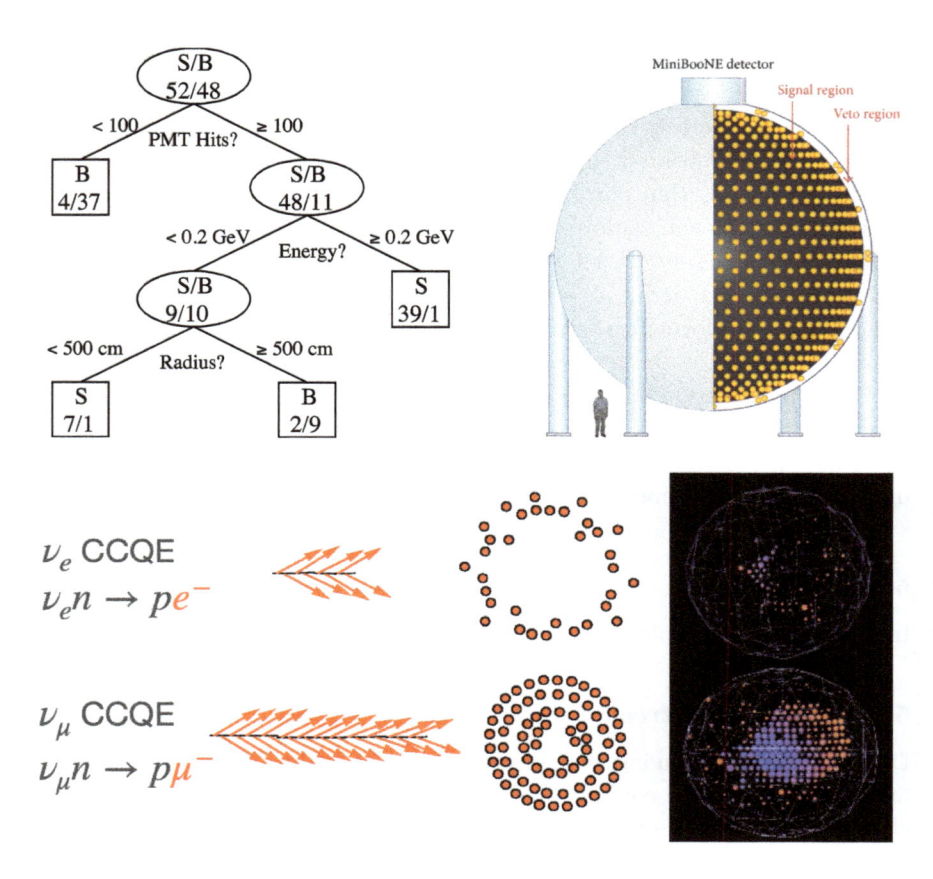

Figure 5.18. A schematic decision tree for event classification in the MiniBooNE experiment [109]. The goal is to differentiate signal $\nu_e n \to p e^-$ charged current quasi-elastic (ν_e CCQE) interactions from background $\nu_\mu n \to p \mu^-$ (ν_μ CCQE) interactions based on the Cherenkov radiation patterns measured by the photomultiplier tubes.

of the query function is a cut in an individual component $x^{(d_i)}$ of x:

$$q_i(x) = \delta[x^{(d_i)} > c_i] \tag{5.28}$$

Every leaf node makes a constant prediction. For a given sample x, prediction begins at the root node, calling the query function for each visited node. If the returned value is 1, the left child node is chosen, while the right child node is chosen otherwise. This process is repeated until a leaf node is reached.

Decision trees express piecewise-constant functions. A given tree creates J *axis-aligned* partitions of the input space $\mathcal{X} = \mathcal{X}^1 \cup \cdots \cup \mathcal{X}^J$, through a sequence of binary splits, where the length of the sequence is the *depth* of the tree. The number of leaf nodes is J. Each partition has a constant prediction b_j. The model can be written as

$$f(x|\theta) = \sum_j b_j \delta[x \in \mathcal{X}^j] \tag{5.29}$$

where $j \in \{1, \ldots, J\}$ indexes each leaf node.

Decision trees can often outperform linear models because they can learn nonlinear decision boundaries, as shown in Fig. 5.19 (upper). However, because most tree-based models consider splits aligned with individual feature components, there are some failure modes. In particular, it can be difficult to learn decision boundaries diagonally across two components, as shown in Fig. 5.19 (lower). Nonetheless, tree-based models are often preferred over other models because they work well with tabular data that

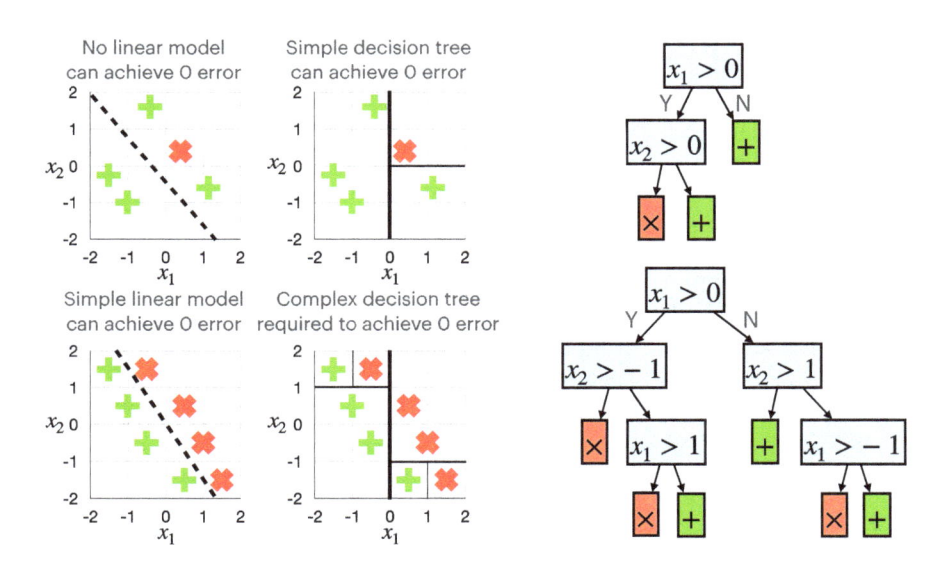

Figure 5.19. Two different cases demonstrating the strengths and weaknesses of linear models and decision trees. A decision tree can learn a nonlinear decision boundary unlike a linear model (upper). A simple linear model can learn a decision boundary diagonally across two feature components, while it requires a more complex decision tree to approximate the same decision boundary (lower).

may comprise a mix of continuous and discrete features, and there is less need for preprocessing.

So far we have described how decision trees look, but how are they constructed in the first place? A common (top-down) approach to building a decision tree starts with a root node and grows the tree with splits based on individual components of x. To decide when a given split is advantageous, we need to use a metric, called an *impurity measure*. Generally, they quantify to what degree a split refines the terminal nodes to be more pure than the parent node. The most widely used measure is the Gini impurity [16] defined as

$$I_{\text{Gini}} = \left(1 - p^2 - (1-p)^2\right) \tag{5.30}$$

where p is the fraction of positive examples ($y = 1$) in the partition. Intuitively, the Gini impurity is the probability of incorrectly classifying a randomly chosen element in the dataset if it were randomly labeled according to the class distribution in the dataset. Other popular impurity measures include (cross-)entropy (also known as information gain) and Bernoulli variance.

Regularization is an important consideration with tree-based models as one can always learn a tree that assigns exactly one training data point to each leaf node, memorizing the training dataset exactly. Regularization methods include restricting the tree building process, based on

- *minimum size*: stop splitting if the resulting children are smaller than a minimum size;
- *maximum depth*: stop splitting if the the resulting children are beyond some maximum tree depth;
- *maximum number of nodes*: stop splitting if the tree already has maximum number of allowable nodes; and
- *minimum reduction in impurity*: stop splitting if resulting children do not reduce impurity by at least $\delta\%$.

Individual trees are known as *weak learners* because they generally perform only slightly better than random guessing. Multiple trees can be combined in various ways via *ensemble methods* to create stronger classifiers. The two main types of tree ensemble methods are *bootstrap aggregation (bagging)* [17], which aims to reduce the variance of low-bias models, and *boosting* [48], which aims to reduce the bias of many low-variance models. The differences between the two methods are illustrated in Fig. 5.20.

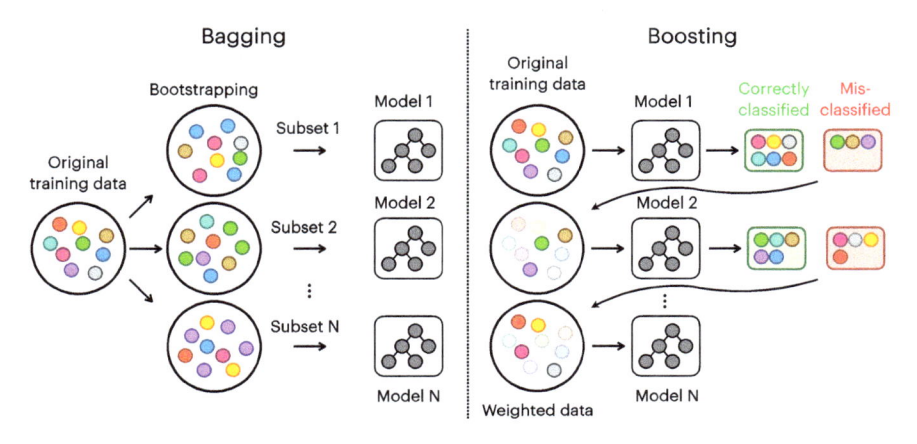

Figure 5.20. Comparison between bagging (left) and boosting (right) ensemble methods for decision trees. In bagging, N models are trained (potentially in parallel) after randomly sampling N subsets from the original training data with replacement. In boosting, N models are trained sequentially by placing higher weight on those events that are misclassified by previous models.

In bagging, the goal is to learn T models and then average the prediction for regression tasks

$$f(x|\theta) = \frac{1}{T} \sum_{t=1}^{T} f_t(x|\theta_t) \quad (5.31)$$

or return the class selected by most trees for classification tasks. Typically, the T training datasets $B_1, \ldots, B_T$, each of size N, are resampled with replacement from the original training dataset S (bootstrap resampling). If the T training datasets were completely independent, then the bias of the average model would be the same as the original model, but the variance would be reduced by a factor of T. With bootstrap resampling, the bias may increase, but reducing the variance often improves performance.

Random forests [63] combine bagging with the selection of random subsets of attributes. Instead of choosing the best split among all attributes, the best split among a random subset of k attributes is chosen. Random forests are more resistant to overfitting their training set.

One of the first boosting algorithms, adaptive boosting (AdaBoost) [47], builds a sequence of trees $f_1, \ldots, f_T$, each trained with reweighted versions of the original training dataset. The weight of an individual training sample is based on the prediction error of the previous iteration. The loss function and training procedure for each iteration are modified to account for the weighted training dataset $\{x_i, y_i, w_i\}, i = 1, \ldots, N$.

The procedure is initiated by setting uniform weights $w^{(t=0)} = 1/N$. For classification, the weighted error of the tth model is

$$E_t = \frac{\sum_{i=1}^{N} w_i^{(t)} \delta[y_i \neq f_t(x_i|\theta_t)]}{\sum_{i=1}^{N} w_i^{(t)}} \quad (5.32)$$

For highly accurate models, this error is small, $E_t \sim 0$, while for highly inaccurate models, this error may be large, e.g., $E_t \sim 0.5$. Unlike in Eq. (5.31), where the weight of each model is 1, we set a different weight β_t for each model depending on the error $\beta_t = \ln[(1-E_t)/E_t]$. For the ensemble prediction, we return the class selected by the trees with the largest sum of weights. Since β_t is larger for more accurate models, we prioritize those in the ensemble prediction.

At each iteration, the weights of the misclassified events are updated as $w^{(t+1)} = w^{(t)} \exp(\beta_t)$ and then normalized so that the sum of all the weights is 1. This reweighted dataset is then used to train the next model $f_{t+1}(x|\theta_{t+1})$. As an example, a mediocre classifier, with a misclassification rate $E_t = 30\%$, would have a corresponding $\beta_t = \ln[(1 - 0.3)/0.3)] = 0.85$. So, misclassified events get their weights multiplied by $\exp(0.85) = 2.3$, and the next tree will consider these events to be about twice as important. Now, consider an excellent classifier with an error rate $E_t = 1\%$ and $\beta_t = \ln[(1 - 0.01)/0.01)] = 4.6$. Misclassified events have their influence boosted by a factor of $\exp(4.6) = 99.5$ and thus contribute significantly to the next tree.

In HEP, a popular framework for training BDTs is the Toolkit for Multivariate Data Analysis (TMVA) [65]. More recently, XGBoost [25], which implements a variant of *gradient boosting* [48], has found widespread use in HEP due to its speed, support for GPU acceleration, and integration with the scientific Python ecosystem. Models built with XGBoost have been successfully applied in many HEP data analyses, including winning first place in the Higgs Boson Machine Learning Challenge, hosted on Kaggle [3].

5.3.2 Neural networks

A feedforward, artificial neural network, also referred to as a multi-layer perceptron, is a collection of units organized into L layers $f = f_L \circ \cdots \circ f_1$. The ℓth layer is a mapping from $d_{\ell-1}$ real-valued inputs to d_ℓ real-valued outputs, $f_\ell: \mathbb{R}^{d_{\ell-1}} \to \mathbb{R}^{d_\ell}$. Each layer is implemented as an affine transformation — a multiplication of the input vector $u \in \mathbb{R}^{d_{\ell-1}}$ by a *weight matrix* $W \in \mathbb{R}^{d_\ell \times d_{\ell-1}}$ and the addition of a bias vector $b \in \mathbb{R}^{d_\ell}$—together with a pointwise nonlinear *activation function* σ:

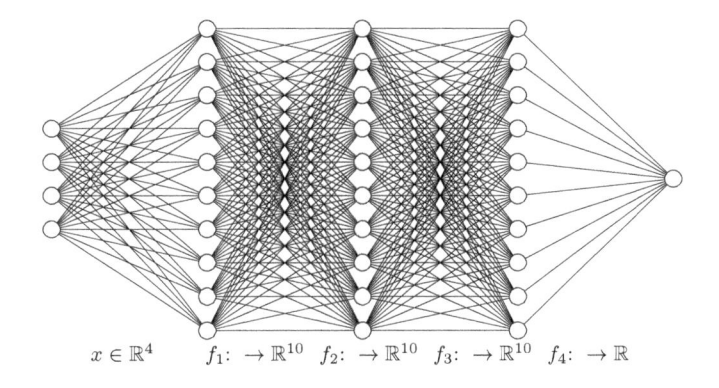

$$f_\ell(u) = \sigma(W_\ell u + b_\ell) \tag{5.33}$$

The purpose of the activation function is to enable learning more complex functions of the input. Without these nonlinearities, the network would be equivalent to a linear model. The parameters of the neural network are the complete set of weights and biases for each layer $\theta = (W_1, \ldots, W_L, b_1, \ldots, b_L)$. An example of a four-layer neural network is shown in Fig. 5.21.

Traditionally, biologically inspired *saturating* activation functions have been used, including the sigmoid function $\text{sigmoid}(u) = 1/(1 + e^{-u})$ and the hyperbolic tangent function $\tanh(u) = (e^u - e^{-u})/(e^u + e^{-u})$. Far from zero input, both sigmoid and tanh saturate at nearly constant values. This can create a problem for gradient-based optimization, especially if the inputs, weights, and biases are not properly scaled so that they take on large positive or negative values. This is known as the "vanishing gradient problem." A popular activation function that partially circumvents this issue is the rectified linear unit (ReLU) [51, 99], $\text{ReLU}(u) = \max(u, 0)$, which is widely used in deep neural networks [59]. However, ReLU suffers a similar saturation problem for negative inputs, known as the "dying ReLU problem," so a variety of alternative solutions have been proposed, including leaky ReLU [91], parameterized ReLU (PReLU) [58], exponential linear unit

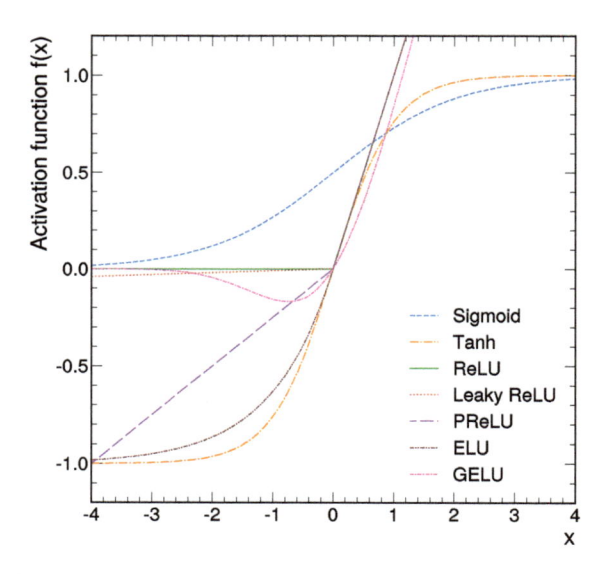

Figure 5.22. Activation functions, including biologically inspired saturating ones, such as sigmoid and tanh, and non-saturating ones, such as ReLU, leaky ReLU, PReLU, ELU, and GELU.

(ELU) [27], and Gaussian error linear unit (GELU) [60]. Visualizations of these different kinds of activation functions are shown in Fig. 5.22.

A softmax function is often used to normalize elements of a discrete vector u, or to interpret the output as a probability over a set of n_C discrete categories as in multi-classification tasks. Given a real-valued input vector $u \in \mathbb{R}^{n_C}$, the softmax function computes an output vector $v \in \mathbb{R}^{n_C}$, whose ith component is given by

$$\mathrm{softmax}(u)_i = v_i = \frac{\exp(u_i)}{\sum_{j=1}^{n_C} \exp(u_j)} \tag{5.34}$$

The output has the property that $v_i \in (0,1)$ and $\sum_i v_i = 1$. The input vector components u_i are often referred to as logits, and the softmax function is commonly used as the last layer in multi-class classifier because it is compatible with the CCE loss.

5.3.2.1 Backpropagation

To train neural networks with gradient descent, we must compute the gradient of the loss function with respect to each parameter. Naively, this requires many individual computations, but by organizing these computations in a specific way and reusing the outputs of previous computations, we can efficiently compute all of the needed gradients. This is known as the *backpropagation* [111] algorithm, and its basis is the chain rule of calculus.

As an explicit example, consider a two-layer neural network. For simplicity, we ignore the bias parameters. It is a composite function, where we can perform the computations layer by layer. Then, to compute the loss function, we only need the output of the neural network and the target y. Writing out these steps explicitly, given an input x,

$$z_1 = W_1 x \tag{5.35}$$

$$u_1 = \sigma(z_1) \tag{5.36}$$

$$z_2 = W_2 u_1 \tag{5.37}$$

$$u_2 = \sigma(z_2) \tag{5.38}$$

$$l = L(y, u_2) \tag{5.39}$$

where W_1 (W_2) is the weight matrix of the first (second) layer, z_1 (z_2) is the pre-activation output of the first (second) layer, σ is the activation function, u_1 (u_2) is the post-activation output of the first (second) layer, and l is the loss function value. These computational steps are called the "forward pass" because we progress through the network in the forward direction.

To compute the gradient of the loss function with respect to all parameters, it is natural to begin from the last layer. So, let us compute the gradient with respect to W_2 in the second layer, denoted $\partial l/\partial W_2$. To do this, we can apply the chain rule to decompose the gradient into three terms:

$$\frac{\partial l}{\partial W_2} = \left(\frac{\partial l}{\partial u_2}\right)\left(\frac{\partial u_2}{\partial z_2}\right)\left(\frac{\partial z_2}{\partial W_2}\right) \tag{5.40}$$

The term $\partial u_2/\partial z_2$ is just the derivative of the nonlinear activation function, which is often easy to compute. We can save the numerical values for each of these separate terms.

Working backward through the network, we can proceed to compute the gradient with respect to W_1 in the first layer with the chain rule:

$$\frac{\partial l}{\partial W_1} = \left(\frac{\partial l}{\partial u_2}\right)\left(\frac{\partial u_2}{\partial z_2}\right)\left(\frac{\partial z_2}{\partial u_1}\right)\left(\frac{\partial u_1}{\partial z_1}\right)\left(\frac{\partial z_1}{\partial W_1}\right) \tag{5.41}$$

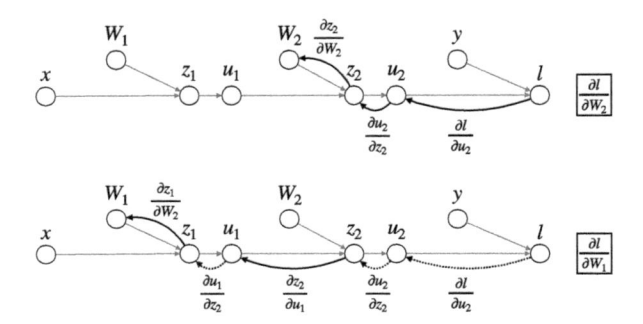

Figure 5.23. Visualizations of the backpropagation algorithm for the computation of $\frac{\partial l}{\partial W_2}$ (upper) and $\frac{\partial l}{\partial W_1}$ (lower). The forward pass is shown with straight gray lines, while the backward pass is shown with curved black lines. For the second computation of $\frac{\partial l}{\partial W_1}$, the reused computations are shown with dotted lines.

Of the five terms, two of them (highlighted in gray) have already been computed in Eq. (5.40). Furthermore, one of the remaining terms is the derivative of the first activation function with respect to its input $\partial u_1/\partial z_1$, which is equal to the previously calculated $\partial u_2/\partial z_2$. So, in order to find the gradient with respect to W_1, we only need to perform two additional computations. This is the essence of the backpropagation algorithm. This process of computing and multiplying gradients is known as the "backward pass." These rules can be extended to arbitrarily deep neural networks, as long as each layer and the loss function are differentiable.

Figure 5.23 shows the computational graph of the network, highlighting the computations needed for $\frac{\partial l}{\partial W_2}$ (upper) and $\frac{\partial l}{\partial W_1}$ (lower). Each node is an input, output, or parameter. The forward pass is shown with straight gray lines, while the backward pass is shown with curved black lines. For the second computation of $\frac{\partial l}{\partial W_1}$, the reused computations are shown with dotted lines.

Modern ML software packages implement automatic differentiation (AD), exploiting the fact that neural networks consist of a sequence of elementary arithmetic operations and functions with known derivatives and repeatedly applying the chain rule to compute the target partial derivative automatically.

5.3.3 Convolutional neural networks

An *inductive bias* expresses assumptions about the data-generating process or the space of solutions, allowing a learning algorithm to prioritize one solution over another [94]. Incorporating an inductive bias into an ML algorithm can enable them to learn more efficiently, for example, with less data or fewer parameters. These models may also generalize better to unseen data.

For image-like data, there are inductive biases that help carry out common tasks, such as classification, regression, and segmentation:

- *Locality*: Nearby areas within an image tend to contain stronger patterns.
- *Translation equivariance*: Only relative positions within an image are relevant.

As an example task, consider classifying galaxy morphologies [5, 38], e.g., spiral, elliptical, or lenticular. For this task, the solution should not depend on the location of the galaxy within an image. Moreover, many of the identifying characteristics of different types of galaxies are localized in small patches of an image.

Convolutional neural networks (CNNs), as shown in Fig. 5.24, incorporate these inductive biases through their design. An input image is described by a tensor $x \in \mathbb{R}^{H \times W \times C}$, where H is the height of the image in pixels and W is the width of the image in pixels, and at each pixel location, there is a vector of C features or *channels*. For natural images, there are typically three channels representing the red, green, and blue color channels. CNNs implement a convolution of the input image and a *filter*, or *kernel*, with height J and width K. The parameters of the filter are learnable and the convolution involves traversing over input and calculating the product of the filter W with a patch of the input, which has the same spatial shape as the filter and is centered at the target pixel. In practice, M filters are combined into a single tensor $W \in \mathbb{R}^{J \times K \times M}$.

We can calculate one element of the output tensor $y \in \mathbb{R}^{V \times U \times M}$ from the input tensor x, filter tensor W, and length-M bias vector b as[3]

$$y[v, u, m] = \left(\sum_{c=1}^{C} \sum_{j=1}^{J} \sum_{k=1}^{K} W[j, k, c, m] \, x[v+j, u+k, c] \right) + b[m] \quad (5.42)$$

For simplicity, we typically assume $J = K$ (square kernel). By repeating the operation over all the input pixels, the result of a kernel convolution is also an image.

---

[3] Note that in practice x is shifted by, e.g., $\left(\frac{J+1}{2}, \frac{K+1}{2}\right)$ in order to be symmetric around (v, u).

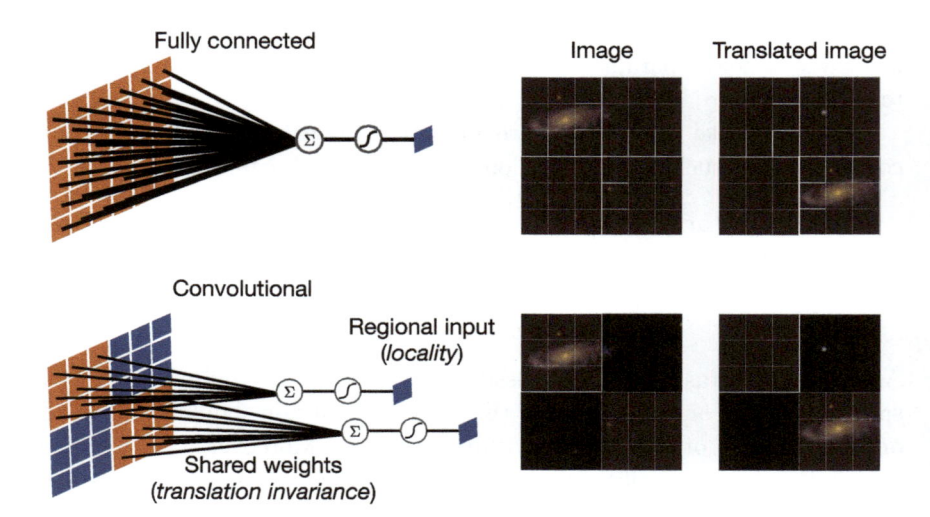

Figure 5.24. Convolutional neural networks incorporate the inductive biases of locality and translation equivariance through their design. A CNN can be interpreted as an MLP with shared weights.

A key feature of convolutions is that they are equivariant to translations: if the input image is shifted $x[i,\ldots] \to x[i-j,\ldots]$, then the output is also shifted by the same amount $y[v,\ldots] \to y[v-j,\ldots]$. Another way of looking at this is to compare this to a fully connected MLP. A fully connected MLP acting on the same image as a fully unrolled vector would generally not have this symmetry. Another way of interpreting a CNN is as a very specific type of MLP with shared weights. CNNs generally have fewer parameters than the corresponding fully connected MLP, which can improve the optimization process. The CNN structure allows for patterns in one part of an image in the training dataset effectively contribute to learning that pattern anywhere in the image.

A kernel convolution involves three hyperparameters: the kernel size (typically an odd number so that the filter has an unambiguous center), stride, and padding. In practice, kernel sizes of 1×1, 3×3, or 5×5 are frequently used. A 1×1 convolution cannot capture correlations among different pixels, but it can increase or decrease the number of features per pixel [59, 86, 119]. The stride is the number of pixels between each target pixel. For example, for a stride of 1, the target pixels are adjacent, whereas for a stride of 2, 1 pixel is skipped along each axis. Padding expands the

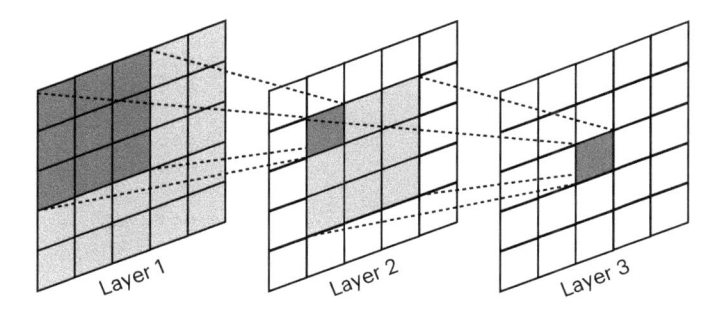

Figure 5.25. Illustration of receptive field in CNNs. Stacking two 3 × 3 kernels will lead to a larger receptive field equivalent to a 5 × 5 kernel.

input image by a specified number of pixels for when the target pixel is near the edge and the filter would extend beyond the input image.

CNNs can identify features with a spatial size larger than the kernel size by stacking multiple convolutional layers. For example, stacking two 3 × 3 kernels will lead to a larger *receptive field* equivalent to a 5 × 5 kernel, as shown in Fig. 5.25. Another approach, known as an inception module, extracts features using kernels of different sizes simultaneously [119, 120].

CNNs often use *pooling* to downsample the image, further extending the receptive field. A pooling operation is a type of aggregation that takes many input pixels and produces one output pixel. The most popular pooling operations are max pooling and average pooling. Max pooling picks the highest activation pixel value within the specified receptive field, while average pooling computes the average pixel value in the receptive field. Average pooling tends to smooth out an image, so sharp features may not be preserved. However, a drawback of max pooling is that all pixels other than the maximum one are ignored. Examples of the two operations are shown in Fig. 5.26.

By globally pooling over the entire image, a single feature vector with no spatial index can be created, giving rise to a potentially translation-invariant CNN. Reducing the image size, either through pooling or a convolution with stride larger than 1, can also be computationally beneficial. The reduction in the spatial size of an image is carried out gradually, typically by a factor of 2. After the spatial size is reduced, the number of channels is typically increased (usually by the same factor of 2). CNNs can consist of dozens or sometimes hundreds of convolutional layers, and their

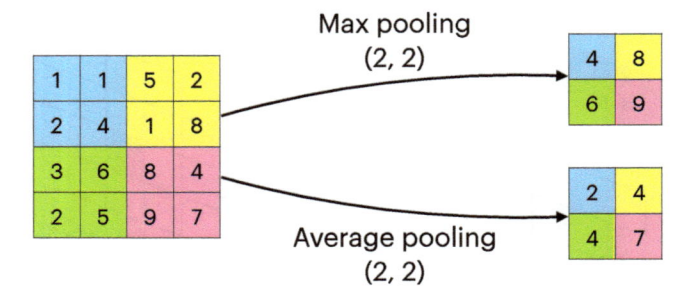

Figure 5.26. Max (upper) and average (lower) pooling in CNNs. Max pooling picks the highest activation pixel value within the specified receptive field, while average pooling computes the average pixel value in the receptive field.

optimization may be challenging due to the vanishing gradient problem. Techniques such as batch normalization [73], which normalize the tensors at each convolution layer, and skip connections [59] can mitigate this and have contributed to the tremendous success of CNNs for image-based tasks.

5.4 Applications

Machine learning has found numerous natural applications in analysis reconstruction in particle physics. At the lowest level, machine learning tools can perform hit reconstruction or track finding in individual detector systems. These tools can also identify objects such as electrons, photons, τ leptons, and jets, using information from various detector systems. Recently, researchers have also explored the use of ML to accelerate or replace computationally intensive detector simulation [4,19]. Finally, ML tools have been widely used to classify entire events as background- or signal-like, both in the final statistical analysis and at the initial trigger decision.

ML tools have found high-profile applications in particle physics. For example, BDTs were instrumental in the discovery of the Higgs boson [1, 23, 24], including in the CMS $H \to \gamma\gamma$, CMS $VH \to b\bar{b}$, and ATLAS $H \to \tau\tau$ analyses. For five key Higgs boson analyses, ML greatly increased the sensitivity of the LHC experiments, improving the discovery p-values by factors ranging from about 2–20, or equivalently, reducing the amount of data that would need to be collected by about 13–56% [106].

In this section, we discuss two representative use cases of ML in HEP, intended to highlight unique aspects of HEP data and requirements: jet tagging and trigger applications.

5.4.1 Jet tagging

Quarks and gluons originating from high energy particle collisions, such as the proton–proton collisions at the LHC, generate a cascade of other particles (mainly other quarks or gluons) that then arrange themselves into hadrons. The stable and unstable hadrons' decay products are observed by large particle detectors, reconstructed by algorithms that combine the information from different detector components and then clustered into *jets*, using physics-motivated sequential recombination algorithms [20–22, 39]. Jet identification, or *tagging*, algorithms are designed to identify the nature of the particle that initiated a given cascade, inferring it from the collective features of the particles generated in the cascade. This is illustrated in Fig. 5.27.

Traditionally, jet tagging was meant to distinguish three classes of jets: light flavor quarks, gluons, or bottom quarks. At the LHC, due to the large collision energy, new jet topologies emerge when heavy particles, e.g., W, Z, or Higgs bosons or top quarks, are produced with large momentum and decay to all-quark final states. In this case, the resulting jets contain

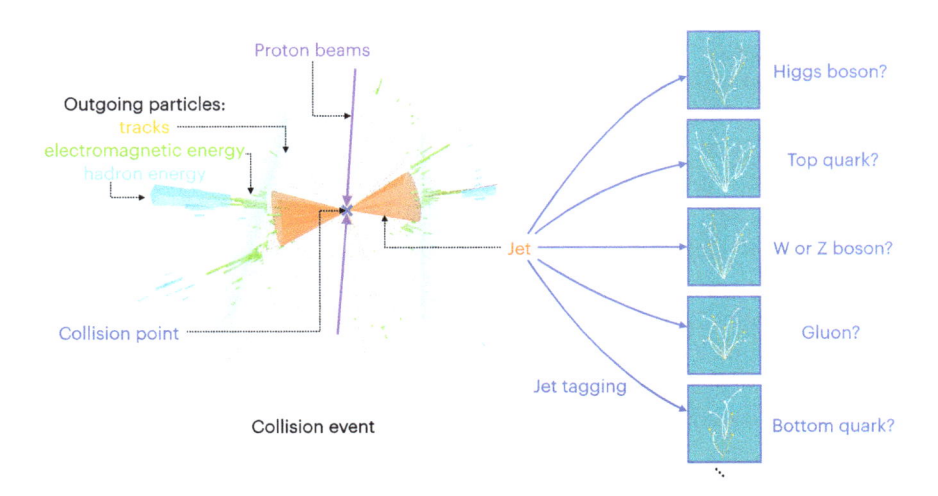

Figure 5.27. A visual representation of a collision event at the LHC and the task of jet tagging. Proton beams (purple arrows) cross at a collision point (blue cross). Outgoing particles make tracks (curved orange lines), energy deposits in the electromagnetic calorimeter (green boxes), and energy deposits in the hadron calorimeter (blue boxes). The orange cone represents a cluster of tracks and energy deposits reconstructed as a jet. The task of jet tagging is to infer, on a statistical basis, the origin of a jet based on its measured characteristics.

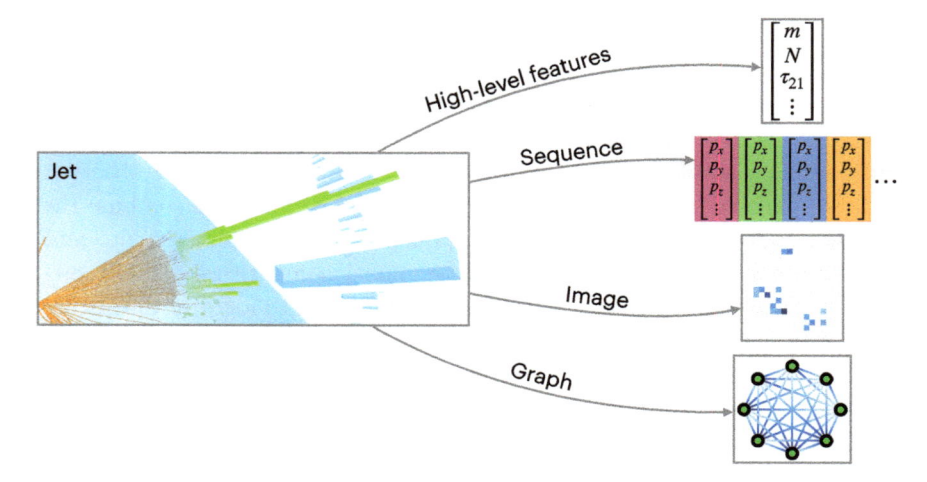

Figure 5.28. Visualization of different jet representations, including high-level features, sequences, images, and graphs.

the overlapping showers of these decay products and can appear similar to showers from single quarks or gluons. These jets are characterized by a large invariant mass and differ from quark and gluon jets in their energy correlations. Several techniques have been proposed to identify these jets by using physics-motivated quantities, collectively referred to as "jet substructure" variables [80].

Machine learning approaches for jet tagging have been extensively investigated using different *representations* of the jet, i.e., ways to encode and preprocess the information, as shown in Fig. 5.28. Different representations are naturally coupled to different kinds of ML models. For example, physics-motivated quantities, also known as high-level features, such as mass, particle multiplicity, or N-subjettiness [121] can be processed with fully connected neural networks or BDTs. A lower-level representation consists of treating the particle features as a sequence, list, or set of inputs. This type of representation can be processed by recurrent neural networks (RNNs) [90], which act on each element in a sequence and contain an internal memory, or deep sets [78], whose output is invariant under permutations of the inputs.

Jets can also be preprocessed into two-dimensional images in the (η, ϕ) plane, in which each pixel value represents the sum of the particle transverse momenta p_T or energies deposited in a given spatial detector cell. Unlike natural images, jet images are typically sparse, with only a small fraction of non-zero pixels. Jet images can be processed by CNNs, albeit potentially with some modifications, such as larger kernel sizes [32] or specialized layers optimized for sparse inputs [40].

Finally, jets can also be represented as graphs, with nodes representing particles and edges representing pairwise relationships between particles. This graph data can be processed by graph neural networks (GNNs), a class of models for reasoning about explicitly structured data [18, 50, 77, 85, 112, 113, 124]. GNNs have been successfully applied to identify Higgs bosons decaying to bottom quarks and several other types of jets [96, 97, 105]. It is also possible to encode symmetries, such as Lorentz symmetry, or other physics-inspired inductive biases in GNN models [10–13, 52].

5.4.2 Trigger applications

In HEP, a significant amount of data processing, including data compression, filtering, and selection, takes place in real time even before the data is written to disk. For example, at the LHC, proton–proton collisions occur at a rate of 40 MHz, but only roughly 1 kHz of this can be saved for offline analysis. Out of this factor of 40 000 rejection, a factor of 400 must occur within a few microseconds of the collision, and the remaining factor of 100 must occur in the next $\sim$100 ms. In addition, resources are often limited and some applications use specialized hardware such as field-programmable gate arrays (FPGAs) and application-specific integrated circuits (ASICs). Developing ML algorithms for low-latency and resource-constrained environments requires specialized techniques.

FPGAs and ASICs are designed for fast parallel processing with low power usage. The most significant difference between FPGAs and ASICs is that FPGAs can be reprogrammed, while ASICs cannot be changed once manufactured. Therefore, FPGA designs are more flexible, typically consume more power, and have slightly larger latencies than the equivalent ASIC designs. ASICs can also be designed to tolerate high levels of radiation through methods like triplication.

FPGAs contain building blocks of logic gates which can be used to construct algorithms by programming the interconnects between the components. The primary building blocks are dedicated arithmetic units or digital signal processors (DSPs), lookup tables (LUTs) for implementing logic, and two different units for storing information: registers or flip-flops (FFs) and block random-access memory (BRAM). FPGAs also contain a large number of input–output (I/O) links to receive input data and transmit output data. A schematic of a generic FPGA is shown in Fig 5.29. Unlike

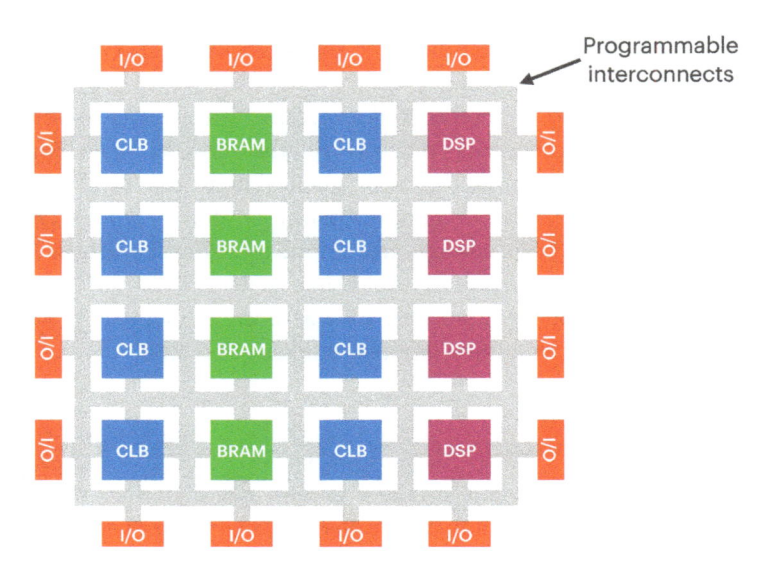

Figure 5.29. Schematic of a generic FPGA. The primary building blocks are digital signal processors (DSPs), lookup tables (LUTs), flip-flops (FFs), block random-access memory (BRAM), and input–output (I/O) links. FFs and LUTs are combined into configurable logic blocks (CLBs).

traditional CPUs, these devices are only capable of running the algorithm(s) that have been programmed. As a result of this specialization and their high clock frequencies (typically hundreds of MHz), algorithms can be executed in $\mathcal{O}(\mathrm{ns})$.

Programming FPGAs requires the use of dedicated hardware description languages (HDLs) such as VHDL or Verilog as well as a design methodology that is aware of the limitations and nature of the relevant device. All components of an FPGA program must be synchronized with the rising and falling edges of the clock, and the relations between components must be thought of in relation to these clock periods. Recently, high-level synthesis (HLS) tools [72, 115, 125], which take algorithms written in untimed (typically C) code decorated with directives and produce equivalent HDL algorithms, have lowered the barrier to entry for using FPGAs and ASICs.

Several tools, including hls4ml [43], FINN [9, 122], Conifer [117], and fwXmachina [68], have been developed to automatically create firmware from ML algorithms, as shown in Fig. 5.30. These tools have been used for applications ranging from jet tagging [75] to muon p_T regression [28],

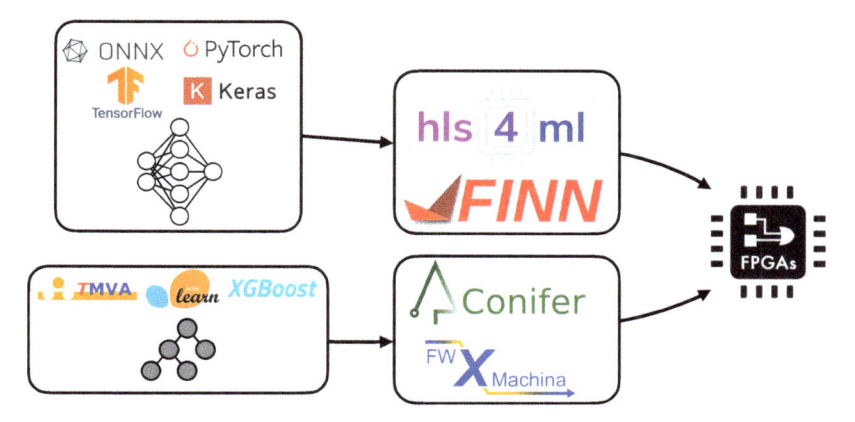

Figure 5.30. Tools like hls4ml [43], FINN [9, 122], Conifer [117], and fwXmachina [68] can translate ML algorithms from libraries like TensorFlow, Keras, PyTorch, ONNX, Scikit-learn, TMVA, or XGBoost into firmware for FPGAs.

on-detector data compression [37], charged particle tracking [45, 69], calorimeter reconstruction [71], and anomaly detection [54]. Hardware-AI co-design principles, including pruning [57], quantization [29], and parallelization [43], are important to consider to produce optimal designs that satisfy strict latency and resource constraints.

5.5 Summary and Outlook

Machine learning (ML) is now an integral part of research in high energy physics (HEP), from analysis to instrumentation, reconstruction, and simulation. Beyond being an essential tool, computational methods like ML are a third fundamental approach for studying physics on the same logical level as theory and experiment. In this chapter, we gave an overview of ML basics, types of models, and advanced techniques like model compression and surveyed some recent ML applications in HEP. There are, of course, a plethora of techniques and tools that we could not cover, many of which can be found in the HEP ML Living Review [61].

The ML in HEP research community benefits tremendously from the proliferation of public datasets and research code on GitHub, open-source software packages, such as TensorFlow [2], Keras [26], PyTorch [102], and JAX [15], commercial hardware for ML training and inference, such as NVIDIA GPUs, and widely available learning resources. This culture of openness advances the fast-paced nature of ML development, in which the latest state-of-the-art methods can be quickly extended and even superseded within months.

Strides in ML and HEP research benefit each other. In one direction, ML has helped revolutionize HEP research by enabling discoveries with less data, model-agnostic searches for exotic new physics, and exploration of final states previously thought impossible. In the other direction, HEP has unique characteristics and challenges, such as the petabyte-scale datasets, enormous data throughput, strict latency and resource constraints, and the physics and symmetry structures underpinning the data, that drive innovation in ML. Despite these benefits, there are valid criticisms of using ML in HEP research, such as the possibility of bias, the need for careful validation and calibration of ML models in data, and the difficulty of reinterpretation of HEP results that heavily rely on ML models.

Beyond the dizzying array of existing HEP applications, there continue to be many new opportunities to apply ML in surprising ways. If advances continue at the current pace, the future is bright for this (still) growing subfield.

References

[1] G. Aad, *et al.* Observation of a new particle in the search for the standard model Higgs Boson with the ATLAS detector at the LHC. *Physics Letters B*, 716:1, 2012. DOI: 10.1016/j.physletb.2012.08.020, arXiv:1207.7214 [hep-ex].

[2] M. Abadi, *et al.* (2015). TensorFlow: Large-scale machine learning on heterogeneous systems. 1603.04467 [cs.DC], https://www.tensorflow.org/.

[3] C. Adam-Bourdarios, G. Cowan, C. Germain-Renaud, I. Guyon, B. Kégl, and D. Rousseau. The Higgs machine learning challenge. *Journal of Physics: Conference Series*, 664(7):072015, 2015. DOI: 10.1088/1742-6596/664/7/072015.

[4] Y. Alanazi, N. Sato, P. Ambrozewicz, A. N. H. Blin, W. Melnitchouk, M. Battaglieri, T. Liu, and Y. Li. A survey of machine learning-based physics event generation. In: *Proceedings of the 13th International Joint Conference on Artificial Intelligence*, 2021, p. 4286. DOI: 10.24963/ijcai.2021/588, arXiv:2106.00643.

[5] AstroDave, AstroTom, C. R., Winton, joycenv, and K. Willett. Galaxy zoo - the galaxy challenge, 2013. https://kaggle.com/competitions/galaxy-zoo-the-galaxy-challenge.

[6] A. Aurisano, A. Radovic, D. Rocco, A. Himmel, M. D. Messier, E. Niner, G. Pawloski, F. Psihas, A. Sousa, and P. Vahle. A convolutional

neural network neutrino event classifier. *JINST*, 11(9):P09001, 2016. DOI: 10.1088/1748-0221/11/09/P09001, arXiv:1604.01444 [hep-ex].

[7] R. Banner, Y. Nahshan, E. Hoffer, and D. Soudry. Post-training 4-bit quantization of convolution networks for rapid-deployment. In: H. Wallach, H. Larochelle, A. Beygelzimer, F. d'Alché Buc, E. Fox, and R. Garnett (Eds.), *Advances in Neural Information Processing Systems*, Vol. 32. Curran Associates, Inc., 2019, p. 7950. arXiv:1810.05723 [cs.CV], https://proceedings.neurips.cc/paper/2019/file/c0a62e133894cdce435bcb4a5df1db2d-Paper.pdf.

[8] D. Blalock, J. J. G. Ortiz, J. Frankle, and J. Guttag. What is the state of neural network pruning? In: I. Dhillon, D. Papailiopoulos, and V. Sze (Eds.), *Proceedings of Machine Learning and Systems*, Vol. 2, 2020, p. 129. arXiv:2003.03033 [cs.LG], https://proceedings.mlsys.org/paper/2020/file/d2ddea18f00665ce8623e36bd4e3c7c5-Paper.pdf.

[9] M. Blott, T. Preusser, N. Fraser, G. Gambardella, K. O'Brien, and Y. Umuroglu. FINN-R: An end-to-end deep-learning framework for fast exploration of quantized neural networks. *ACM Transactions on Reconfigurable Technology and Systems*, 11:3, 2018. DOI: 10.1145/3242897, arXiv:1809.04570 [cs.AR].

[10] A. Bogatskiy, B. Anderson, J. T. Offermann, M. Roussi, D. W. Miller, and R. Kondor. Lorentz group equivariant neural network for particle physics. In: H. Daume III and A. Singh (Eds.), *Proceedings of the 37th International Conference on Machine Learning*, Vol. 119, 2020, p. 992. arXiv:2006.04780 [hep-ph], https://proceedings.mlr.press/v119/bogatskiy20a.html.

[11] A. Bogatskiy, T. Hoffman, D. W. Miller, and J. T. Offermann. PELICAN: Permutation equivariant and Lorentz invariant or covariant aggregator network for particle physics. In: *Machine Learning and the Physical Sciences Workshop at Neural Informance Processing Systems*, 2022. arXiv:2211.00454 [hep-ph], https://ml4physicalsciences.github.io/2022/files/NeurIPS_ML4PS_2022_132.pdf.

[12] A. Bogatskiy, T. Hoffman, D. W. Miller, J. T. Offermann, and X. Liu. Explainable equivariant neural networks for particle physics: PELICAN, 2023. arXiv:2307.16506 [hep-ph].

[13] Bogatskiy, A. *et al.* Symmetry group equivariant architectures for physics. In: *Snowmass 2021*, 2022. arXiv:2203.06153 [cs.LG].

[14] D. Bourilkov. Machine and deep learning applications in particle physics. *International Journal of Modern Physics A*, 34:1930019, 2020. DOI: 10.1142/S0217751X19300199, arXiv:1912.08245 [physics.data-an].

[15] J. Bradbury, R. Frostig, P. Hawkins, M. J. Johnson, C. Leary, D. Maclaurin, G. Necula, A. Paszke, J. VanderPlas, S. Wanderman-Milne, and Q. Zhang. JAX: Composable transformations of Python+NumPy programs, 2018. http://github.com/google/jax.

[16] L. Breiman. *Classification and Regression Trees*, 1st edn. Routledge, 1984. DOI: 10.1201/9781315139470.

[17] L. Breiman. Bagging predictors. *Machine Learning*, 24(2):123, 1996. DOI: 10.1007/BF00058655.
[18] M. M. Bronstein, J. Bruna, Y. LeCun, A. Szlam, and P. Vandergheynst. Geometric deep learning: Going beyond Euclidean data. *IEEE Signal Processing Magazine*, 34:18, 2017. DOI: 10.1109/MSP.2017.2693418, arXiv:1611.08097 [cs.CV].
[19] A. Butter and T. Plehn. Generative networks for LHC events. In: P. Calafiura, D. Rousseau, and K. Terao (Eds.), *Artificial Intelligence for High Energy Physics*. World Scientific, 2022, p. 191. DOI: 10.1142/9789811234033_0007, arXiv:2008.08558.
[20] M. Cacciari, G. P. Salam, and G. Soyez. The anti-k_t jet clustering algorithm. *JHEP*, 04:063, 2008. DOI: 10.1088/1126-6708/2008/04/063, arXiv:0802.1189 [hep-ph].
[21] M. Cacciari, G. P. Salam, and G. Soyez. FastJet user manual. *The European Physical Journal C*, 72:1896, 2012. DOI: 10.1140/epjc/s10052-012-1896-2, arXiv:1111.6097 [hep-ph].
[22] S. Catani, Y. L. Dokshitzer, M. H. Seymour, and B. R. Webber. Longitudinally invariant K_t clustering algorithms for hadron hadron collisions. *Nuclear Physics B*, 406:187–224, 1993. DOI: 10.1016/0550-3213(93)90166-M.
[23] S. Chatrchyan, *et al.* Observation of a new boson at a mass of 125 GeV with the CMS experiment at the LHC. *Physics Letters B*, 716:30, 2012. DOI: 10.1016/j.physletb.2012.08.021, arXiv:1207.7235 [hep-ex].
[24] S. Chatrchyan, *et al.* Observation of a new boson with mass near 125 GeV in pp collisions at $\sqrt{s} = 7$ and 8 TeV. *JHEP*, 06:81, 2013. DOI: 10.1007/JHEP06(2013)081, arXiv:1303.4571 [hep-ex].
[25] T. Chen and C. Guestrin. Xgboost: A scalable tree boosting system. In: *Proceedings of the 22nd ACM SIGKDD International Conference on Knowledge Discovery and Data Mining*. Association for Computing Machinery, New York, 2016, p. 785. DOI: 10.1145/2939672.2939785.
[26] F. Chollet, *et al.* Keras, 2015. https://keras.io.
[27] D. Clevert, T. Unterthiner, and S. Hochreiter. Fast and accurate deep network learning by exponential linear units (ELUs). In: Y. Bengio and Y. LeCun (Eds.), *4th International Conference on Learning Representations (ICLR), Conference Track Proceedings*, 2016. arXiv:1511.07289 [cs.LG].
[28] CMS Collaboration. The phase-2 upgrade of the CMS Level-1 trigger. CMS Technical Design Report CERN-LHCC-2020-004. CMS-TDR-021, 2020. https://cds.cern.ch/record/2714892.
[29] C. N. Coelho, A. Kuusela, S. Li, H. Zhuang, J. Ngadiuba, T. K. Aarrestad, V. Loncar, M. Pierini, A. A. Pol, and S. Summers. Automatic heterogeneous quantization of deep neural networks for low-latency inference on the edge for particle detectors. *Nature Machine Intelligence*, 2021. DOI: 10.1038/s42256-021-00356-5, arXiv:2006.10159.

[30] M. Courbariaux, Y. Bengio, and J.-P. David. BinaryConnect: Training deep neural networks with binary weights during propagations. In: C. Cortes, N. D. Lawrence, D. D. Lee, M. Sugiyama, and R. Garnett (Eds.), *Advances in Neural Information Processing Systems*, Vol. 28. Curran Associates, Inc., 2015, p. 3123. 1511.00363, https://proceedings.neurips.cc/paper/2015/file/3e15cc11f979ed25912dff5b0669f2cd-Paper.pdf.

[31] I. Csabai, F. Czako, and Z. Fodor. Quark and gluon jet separation using neural networks. *Physical Review D*, 44:1905, 1991. DOI: 10.1103/PhysRevD.44.R1905.

[32] L. de Oliveira, M. Kagan, L. Mackey, B. Nachman, and A. Schwartzman. Jet-images — deep learning edition. *Journal of High Energy Physics*, 7:69, 2016. DOI: 10.1007/JHEP07(2016)069, arXiv:1511.05190 [hep-ph].

[33] B. H. Denby. Neural networks and cellular automata in experimental high-energy physics. *Computer Physics Communications*, 49:429, 1988. DOI: 10.1016/0010-4655(88)90004-5.

[34] B. H. Denby. The use of neural networks in high-energy physics. *Neural Computation*, 5:505, 1993. DOI: 10.1162/neco.1993.5.4.505.

[35] B. H. Denby, T. Lindblad, C. S. Lindsey, G. Szekely, J. Molnar, A. Eide, S. R. Amendolia, and A. Spaziani. Investigation of a VLSI neutral network chip as part of a secondary vertex trigger. *Nuclear Instruments and Methods in Physics Research Section A: Accelerators, Spectrometers, Detectors and Associated Equipment*, 335:296, 1993. DOI: 10.1016/0168-9002(93)90284-O.

[36] G. Di Guglielmo, et al. Compressing deep neural networks on FPGAs to binary and ternary precision with `hls4ml`. *Machine Learning: Science and Technology*, 2(1):015001, 2020. DOI: 10.1088/2632-2153/aba042, 2003.06308.

[37] G. Di Guglielmo, et al. A reconfigurable neural network ASIC for detector front-end data compression at the HL-LHC. *IEEE Transactions on Nuclear Science*, 68(8):2179, 2021. DOI: 10.1109/TNS.2021.3087100, arXiv:2105.01683 [physics.ins-det].

[38] S. Dieleman, K. W. Willett, and J. Dambre. Rotation-invariant convolutional neural networks for galaxy morphology prediction. *MNRAS*, 450(2):1441, 2015. DOI: 10.1093/mnras/stv632, arXiv:1503.07077 [astro-ph.IM].

[39] Y. L. Dokshitzer, G. D. Leder, S. Moretti, and B. R. Webber. Better jet clustering algorithms. *JHEP*, 8:1, 1997. DOI: 10.1088/1126-6708/1997/08/001, arXiv:hep-ph/9707323 [hep-ph].

[40] L. Dominé and K. Terao. Scalable deep convolutional neural networks for sparse, locally dense liquid argon time projection chamber data. *Physical Review D*, 102(1):012005, 2020. DOI: 10.1103/PhysRevD.102.012005, arXiv:1903.05663 [hep-ex].

[41] Z. Dong, Z. Yao, Y. Cai, D. Arfeen, A. Gholami, M. W. Mahoney, and K. Keutzer. HAWQ-V2: Hessian aware trace-weighted quantization

of neural networks. In: H. Larochelle, M. Ranzato, R. Hadsell, M. F. Balcan, and H. Lin (Eds.), *Advances in Neural Information Processing Systems*, Vol. 33. Curran Associates, Inc., 2020, p. 18518. arXiv:1911.03852 [cs.CV], https://proceedings.neurips.cc/paper/2020/file/d77c703536718b95308130ff2e5cf9ee-Paper.pdf.

[42] Z. Dong, Z. Yao, A. Gholami, M. Mahoney, and K. Keutzer. HAWQ: Hessian aware quantization of neural networks with mixed-precision. In: *2019 IEEE/CVF International Conference on Computer Vision*, Seoul, South Korea, October 27, 2019, p. 293. DOI: 10.1109/ICCV.2019.00038, 1905.03696.

[43] J. Duarte, et al. Fast inference of deep neural networks in FPGAs for particle physics. *Journal of Instrumentation*, 13:P07027, 2018. DOI: 10.1088/1748-0221/13/07/P07027, arXiv:1804.06913 [physics.ins-det].

[44] J. Duchi, E. Hazan, and Y. Singer. Adaptive subgradient methods for online learning and stochastic optimization. *Journal of Machine Learning Research*, 12(61):2121, 2011. http://jmlr.org/papers/v12/duchi11a.html.

[45] A. Elabd, et al. Graph neural networks for charged particle tracking on FPGAs. *Frontiers in Big Data*, 5, 2022. DOI: 10.3389/fdata.2022.828666, arXiv:2112.02048 [physics.ins-det].

[46] J. Frankle and M. Carbin. The lottery ticket hypothesis: Training pruned neural networks. In: *7th International Conference on Learning Representations*, 2019. arXiv:1803.03635, https://openreview.net/forum?id=rJl-b3RcF7.

[47] Y. Freund and R. E. Schapire. A decision-theoretic generalization of on-line learning and an application to boosting. *Journal of Computer and System Sciences*, 55(1):119, 1997. DOI: 10.1006/jcss.1997.1504.

[48] J. H. Friedman. Greedy function approximation: A gradient boosting machine. *Annals of Statistics*, 29(5):1189, 2001. DOI: 10.1214/aos/1013203451.

[49] G. Cowan. Chapter 40: Statistics. In: *Particle Data Group*. R. L. Workman, et al. Review of particle physics. *PTEP*, 2022:083C01, 2022. DOI: 10.1093/ptep/ptac097, https://pdg.lbl.gov/2023/reviews/rpp2022-rev-statistics.pdf.

[50] J. Gilmer, S. S. Schoenholz, P. F. Riley, O. Vinyals, and G. E. Dahl. Neural message passing for quantum chemistry. In: *Proceedings of the 34th International Conference on Machine Learning*, Proceedings of Machine Learning Research, Vol. 70, 2017, p. 1263. arXiv:1704.01212 [cs.LG], http://proceedings.mlr.press/v70/gilmer17a.html.

[51] X. Glorot, A. Bordes, and Y. Bengio. Deep sparse rectifier neural networks. In: G. Gordon, D. Dunson, and M. Dudík (Eds.), *Proceedings of the 14th International Conference on Artificial Intelligence and Statistics (AISTATS)*, Vol. 15, 2011, p. 315. http://proceedings.mlr.press/v15/glorot11a.html.

[52] S. Gong, Q. Meng, J. Zhang, H. Qu, C. Li, S. Qian, W. Du, Z.-M. Ma, and T.-Y. Liu. An efficient Lorentz equivariant graph neural network for jet tagging. *JHEP*, 7:30, 2022. DOI: 10.1007/JHEP07(2022)030, arXiv:2201.08187 [hep-ph].

[53] M. C. Gonzalez-Garcia and M. Yokoyama. Neutrino masses, mixing, and oscillations. In: *Review of Particle Physics*, Vol. 2022, 2022, p. 083C01. DOI: 10.1093/ptep/ptac097, https://pdg.lbl.gov/2023/reviews/rpp2022-rev-neutrino-mixing.pdf.

[54] E. Govorkova, et al. Autoencoders on field-programmable gate arrays for real-time, unsupervised new physics detection at 40 MHz at the Large Hadron Collider. *Nature Machine Intelligence*, 4:154–161, 2022. DOI: 10.1038/s42256-022-00441-3, arXiv:2108.03986 [physics.ins-det].

[55] P. Goyal, P. Dollár, R. Girshick, P. Noordhuis, L. Wesolowski, A. Kyrola, A. Tulloch, Y. Jia, and K. He. Accurate, large minibatch sgd: Training imagenet in 1 hour, 2018. arXiv:1706.02677 [cs.CV].

[56] D. Guest, K. Cranmer, and D. Whiteson. Deep learning and its application to LHC physics. *Annual Review of Nuclear and Particle Science*, 68: 161, 2018. DOI: 10.1146/annurev-nucl-101917-021019, arXiv:1806.11484 [hep-ex].

[57] S. Han, H. Mao, and W. J. Dally. Deep compression: Compressing deep neural networks with pruning, trained quantization and Huffman coding. In: Y. Bengio and Y. LeCun (Eds.), *4th International Conference on Learning Representations*, San Juan, Puerto Rico, 2 May 2016. arXiv:1510.00149 [cs.CV].

[58] K. He, X. Zhang, S. Ren, and J. Sun. Delving deep into rectifiers: Surpassing human-level performance on imagenet classification. In: *2015 IEEE International Conference on Computer Vision (ICCV)*, 2015, p. 1026. DOI: 10.1109/ICCV.2015.123, 1502.01852.

[59] K. He, X. Zhang, S. Ren, and J. Sun. Deep residual learning for image recognition. In: *2016 IEEE Conference on Computer Vision and Pattern Recognition (CVPR)*, 2016, pp. 770–778. DOI: 10.1109/CVPR.2016.90.

[60] D. Hendrycks and K. Gimpel. Gaussian error linear units (GELUs), 2023. arXiv:1606.08415 [cs.LG].

[61] HEP ML Community. A living review of machine learning for particle physics, 2021. arXiv:2102.02770 [hep-ph], https://iml-wg.github.io/HEPML-LivingReview.

[62] G. Hinton. Coursera neural networks for machine learning lecture 6, 2018. https://www.cs.toronto.edu/~tijmen/csc321/slides/lecture_slides_lec6.pdf.

[63] T. K. Ho. Random decision forests. In: *Proceedings of 3rd International Conference on Document Analysis and Recognition*, Vol. 1, 1995, p. 278. DOI: 10.1109/ICDAR.1995.598994.

[64] S. Hochreiter and J. Schmidhuber. Long short-term memory. *Neural Computation*, 9(8):1735, 1997. DOI: 10.1162/neco.1997.9.8.1735, https://doi.org/10.1162/neco.1997.9.8.1735.

[65] A. Hocker, et al. TMVA - Toolkit for multivariate data analysis, 2007. arXiv:physics/0703039.

[66] E. Hoffer, I. Hubara, and D. Soudry. Train longer, generalize better: Closing the generalization gap in large batch training of neural networks, 2018. arXiv:1705.08741 [stat.ML].

[67] T. Hofmann, B. Schölkopf, and A. J. Smola. Kernel methods in machine learning. *Annals of Statistics*, 36(3), 2008. DOI: 10.1214/009053607000000677, math/0701907.

[68] T. M. Hong, B. Carlson, B. Eubanks, S. Racz, S. Roche, J. Stelzer, and D. Stumpp. Nanosecond machine learning event classification with boosted decision trees in FPGA for high energy physics. *JINST*, 16(8):P08016, 2021. DOI: 10.1088/1748-0221/16/08/P08016, arXiv:2104.03408 [hep-ex].

[69] S.-Y. Huang, Y.-C. Yang, Y.-R. Su, B.-C. Lai, J. Duarte, S. Hauck, S.-C. Hsu, J.-X. Hu, and M. S. Neubauer. Low latency edge classification GNN for particle trajectory tracking on FPGAs. In: *33rd International Conference on Field-Programmable Logic and Applications*, 2023. arXiv:2306.11330 [cs.AR].

[70] I. Hubara, M. Courbariaux, D. Soudry, R. El-Yaniv, and Y. Bengio. Quantized neural networks: Training neural networks with low precision weights and activations. *Journal of Machine Learning Research*, 18(187):1, 2018. 1609.07061, http://jmlr.org/papers/v18/16-456.html.

[71] Y. Iiyama, et al. Distance-weighted graph neural networks on FPGAs for real-time particle reconstruction in high energy physics. *Frontiers in Big Data*, 3:44, 2021. DOI: 10.3389/fdata.2020.598927, arXiv:2008.03601 [hep-ex].

[72] Intel. Intel high level synthesis compiler, 2023. https://www.intel.com/content/www/us/en/software/programmable/quartus-prime/hls-compiler.html.

[73] S. Ioffe and C. Szegedy. Batch normalization: Accelerating deep network training by reducing internal covariate shift. In: F. Bach and D. Blei (Eds.), *32nd International Conference on Machine Learning*, Vol. 37. PMLR, Lille, France, 2015, p. 448. arXiv:1502.03167, http://proceedings.mlr.press/v37/ioffe15.html.

[74] N. S. Keskar, D. Mudigere, J. Nocedal, M. Smelyanskiy, and P. T. P. Tang. On large-batch training for deep learning: Generalization gap and sharp minima, 2017. arXiv:1609.04836 [cs.LG].

[75] E. E. Khoda, et al. Ultra-low latency recurrent neural network inference on FPGAs for physics applications with hls4ml. *Machine Learning: Science and Technology*, 4(2):025004, 2023. DOI: 10.1088/2632-2153/acc0d7, arXiv:2207.00559 [cs.LG].

[76] D. P. Kingma and J. Ba. Adam: A method for stochastic optimization. In: Y. Bengio and Y. LeCun (Eds.), *3rd International Conference on Learning Representations (ICLR), Conference Track Proceedings*, 2015. arXiv:1412.6980 [cs.LG].

[77] T. Kipf, E. Fetaya, K.-C. Wang, M. Welling, and R. Zemel. Neural relational inference for interacting systems. In: J. Dy and A. Krause (Eds.), *Proceedings of the 35th International Conference on Machine Learning. Proceedings of Machine Learning Research*, Vol. 80. PMLR, Stockholmsmässan, Stockholm, Sweden, 2018, p. 2688. arXiv:1802.04687 [stat.ML], http://proceedings.mlr.press/v80/kipf18a.html.

[78] P. T. Komiske, E. M. Metodiev, and J. Thaler. Energy flow networks: Deep sets for particle jets. *Journal of High Energy Physics*, 1:121, 2019. DOI: 10.1007/JHEP01(2019)121, arXiv:1810.05165 [hep-ph].

[79] A. Krizhevsky, I. Sutskever, and G. E. Hinton. Imagenet classification with deep convolutional neural networks. In: *Proceedings of the 25th International Conference on Neural Information Processing Systems — Volume 1*. NIPS'12. Curran Associates Inc., Red Hook, NY, USA, 2012, p. 1097.

[80] A. J. Larkoski, I. Moult, and B. Nachman. Jet substructure at the large hadron collider: A review of recent advances in theory and machine learning. *Physics Reports*, 841:1, 2020. DOI: 10.1016/j.physrep.2019.11.001, arXiv:1709.04464 [hep-ph].

[81] Y. LeCun, J. S. Denker, and S. A. Solla. Optimal brain damage. In: D. S. Touretzky (Ed.), *Advances in Neural Information Processing Systems*, Vol. 2. Morgan-Kaufmann, 1990, p. 598. http://papers.nips.cc/paper/250-optimal-brain-damage.

[82] Y. A. LeCun, L. Bottou, G. B. Orr, and K.-R. Müller. *Efficient BackProp*. Springer, Berlin, 2012, pp. 9–48. DOI: 10.1007/978-3-642-35289-8_3.

[83] F. Li and B. Liu. Ternary weight networks, 2016. arXiv:1605.04711.

[84] L. Li, K. Jamieson, A. Rostamizadeh, E. Gonina, J. Ben-tzur, M. Hardt, B. Recht, and A. Talwalkar. A system for massively parallel hyperparameter tuning. In: I. Dhillon, D. Papailiopoulos, and V. Sze (Eds.), *Proceedings of Machine Learning and Systems*, Vol. 2, 2020, p. 230. 1810.05934, https://proceedings.mlsys.org/paper/2020/file/f4b9ec30ad9f68f89b29639786cb62ef-Paper.pdf.

[85] Y. Li, O. Vinyals, C. Dyer, R. Pascanu, and P. Battaglia. Learning deep generative models of graphs. In: *6th International Conference on Learning Representations, Workshop Track*, 2018. arXiv:1803.03324 [cs.LG], https://openreview.net/forum?id=Hy1d-ebAb.

[86] M. Lin, Q. Chen, and S. Yan. Network in network, 2014. arXiv:1312.4400 [cs.NE].

[87] C. S. Lindsey, B. H. Denby, and H. Haggerty. Drift chamber tracking with neural networks. *IEEE Transactions on Nuclear Science*, 40:607, 1993. DOI: 10.1109/23.256626.

[88] L. Lonnblad, C. Peterson, and T. Rognvaldsson. Finding gluon jets with a neural trigger. *Physical Review Letters*, 65:1321, 1990. DOI: 10.1103/PhysRevLett.65.1321.

[89] L. Lonnblad, C. Peterson, and T. Rognvaldsson. Using neural networks to identify jets. *Nuclear Physics B*, 349:675–702, 1991. DOI: 10.1016/0550-3213(91)90392-B.

[90] G. Louppe, K. Cho, C. Becot, and K. Cranmer. QCD-aware recursive neural networks for jet physics. *Journal of High Energy Physics*, 1:57, 2019. DOI: 10.1007/JHEP01(2019)057, arXiv:1702.00748 [hep-ph].

[91] A. Maas, A. Hannun, and A. Ng. Rectifier nonlinearities improve neural network acoustic models. In: *Proceedings of the International Conference on Machine Learning*, Atlanta, Georgia, 2013.

[92] E. Meller, A. Finkelstein, U. Almog, and M. Grobman. Same, same but different: Recovering neural network quantization error through weight factorization. In: K. Chaudhuri and R. Salakhutdinov (Eds.), *Proceedings of the 36th International Conference on Machine Learning*, Long Beach, CA, USA, Vol. 97. PMLR, 9 June 2019, p. 4486. arXiv:1902.01917 [cs.LG], http://proceedings.mlr.press/v97/meller19a.html.

[93] P. Micikevicius, S. Narang, J. Alben, G. F. Diamos, E. Elsen, D. García, B. Ginsburg, M. Houston, O. Kuchaiev, G. Venkatesh, and H. Wu. Mixed precision training. In: *6th International Conference on Learning Representations*, Vancouver, BC, Canada, 30 April 2018. 1710.03740, https://openreview.net/forum?id=r1gs9JgRZ, https://openreview.net/forum?id=r1gs9JgRZ.

[94] T. M. Mitchell. The need for biases in learning generalizations. Technical Report CBM-TR-117, Rutgers University, New Brunswick, NJ, 1980. https://www.cs.cmu.edu/~tom/pubs/NeedForBias_1980.pdf.

[95] B. Moons, K. Goetschalckx, N. V. Berckelaer, and M. Verhelst. Minimum energy quantized neural networks. In: M. B. Matthews (Ed.), *2017 51st Asilomar Conference on Signals, Systems, and Computers*, Pacific Grove, CA, USA, 29 October 2017, p. 1921. DOI: 10.1109/ACSSC.2017.8335699, arXiv:1711.00215.

[96] E. A. Moreno, et al. Interaction networks for the identification of boosted $H \to b\bar{b}$ decays. *Physical Review D*, 102:012010, 2020. DOI: 10.1103/PhysRevD.102.012010, arXiv:1909.12285 [hep-ex].

[97] E. A. Moreno, et al. JEDI-net: A jet identification algorithm based on interaction networks. *The European Physical Journal C*, 80:58, 2020. DOI: 10.1140/epjc/s10052-020-7608-4, arXiv:1908.05318 [hep-ex].

[98] M. Nagel, M. van Baalen, T. Blankevoort, and M. Welling. Data-free quantization through weight equalization and bias correction. In: *2019 IEEE/CVF International Conference on Computer Vision*, Seoul, South Korea, 27 October 2019, p. 1325. DOI: 10.1109/ICCV.2019.00141, arXiv:1906.04721 [cs.LG].

[99] V. Nair and G. E. Hinton. Rectified linear units improve restricted Boltzmann machines. In: *Proceedings of the 27th International Conference on Machine Learning (ICML)*, 2010, p. 807. https://icml.cc/Conferences/2010/papers/432.pdf.

[100] Y. E. Nesterov. A method of solving a convex programming problem with convergence rate $O(\frac{1}{k^2})$. *Doklady Akademii Nauk SSSR*, 269:543, 1983.

[101] R. Ormiston, T. Nguyen, M. Coughlin, R. X. Adhikari, and E. Katsavounidis. Noise reduction in gravitational-wave data via deep learning. *Physical*

Review Research, 2(3):033066, 2020. DOI: 10.1103/PhysRevResearch.2.033066, arXiv:2005.06534 [astro-ph.IM].

[102] A. Paszke, S. Gross, F. Massa, A. Lerer, J. Bradbury, G. Chanan, T. Killeen, Z. Lin, N. Gimelshein, L. Antiga, A. Desmaison, A. Kopf, E. Yang, Z. DeVito, M. Raison, A. Tejani, S. Chilamkurthy, B. Steiner, L. Fang, J. Bai, and S. Chintala. Pytorch: An imperative style, high-performance deep learning library. In: H. Wallach, H. Larochelle, A. Beygelzimer, F. d'Alché-Buc, E. Fox, and R. Garnett (Eds.), *Advances in Neural Information Processing Systems*, Vol. 32. Curran Associates, Inc., 2019, p. 8024. arXiv:1912.01703, http://papers.neurips.cc/paper/9015-pytorch-an-imperative-style-high-performance-deep-learning-library.pdf.

[103] J. Pata, J. Duarte, J.-R. Vlimant, M. Pierini, and M. Spiropulu. MLPF: Efficient machine-learned particle-flow reconstruction using graph neural networks. *The European Physical Journal C*, 81(5):381, 2021. DOI: 10.1140/epjc/s10052-021-09158-w, arXiv:2101.08578 [physics.data-an].

[104] J. Pata, E. Wulff, F. Mokhtar, D. Southwick, M. Zhang, M. Girone, and J. Duarte. Improved particle-flow event reconstruction with scalable neural networks for current and future particle detectors, 2023. arXiv:2309.06782 [physics.data-an].

[105] H. Qu and L. Gouskos. ParticleNet: Jet tagging via particle clouds. *Physical Review D*, 101:056019, 2020. DOI: 10.1103/PhysRevD.101.056019, arXiv:1902.08570 [hep-ph].

[106] A. Radovic, M. Williams, D. Rousseau, M. Kagan, D. Bonacorsi, A. Himmel, A. Aurisano, K. Terao, and T. Wongjirad. Machine learning at the energy and intensity frontiers of particle physics. *Nature*, 560:41, 2018. DOI: 10.1038/s41586-018-0361-2.

[107] M. Rastegari, V. Ordonez, J. Redmon, and A. Farhadi. XNOR-Net: ImageNet classification using binary convolutional neural networks. In: *14th European Conference on Computer Vision (ECCV)*. Springer International Publishing, Cham, Switzerland, 2016, p. 525. DOI: 10.1007/978-3-319-46493-0_32, arXiv:1603.05279.

[108] A. Renda, J. Frankle, and M. Carbin. Comparing rewinding and fine-tuning in neural network pruning. In: *8th International Conference on Learning Representations*, Addis Ababa, Ethiopia, 26 April 2020. arXiv:2003.02389 [cs.LG], https://openreview.net/forum?id=S1gSj0NKvB, https://openreview.net/forum?id=S1gSj0NKvB.

[109] B. P. Roe, H.-J. Yang, J. Zhu, Y. Liu, I. Stancu, and G. McGregor. Boosted decision trees as an alternative to artificial neural networks for particle identification. *Nuclear Instruments and Methods in Physics Research Section A: Accelerators, Spectrometers, Detectors and Associated Equipment*, 543(2):577, 2005. DOI: 10.1016/j.nima.2004.12.018, arXiv:physics/0408124.

[110] D. E. Rumelhart, G. E. Hinton, and R. J. Williams. Learning representations by back-propagating errors. *Nature*, 323(6088):533, 1986. DOI: 10.1038/323533a0.

[111] D. E. Rumelhart, G. E. Hinton, and R. J. Williams. Learning representations by back-propagating errors. *Nature*, 323(6088):533, 1986. DOI: 10.1038/323533a0.

[112] A. Santoro, D. Raposo, D. G. T. Barrett, M. Malinowski, R. Pascanu, P. Battaglia, and T. Lillicrap. A simple neural network module for relational reasoning. In: I. Guyon, U. V. Luxburg, S. Bengio, H. Wallach, R. Fergus, S. Vishwanathan, and R. Garnett (Eds.), *Advances in Neural Information Processing Systems 30*, Vol. 30. Curran Associates, Inc., 2017, p. 4967. arXiv:1706.01427 [cs.CL], https://papers.nips.cc/paper/2017/hash/e6acf4b0f69f6f6e60e9a815938aa1ff-Abstract.html.

[113] F. Scarselli, M. Gori, A. C. Tsoi, M. Hagenbuchner, and G. Monfardini. The graph neural network model. *IEEE Transactions on Neural Networks and Learning Systems*, 20(1):61, 2009. DOI: 10.1109/TNN.2008.2005605.

[114] B. Scholkopf and A. Smola. *Learning with Kernels: Support Vector Machines, Regularization, Optimization, and Beyond*. Adaptive Computation and Machine Learning Series. MIT Press, 2018.

[115] Siemens. Catapult high-level synthesis and verification, 2023. https://eda.sw.siemens.com/en-US/ic/catapult-high-level-synthesis/.

[116] N. Srivastava, G. Hinton, A. Krizhevsky, I. Sutskever, and R. Salakhutdinov. Dropout: A simple way to prevent neural networks from overfitting. *Journal of Machine Learning Research*, 15:1929, 2014. http://jmlr.org/papers/v15/srivastava14a.html.

[117] S. Summers, *et al.* Fast inference of boosted decision trees in FPGAs for particle physics. *Journal of Instrumentation*, 15:P05026, 2020. DOI: 10.1088/1748-0221/15/05/p05026, arXiv:2002.02534 [physics.comp-ph].

[118] I. Sutskever, J. Martens, G. Dahl, and G. Hinton. On the importance of initialization and momentum in deep learning. In: S. Dasgupta and D. McAllester (Eds.), *Proceedings of the 30th International Conference on Machine Learning*, Proceedings of Machine Learning Research, Vol. 28. PMLR, Atlanta, Georgia, USA, 2013, p. 1139. https://proceedings.mlr.press/v28/sutskever13.html.

[119] C. Szegedy, W. Liu, Y. Jia, P. Sermanet, S. Reed, D. Anguelov, D. Erhan, V. Vanhoucke, and A. Rabinovich. Going deeper with convolutions. In: *2015 IEEE Conference on Computer Vision and Pattern Recognition (CVPR)*, 2015, p. 1. DOI: 10.1109/CVPR.2015.7298594, 1409.4842.

[120] C. Szegedy, V. Vanhoucke, S. Ioffe, J. Shlens, and Z. Wojna. Rethinking the inception architecture for computer vision. In: *2016 IEEE Conference on Computer Vision and Pattern Recognition (CVPR)*, 2016, p. 2818. DOI: 10.1109/CVPR.2016.308, 1512.00567.

[121] J. Thaler and K. Van Tilburg. Identifying boosted objects with N-subjettiness. *JHEP*, 03:015, 2011. DOI: 10.1007/JHEP03(2011)015, arXiv:1011.2268 [hep-ph].

[122] Y. Umuroglu, N. J. Fraser, G. Gambardella, M. Blott, P. Leong, M. Jahre, and K. Vissers. FINN: A framework for fast, scalable binarized neural network inference. In: *Proceedings of the 2017 ACM/SIGDA International*

Symposium on Field-Programmable Gate Arrays. ACM, New York, NY, USA, 2017, p. 65. DOI: 10.1145/3020078.3021744, arXiv:1612.07119.

[123] N. Wang, J. Choi, D. Brand, C.-Y. Chen, and K. Gopalakrishnan. Training deep neural networks with 8-bit floating point numbers. In: S. Bengio, H. Wallach, H. Larochelle, K. Grauman, N. Cesa-Bianchi, and R. Garnett (Eds.), *Advances in Neural Information Processing Systems*, Vol. 31. Curran Associates, Inc., 2018, p. 7675. arXiv:1812.08011 [cs.LG], https://proceedings.neurips.cc/paper/2018/file/335d3d1cd7ef05ec77714a215134914c-Paper.pdf.

[124] X. Wang, R. Girshick, A. Gupta, and K. He. Non-local neural networks. In: *2018 IEEE/CVF Conference on Computer Vision and Pattern Recognition*, 2018, p. 7794. DOI: 10.1109/CVPR.2018.00813, arXiv:1711.07971 [cs.CV].

[125] Xilinx. Vitis unified software platform overview, 2023. https://www.xilinx.com/products/design-tools/vitis/vitis-platform.html.

[126] Y. You, I. Gitman, and B. Ginsburg. Large batch training of convolutional networks, 2017. arXiv:1708.03888 [cs.CV].

[127] M. D. Zeiler. Adadelta: An adaptive learning rate method, 2012. arXiv:1212.5701 [cs.LG].

[128] D. Zhang, J. Yang, D. Ye, and G. Hua. LQ-nets: Learned quantization for highly accurate and compact deep neural networks. In: V. Ferrari, M. Hebert, C. Sminchisescu, and Y. Weiss (Eds.), *Proceedings of the European Conference on Computer Vision*, Munich, Germany, 8 September 2018, p. 373. DOI: 10.1007/978-3-030-01237-3_23, arXiv:1807.10029 [cs.CV].

[129] R. Zhao, Y. Hu, J. Dotzel, C. D. Sa, and Z. Zhang. Improving neural network quantization without retraining using outlier channel splitting. In: K. Chaudhuri and R. Salakhutdinov (Eds.), *Proceedings of the 36th International Conference on Machine Learning*, 9 June 2019, Long Beach, CA, USA, Vol. 97. PMLR, 2019, p. 7543. arXiv:1901.09504 [cs.LG], http://proceedings.mlr.press/v97/zhao19c.html.

[130] H. Zhou, J. Lan, R. Liu, and J. Yosinski. Deconstructing lottery tickets: Zeros, signs, and the supermask. In: H. Wallach, H. Larochelle, A. Beygelzimer, F. d'Alché Buc, E. Fox, and R. Garnett (Eds.), *Advances in Neural Information Processing Systems*, Vol. 32. Curran Associates, Inc., 2019, p. 3597. arXiv:1905.01067 [cs.LG], https://proceedings.neurips.cc/paper/2019/file/1113d7a76ffceca1bb350bfe145467c6-Paper.

[131] S. Zhou, Y. Wu, Z. Ni, X. Zhou, H. Wen, and Y. Zou. DoReFa-Net: Training low bitwidth convolutional neural networks with low bitwidth gradients, 2016. 1606.06160.

[132] B. Zhuang, C. Shen, M. Tan, L. Liu, and I. Reid. Towards effective low-bitwidth convolutional neural networks. In: *2018 IEEE/CVF Conference on Computer Vision and Pattern Recognition*, Salt Lake City, UT, USA, 18 June 2018, p. 7920. DOI: 10.1109/CVPR.2018.00826, arXiv:1711.00205 [cs.CV].

Chapter 6

Jets at Colliders

Simone Marzani

Dipartimento di Fisica, Università di Genova and INFN, Sezione di Genova,
Via Dodecaneso 33, 16146, Italy
simone.marzani@ge.infn.it

6.1 A Brief Introduction

Collisions at very high energies produce a plethora of particles that are collected by detectors surrounding the interaction point. In particular, because of the conspicuous magnitude of the coupling α_s, strongly interacting particles are abundantly produced in every such collision. This occurs for both lepton (e.g., e^+e^-) colliders and for experiments in which at least one hadron is brought to collision, such as, for instance, proton–proton (pp) collisions at the CERN Large Hadron Collider (LHC) or lepton–proton (ep) or, more generically, lepton–hadron (eh) collisions, such as the ones that will be investigated by the future BNL Electron Ion Collider (EIC).

Studies of hadronic final states in e^+e^- collisions have been instrumental to establish Quantum-Chromo Dynamics (QCD) as the theory of strong interactions. This is because the initial-state leptons carry no color charge and, consequently, QCD radiation can only be produced by the final state. More complex environments are found in ep and pp collisions because QCD radiation can also originate from the hadronic initial states. In this context, past ep experiments allowed us to reach a deep understanding of the structure of the proton in terms of parton distribution functions.

2024 © The Author(s). This is an Open Access chapter published by World Scientific Publishing Company, licensed under the terms of the Creative Commons Attribution 4.0 International License (CC BY 4.0). https://doi.org/10.1142/9789819801107_0006

The successful physics program of the LHC, including the study of strong interactions at unprecedented energies, builds upon the knowledge acquired at previous particle colliders. Even more challenging is the study of collisions involving heavy ions, which allow us to probe new regions of the QCD phase diagrams, such as the color glass condensate and the quark-gluon plasma.

Studies of strong interactions in particle collisions come with enormous theoretical and experimental challenges. From the theory point of you, we can exploit a fundamental property of QCD, called asymptotic freedom, to perform perturbative calculations. In this framework, valid at high energies, i.e., far above the characteristic energy scale of hadron formation, typically denote by Λ or taken to be of the order of hadron masses, i.e., 1 GeV, the theory is weakly coupled and quarks and gluons, collectively referred to as partons, are good degrees of freedom. Thus, at high energy, QCD processes can be described in terms of scattering and production of these states.

Quarks and gluons cannot be directly detected in experimental apparatuses. We can imagine highly energetic quarks and gluons, which are produced in the collision, or from the decay of a high-mass intermediate particle, starting radiating further partons, thus reducing their energy. This process of successive splittings, usually referred to as *parton shower*, continues until one reaches the characteristic scale of hadron formation Λ. In this regime, QCD is no longer perturbative and, because of confinement, quarks and gluons form hadrons. Although some first-principle understanding of the hadronization process does exist, we often rely on phenomenological models implemented in Monte Carlo event generators to describe the transition from partons to hadrons.

One peculiar feature of parton showers is that, because of the structure of QCD matrix elements, QCD splittings preferentially happen at small angles, giving rise to a series of collimated quarks and gluons. This characteristic is not washed out by the hadronization process and hence hadrons resulting from high-energy interactions are not uniformly distributed in the detector but rather appear in a few collimated sprays that are named *jets*. This peculiar feature can be exploited to perform meaningful comparisons between theoretical calculations and experimental data. This is extremely useful because calculations in perturbative QCD feature a few final-state partons in fixed-order calculations, or a few tens of partons after the showering process, while a hadron-level event contains hundreds, if not thousands, of particles, the dynamics of which would be very difficult to individually determine. In some sense, jets constitute a portal between theory land and the real world.

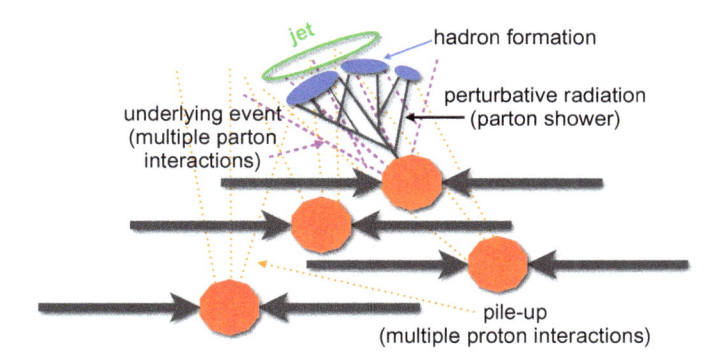

Figure 6.1. A cartoon representing jet formation in proton–proton collisions, such as the ones happening at the LHC. On top of highly energetic phenomena, which we can describe using perturbative field theory, jet formation is affected by soft, and hence non-perturbative, QCD effects, such as hadronization, the underlying event and pile-up.

Despite the remarkably successful application of perturbative calculations to describe collider phenomenology, we should bear in mind that actual collision events are much more complicated, as depicted in Fig. 6.1. Every time those two protons collide, multiple (semi-hard) partonic interactions can happen, giving rise to more hadronic activity, denoted by the term *underlying event*. Furthermore, in actual colliders, bunches of protons are brought to collisions and so multiple proton–proton interactions per bunch crossing can happen. This produces rather uniform soft radiation, usually referred to as *pile-up*. This is an unwanted consequence of the desire for higher and higher luminosity, which is necessary in order to probe rare events and pile-up mitigation is a very active area of research [1].

6.2 The Concept of Jets

The parton-shower picture described above, which may appear hand-wavy, finds its foundation on the factorization properties of QCD. However, it does simplify several aspects because it is essentially based on a semiclassical approximation of quantum field theory. If higher-order corrections are included, the concept of parton becomes ill-defined because both real emissions and virtual contributions must be taken into account. We discuss some of the issues we encounter when doing higher-order calculations in Section 6.2.1.1.

From a more practical point of view, we immediately realize that the concept of jet is somewhat ambiguous. Assigning two particles (or two partons in perturbative calculations) to the same jet, or to different ones, has some degree of arbitrariness because it depends on what we mean by two objects being collimated. In a more precise way, when talking about jets, we must introduce a resolution scale that allows us to separate objects in an event. This concept can be formalized by saying that we have to introduce a *jet definition*, i.e., a procedure that dictates how to reconstruct jets from the set of final-state hadrons (or partons) in a collision event. Jet definitions usually contain two parts:

- The *jet algorithm* is the set of rules that we must follow in order to map the set of final-state particles into jets. Most jet algorithms can be applied in an *inclusive* way, whereby the number of resulting jets is not fixed *a priori*, or in an *exclusive* mode, whereby an event is mapped into a specified number of jets. Jet algorithms feature free resolution parameters that are set by the user according to the physics case they are interested in. For example, a parameter that is present in most jet definitions for LHC studies is the jet radius, which sets the jet resolution scale in the azimuth-rapidity plane.
- The *recombination scheme* specifies how the kinematic properties of a jet, e.g., the jet four-momentum or its axis, are derived from the kinematics of the jet constituents. In most applications, the so-called E-*scheme* is employed. In this approach, the jet momentum is simply the vectorial sum of the four momenta of its constituents and the jet axis is aligned with the jet momentum. Although this choice does appear as the most natural one, specific applications may require different recipes. For instance, in the context of jet substructure studies, the so-called *Winner-Take-All* (WTA) [2] scheme is sometimes employed. In this scheme, the result of the recombination of two particles has the rapidity, azimuth, and mass of the particle with the larger transverse momentum, while the transverse momenta themselves are summed up. As a consequence, in the WTA scheme, the jet axis always lies along the direction of the hardest particle in the jet.

The design and the implementation of jet definitions are still an area of active research and a detailed discussion of the several algorithms that have been proposed in past few decades goes beyond the scope of this chapter.[1] Here, we limit ourselves to discuss and highlight, from both theoretical

---

[1] For an extensive review on jet definitions, we highly recommend the reading of Ref. [3].

and experimental viewpoints, the features of two main categories of jet definitions: the ones that feature *cone algorithms* and the ones based on *sequential recombination*. Before doing so, let us discuss the basic properties that jet definitions should respect.

6.2.1 What experimenters want... what theorists want...

Jets live at the boundary between theoretical and experimental high-energy physics. Thus, their definition should be meaningful both when applied to observable particles without considering detector effects (e.g., truth level) and also when applied to real data, which is to say detector signals. These signals include things like tracks left by charged particles or energy deposits in calorimeter cells. At the same time, the very same jet definitions should be used by theorists when performing perturbative calculations involving quarks, gluons, loops, and all that. In the 1990s, a group of theorists and Tevatron experimentalists formulated what is known as the Snowmass accord [5]. To date, this document represents the minimal set of fundamental criteria that any jet algorithm should satisfy:

(1) simple to implement in an experimental analysis;
(2) simple to implement in theoretical calculations;
(3) defined at any order of perturbation theory;
(4) yields finite cross-sections at any order of perturbation theory;
(5) yields cross-sections and distributions that are relatively insensitive to hadronization.

The first point of the list is the main demand that arises from experimental considerations. The information gathered from the various detector components, such as the trackers, the electromagnetic and hadronic calorimeters, and the muon spectrometer allows us to obtain a good picture of the types of particles that are produced in a given collision. However, jet reconstruction, often referred to as *clustering*, is typically performed at an early stage, when particle identification is still incomplete. In the first two runs of the LHC, ATLAS and CMS used different strategies to define jets. The former predominantly exploited topological clusters, so-called topoclusters, which are based on information obtained from the calorimeters, while the latter used so-called particle flow objects, which combine information from the tracker and the calorimeter to build a coherent single object. All major experimental collaborations have

Time to cluster N particles

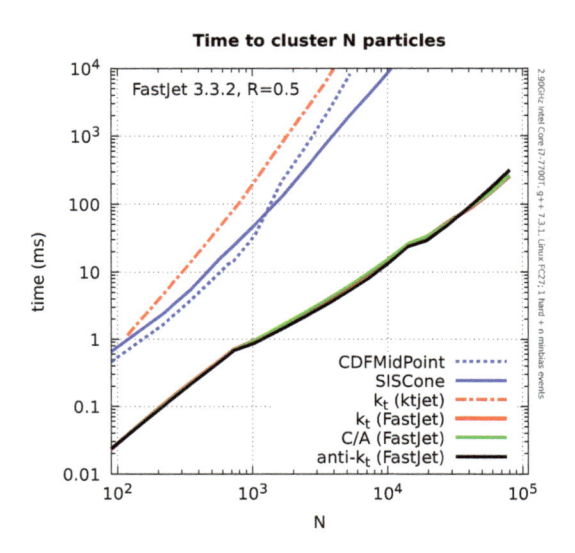

Figure 6.2. In this plot, the average clustering time for a set of representative algorithms is shown as a function of the event multiplicity N. Curves are obtained with either the algorithm original implementation or with the FastJet.

Source: Figure taken from Ref. [4].

dedicated groups actively working on the performance of jet definitions. For instance, the ATLAS collaboration has introduced for LHC Run 3 new Unified Flow Objects (UFOs) that aim to maximize performance across many orders of magnitude in the jet transverse momentum by combing the virtues of calorimetric and particle-flow approaches [6].

Once the inputs have been defined, jets must be reconstructed. Currently, the standard computer program for doing this step is FastJet[2] [7,8], used by both the experimental and theoretical communities. FastJet employs different strategies, including ideas from computational geometry, in order to speed up jet reconstruction. To illustrate this point, the plot in Fig. 6.2 shows the average time it takes to cluster an event with N particles into jets, for a few representative algorithms. There is a noticeable difference between the original ktjet implementation [9] of the k_t algorithm, which was deemed too slow, and the FastJet implementation which is faster by 2–3 orders of magnitude in the region relevant for phenomenology.

---

[2] See also http://fastjet.fr.

Conditions (2), (3), and (4) come from the theorists. We have already discussed the second one, namely, one should be able to use quarks and gluons as inputs to the jet algorithms. Conditions (3) and (4) have instead to do with *InfraRed and Collinear (IRC) safety*, a concept so important that deserves a separate discussion. We dive into this topic in Section 6.2.1.1, but before doing that, let us briefly comment on condition (5). Admittedly, this point is less precise and somewhat more subjective. Since jets are supposed to capture the "hard partons in an event," we should hope that observables built from jet quantities are as little sensitive as possible to non-perturbative effects like hadronization, the underlying event, and pile-up. Furthermore, jets should not be too sensitive to detector effects so that corrections deriving from moving from detector-level to particle-level quantities, the so-called unfolding procedure, remain under control.

6.2.1.1   *A detour about IRC safety*

Following the Snowmass accord, we work with jet algorithms that are defined and yield finite cross-sections at any order of perturbation theory. In order to better understand the origin of this request, let us work through a simple example that initially does not involve jets. We consider the calculation of the total cross-section for the production of hadrons in e^+e^- collisions. In this discussion, we are going to mostly quote results of perturbative calculations and interpret them with physical arguments. We encourage the interested readers to actually perform such calculations, following one of the many beautiful textbooks about high-energy applications of perturbative quantum field theory.

As we have already mentioned, hadrons are bound states that cannot be described in perturbation theory. However, hadron formation happens at an energy scale that is much smaller than the scale of the hard interaction. For instance, at LEP1, leptons were brought to collision at an energy Q equal to the Z boson mass, which is two orders of magnitude bigger than the hadron formation scale Λ. We can separate, we say factorize, the production cross-section as follows:

$$d\sigma_{e^+e^- \to \text{hadrons}} = \sum_{\{i\}} d\sigma_{e^+e^- \to \{i\}} \times dF_{\{i\} \to \text{hadrons}} + \mathcal{O}\left(\frac{\Lambda^2}{Q^2}\right), \qquad (6.1)$$

where the $\{i\}$ sum runs over all partonic state that are possible at a given perturbative order. Thus, up to power corrections that are small at very high-energy colliders, we can separate a partonic cross-section, which we can compute in perturbation theory, from a non-perturbative contribution that describes the fragmentation of partons into hadrons. Theorists usually focus on the former, computing higher and higher orders in the perturbative expansion. The calculation of the lowest order contribution is particularly straightforward. We only have to consider two Feynman diagrams, corresponding to the processes:

$$e^+e^- \to Z/\gamma^* \to q\bar{q}. \tag{6.2}$$

Note that the cross-section for this process at leading order (LO), or Born-level, only involves electroweak couplings. Its expression is a bit cumbersome because it involves the photon contribution, the Z one, and their interference. At energies much lower than the Z mass, but still larger enough than Λ, so that we can trust our factorized formula in Eq. (6.1), the photon contribution dominates and the inclusive, i.e., after integration over the phase space, Born cross-section has a particularly simple form:

$$\sigma_0^{\gamma*} = \frac{4\pi\alpha^2}{3Q^2} N_C \sum_f Q_f^2, \tag{6.3}$$

where α is the fine-structure constant, the sum is over the quark flavors that are accessible at the energy Q considered here, Q_f is the fractional quark electric charge, and $N_C = 3$ is the number of colors in QCD.

We are now interested in the next-to-leading order (NLO) corrections, i.e., the $\mathcal{O}(\alpha_s)$ contributions, to the partonic cross-section. We have to consider two types of contributions. First, we can dress the LO diagram with loops involving quarks and gluons. At $\mathcal{O}(\alpha_s)$, we have only one such diagram, which is depicted in Fig. 6.3(2). Second, we should remember that we are ultimately interested in the inclusive cross-section for the production of hadrons, and according to Eq. (6.1), we must consider all possible partonic states $\{i\}$. At $\mathcal{O}(\alpha_s)$, this means that we should also consider the emission of a real gluon, as shown in Figs. 6.3(3) and 6.3(4):

$$\sigma_{\text{NLO}} = \int d\Phi_2(k_1, k_2) \mid \mathcal{M}_0 + \mathcal{M}_{\text{loop}} \mid^2 + \int d\Phi_3(k_1, k_2, k_3) \mid \mathcal{M}_{\text{real}} \mid^2, \tag{6.4}$$

where $d\Phi_n$ is the n-body Lorentz-invariant phase space. It goes beyond the scope of this presentation to describe the details of the calculation. Here, we simply state that both the integral over the loop momentum in the virtual amplitude and the one over the phase space of the real gluon

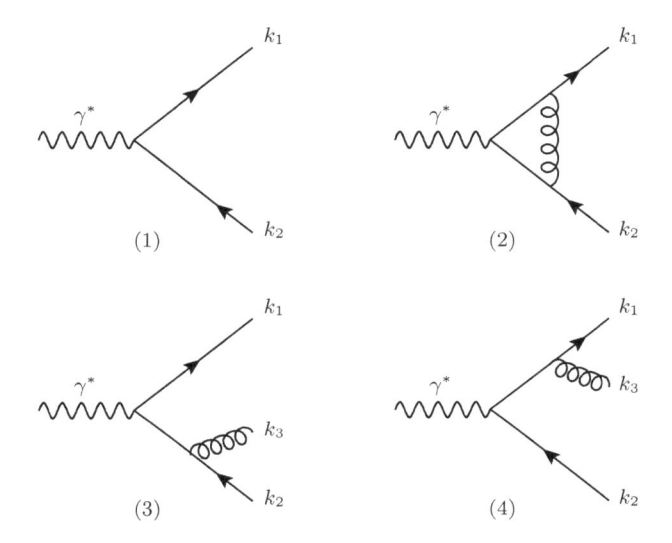

Figure 6.3. Feynman diagrams contributing to the cross-section of $e^+e^- \to q\bar{q}$ up to NLO. Diagram (1) gives the Born-level contribution, (2) the one-loop correction, and (3) and (4) describe the real-emission contribution.

are divergent. In order to understand the origin of these singularities, it is convenient to inspect the kinematics of the real emission. We find that the real emission contribution is singular when the gluon is either soft, i.e., with vanishing energy, or its three momentum becomes collinear to the directions of either the quark or the antiquark. This is a very general feature of massless gauge theories: infrared and collinear singularities arise when massless gauge bosons become soft or when two massless particles become collinear. It is interesting to note that in these singular limits, the kinematics of the three-body final state reduces to one of the two-body final states, i.e., one of the Born contribution and of the loop correction. This makes sense because we cannot resolve infinitely soft particles or two particles that are too close in angle. Thus, it is at least conceivable that the singular behavior of the real contribution may conspire with one of the loop diagrams, giving a finite result.[3] It is useful to rewrite the cross-section

---

[3] Loop amplitudes can also exhibit singularities in the ultra-violet, which can be dealt with the renormalization procedure.

separating out the divergent contributions:

$$\sigma_{\text{NLO}} = \int d\Phi_2(k_1,k_2) |\mathcal{M}_0 + \mathcal{M}_{\text{loop-finite}}|^2 + \int d\Phi_3(k_1,k_2,k_3) |\mathcal{M}_{\text{real-hard}}|^2$$
$$+ \int d\Phi_2(k_1,k_2) \left[2\text{Re}\,\mathcal{M}_0^* \mathcal{M}_{\text{loop-div}} + \int d\Phi_1(k_3) \,|\mathcal{M}_{\text{real-IRC}}|^2 \right]$$
$$+ \mathcal{O}\left(\alpha_s^2\right), \tag{6.5}$$

where we have exploited the factorization properties of phase-space integrals. The explicit computation of the problematic contributions reveals that

$$2\text{Re}\,\mathcal{M}_0^* \mathcal{M}_{\text{loop-div}} = - \int d\Phi_1(k_3)\,|\mathcal{M}_{\text{real-IRC}}|^2, \tag{6.6}$$

Thus, IRC singularities cancel and the cross-section that describes the process $e^+e^- \to$ hadrons can be safely computed in perturbation theory by considering the corresponding partonic process. The cross-section up to NLO reads

$$\sigma_{\text{NLO}} = \sigma_0 \left(1 + \frac{\alpha_s}{\pi}\right), \tag{6.7}$$

where σ_0 is the generalization of Eq. (6.3) that also includes the Z contribution and the Z/γ^* interference.

This important result is a manifestation of rather general theorems: the Bloch–Nordsieck [10] and Kinoshita–Lee–Nauenberg [11, 12] theorems state that observable transition probabilities are free of IRC singularities. However, as it stands, it leads to rather boring phenomenology because it holds for the inclusive cross-section. It is therefore interesting to investigate whether it can be generalized to more exclusive processes, such as the production of jets. In order to study this, we introduce a measurement function $J_r(\{k_i\})$ that takes as inputs the momenta of the final-state partons k_i and maps them into a set of jet momenta, with some resolution parameters r. More generally, we can consider measurement functions J_r that define physical observables, also characterized by one or more resolution scales r, with jets being a particular example. Let us go back to our e^+e^- example at $\mathcal{O}(\alpha_s)$ and consider the map J_r that produces two jets. Following the discussion about the inclusive cross-section, we write the

2-jet cross-section separating out the divergent contributions:

$$\sigma_{2\text{ jets}} = \int d\Phi_2(k_1, k_2) \mid \mathcal{M}_0 + \mathcal{M}_{\text{loop-finite}} \mid^2 J_r(k_1, k_2)$$

$$+ \int d\Phi_3(k_1, k_2, k_3) \mid \mathcal{M}_{\text{real-hard}} \mid^2 J_r(k_1, k_2, k_3) \qquad (6.8)$$

$$+ \int d\Phi_2(k_1, k_2) \left[2\text{Re}\, \mathcal{M}_0^* \mathcal{M}_{\text{loop-div}}\, J_r(k_1, k_2) \right.$$

$$\left. + \int d\Phi_1(k_3) \mid \mathcal{M}_{\text{real-IRC}} \mid^2 J_r(k_1, k_2, k_3) \right].$$

Thus, thanks to Eq. (6.6), we obtain a finite 2-jet cross-section, provided that the 3-particle measurement function reduces to the 2-particle one, in the limit in which k_3 becomes soft and/or collinear to the fermions' directions. If the measurement function has this property, we say that the observable (or the jet algorithm) is Infra-Red and Collinear (IRC) safe and its cross-section can be computed in perturbation theory. Clearly, not all possible measurement functions J_r are IRC safe. For instance, a measurement function that simply counts the number of partons, irrespectively of their momenta, does not respect this criterion. Indeed, particle multiplicity, i.e., an observable that simply counts the number of particles in a region of phase space, is not IRC safe.

Different definitions of IRC safety exist in the literature. Here, we have adopted the one in Ref. [13] that ensures cancelation of IRC singularities to any order in perturbation theory:

$$J_r(k_1 \ldots, k_i, k_j, \ldots, k_n) \longrightarrow J_r(k_1 \ldots, k_i + k_j, \ldots, k_n) \qquad \text{if } k_i \parallel k_j,$$
(6.9)

$$J_r(k_1 \ldots, k_i, \ldots, k_n) \longrightarrow J_r(k_1 \ldots, k_{i-1}, \ldots, k_{i+1} \ldots, k_n) \qquad \text{if } k_i \to 0.$$
(6.10)

IRC safe properties of jet cross-sections and related variables, such as event shapes and energy correlation functions, were first studied in Refs. [14–16]. We note that in the case of inclusive observables, for which $J_r = 1$, the cancelation between the soft and collinear contributions in Eq. (6.8) is complete and, consequently, the total cross-section remains unchanged by the emission of soft and collinear particles, as it should. In case of exclusive (but IRC safe) measurements, including jet definitions, although the singularities cancel, the kinematic dependence of the observable can cause an unbalance between real and virtual contributions, which manifests itself with the appearance of potentially large logarithmic corrections to any orders in perturbation theory. There exist techniques to resum these large logarithmic corrections to all perturbative orders. In this context, the concept of recursive IRC safety is particularly useful [17]. Finally, we also mention that recent work [18–21] has introduced the concept of Sudakov safety, which enables to extend the reach of (resummed) perturbation theory beyond the IRC domain.

6.2.2 Cone algorithms

Cone algorithms were first introduced in a famous paper by Sterman and Weinberg [13]. They are based on the idea that jets represent dominant flows of energy in a collision event. According to this definition, a 2-jet event in e^+e^- collisions is such that all, but a fraction ε of the total energy is contained into two cones of opening angle δ. Considering the $\mathcal{O}(\alpha_s)$ calculation in Eq. (6.8), we have that the two-parton measurement function is equal to unity, $J_{\varepsilon,\delta}(k_1, k_2) = 1$, because if we only have two partons in the final states, they must be hard and well separated in angle. If instead we have three partons, the 2-jet condition becomes[4]

$$J_{\varepsilon,\delta}(k_1, k_2, k_3) = \Theta\left(\min(\theta_{12}, \theta_{13}, \theta_{23}) < \delta\right)$$
$$+ \Theta\left(\min(\theta_{12}, \theta_{13}, \theta_{23}) > \delta\right) \Theta\left(\min(E_1, E_2, E_3) < \varepsilon\right), \quad (6.11)$$

where we have introduced the angles θ_{ij} between the directions of motion of particle i and j and their energies E_i. The first Θ function says that if the angle between the three momenta of the closest pair of parton is below δ, then the two partons belong to the same jet and so the event has two jets. The second set of constraints tells us that a configuration in which the three partons are well separated in angle, but the energy of the softest particle is below threshold, leads to two jets. In the limit where two directions become collinear, the second line of Eq. (6.11) is never satisfied, while the first one becomes $\Theta(0 < \delta) = 1$. Similarly, in the soft limit, the energy constraints

---

[4]We introduce the following notation for the Heaviside step function: $\Theta(a > b) = 1$, if $a > b$, and $\Theta(a > b) = 0$, if $a < b$.

are always satisfied and we obtain

$$\Theta\left(\min(\theta_{12}, \theta_{13}, \theta_{23}) < \delta\right) + \Theta\left(\min(\theta_{12}, \theta_{13}, \theta_{23}) > \delta\right) = 1.$$

Thus, Sterman–Weinberg cones are IRC safe, at least to $\mathcal{O}\left(\alpha_s\right)$.

In realistic hadron-collider environments, cone algorithms rely on the concept of a *stable cone*, i.e., the sum of all particles' momenta in the cone should point in the direction of the center of the cone. In order to find stable cones, the JetClu [22] and (various) midpoint-type [23, 24] cone algorithms use a procedure that starts with a given set of seed particles. Taking each of them as a candidate cone center, one calculates the cone contents, finds a new center based on the four-vector sum of the cone contents, and iterates until a stable cone is found. However, stable cones in a given event can overlap, meaning particles can belong to more than one cone. The most common approach is to run a split–merge procedure once the stable cones have been found. This iteratively takes the most overlapping stable cones and either merges them or splits them depending on their overlapping fraction. The procedure is repeated until one is left with non-overlapping objects that can be identified as jets.

Cone algorithms were widely used by the Tevatron experiments. For instance, the JetClu algorithm, used during Run I at the Tevatron, takes the set of particles as seeds, optionally above a given threshold in transverse momentum. This can be shown to be IRC unsafe for configuration for which two hard particles are within a distance smaller than twice the cone radius, rendering JetClu unsatisfactory for theoretical calculations. Midpoint-type algorithms, used for Run II of the Tevatron, added to the list of seeds the intermediate points between any pair of stable cones found by JetClu. This is still infrared unsafe, this time when 3 hard particles are in the same vicinity, i.e., one order later in the perturbative expansion than the JetClu algorithm. This IRC issue was solved by the introduction of the SISCone [25] algorithm, which provably finds all possible stable cones in an event, making the stable cone search IRC safe.

6.2.3 *Sequential recombination algorithms*

Due to the aforementioned problems related to IRC safety, the use of cone algorithms in modern high-energy physics experiments has dwindled in favor of approaches that form jets by successive pairwise combinations of more elementary objects. These sequential recombination algorithms are based on the idea that, from a perturbative QCD viewpoint, jets are the product of successive parton branchings, as we discussed at the beginning of this chapter. Thus, if jets are supposed to capture the properties of the very energetic partons produced in the hard collision, jet algorithms attempt to invert the parton shower process by successively recombining pairs of particles, which are close to each other, according to some user-defined (and physics-inspired) metric, into objects that can be taken as proxies to the hard partons. The metric used in this process determines the type of algorithm.

6.2.3.1 JADE algorithm

A natural choice for the distance metric is the invariant mass of the pair under examination $m_{ij}^2 = (p_i + p_j)^2$. This is clearly a Lorentz-invariant measure that reflects important features of QCD, namely, collinear splittings and soft emissions, which both produce small invariant masses, are favored. The sequential recombination algorithm that exploits this distance measure was first introduced by the JADE collaboration at the PETRA e^+e^- collider and it is therefore called the JADE algorithm [26, 27]. It is formulated as follows:

(1) Take the particles in the event as the initial list of objects.
(2) For each pair of particles i, j work out the distance

$$y_{ij} = \frac{2E_i E_j (1 - \cos\theta_{ij})}{Q^2}, \tag{6.12}$$

where Q is the total energy. If particles i and j are massless, then y_{ij} is the just their squared invariant mass, normalized to the square of the total energy.
(3) Find the minimum $y_{\min}$ of all the y_{ij}.
(4) If $y_{\min}$ is below some *jet resolution threshold* y_{cut}, then recombine i and j into a single new particle (or "pseudojet") and repeat from step 2.
(5) Otherwise, declare all remaining particles to be jets and terminate the iteration.

The parameter y_{cut} plays the role of the resolution variable of the algorithm. In particular, as y_{cut} grows smaller, softer and/or more collinear radiation is resolved into separate jets. Thus, the number of jets found by the JADE algorithm is controlled by a single parameter rather than the two parameters (ε and δ) of Sterman–Weinberg cones.

The JADE algorithm is IRC safe because soft particles are recombined at the beginning of the clustering, as they produce small invariant masses with any other particle, as do pairs of collinear particles. However, the presence of the product $E_i E_j$ in the distance measure means that two very soft particles moving in opposite directions may be recombined into a single particle in the early stages of the clustering, which is at odds with the intuitive picture of a jet as a stream of collimated particles. This peculiar behavior is reflected in a rather intricate structure of higher-order corrections for the distributions of the JADE resolution scale [28–30]. In a modern language, it is possible to show that despite being IRC safe, the JADE algorithm lacks recursive IRC safety [17].

6.2.3.2 Generalized k_t algorithm

Due to the unwanted features of the JADE algorithm, sequential recombination algorithms with alternative metrics have been suggested since the early 1990s. Here, instead of a historical discussion, we group these algorithms into a one-parameter family, the *generalized k_t algorithm* [8], discussing the most common examples. We present the algorithm in its incarnation for hadron–hadron collisions, although it can also be applied to e^+e^-, with small modifications.[5] The algorithm proceeds as follows:

(1) Take the particles in the event as the initial list of objects.
(2) From the list of objects, build two sets of distances: a pairwise distance

$$d_{ij} = \min(p_{t,i}^{2p}, p_{t,j}^{2p}) \Delta R_{ij}^2, \qquad (6.13)$$

where p is a free parameter and $\Delta R_{ij} = \sqrt{(y_i - y_j)^2 + (\phi_i - \phi_j)^2}$ is the geometric distance in the rapidity-azimuthal angle plane, and a "beam distance":

$$d_{iB} = p_{t,i}^{2p} R^2, \qquad (6.14)$$

with R the algorithm resolution parameter, often called the jet radius.

---

[5]At hadron colliders, we typically express the kinematics in terms of transverse momentum, rapidity, and azimuth, while, as we have already seen, in lepton–lepton colliders, energy and (polar) angle are preferred.

(3) Find the minimum of all d_{ij} and d_{iB}.
(4) If the smallest distance is a d_{ij}, then objects i and j are removed from the list and recombined into a pseudo-jet which is itself added to the list.
(5) If the smallest is a d_{iB}, object i is called a jet and removed from the list.
(6) Go back to step 2 until all the list of objects is empty.

In all cases, we see that if two objects are close in the rapidity-azimuth plane, as would be the case after a collinear splitting, the distance d_{ij} becomes small and the two objects are more likely to recombine. Similarly, when $\Delta R_{ij} > R$, the beam distance becomes smaller than the inter-particle distance and objects are no longer recombined, making R a typical measure of the size of the jet. Indeed, if we only have two particles, any member of the generalized k_t family will cluster them together if their distance in the rapidity-azimuth plane is less than R, irrespective of the value of the parameter p:

$$\min(p_{t,i}^{2p}, p_{t,j}^{2p}) \Delta R_{ij}^2 < \min\left(p_{t,i}^{2p} R^2, p_{t,j}^{2p} R^2\right) \quad \Rightarrow \quad \Delta R_{ij} < R. \qquad (6.15)$$

The situation changes if we consider three or more particles and indeed the shape of realistic jets strongly depends on the value of the parameter p, as we are about to discuss.

k_t algorithm: The first solution to alleviate the issues related to the JADE algorithm, while preserving the idea of clustering soft particles first, was the so-called k_t algorithm [31, 32], which corresponds to taking $p = 1$ above. According to this metric, emissions with small transverse momentum are close and therefore are recombined early in the clustering, in accordance with the parton-shower picture. However, the presence of the "minimum" in the distance measure, instead of the product, guarantees that two soft objects geometrically far apart are not recombined, thus avoiding the issues encountered with JADE. It should be noted that, while physically motivated, the k_t distance enhances sensitivity to all sorts of low-energy, non-perturbative, effects, such as the underlying event and pile-up, and for this reason, k_t jets are seldom used in hadron–hadron collisions.

Cambridge/Aachen algorithm: Another specific incarnation is the Cambridge/Aachen algorithm [33, 34], which is obtained by setting $p = 0$ above. With this choice, the metric measures a purely geometrical distance in the rapidity-azimuth plane and particles close in angles are recombined first. This choice is physically motivated because of the collinear enhancement of QCD splittings and it suffers less from the contamination due to soft backgrounds than the k_t algorithm does.

Anti-k_t algorithm: In the context of LHC physics, jets are almost always reconstructed with the anti-k_t algorithm [35], which corresponds to the generalized k_t algorithm with $p = -1$. This choice seems at first rather unnatural because it is at odds with the picture emerging from the QCD parton shower. However, its primary advantage consists in the fact that the anti-k_t metric favors clusterings between hard particles. Thus, anti-k_t jets grow by successively aggregating soft particles around a hard core, until the jet has reached a (geometrical) distance R away from its axis. Since two soft particles are always far away with the anti-k_t metric, anti-k_t jets have very little sensitivity to soft radiation and they appear to have circular shapes in the azimuth-rapidity plane. Indeed, anti-k_t behaves as a rigid cone in the soft limit, which simplifies all-order calculations of jet properties. From an experimental point of view, the resilience against soft radiation implies that anti-k_t jets are easier to calibrate. This is the main reason why it was adopted as the default jet clustering algorithm by all the LHC experiments.

6.2.4 *Sensitivity to soft physics*

The effect of soft radiation on jets clustered with different algorithms is shown in Fig. 6.4. The three-dimensional plots show calorimeter cells in the azimuth-rapidity plane, with the vertical axis measuring the transverse momentum carried by the particles in each cell. The shaded regions correspond to the active catchment area of each jet [36], which is obtained by adding infinitely soft particles (usually called ghosts) that are clustered with the hard jets, thus determining their boundaries. Anti-k_t jets have sharp and round boundaries, demonstrating resilience against soft physics. In actual experimental situations, this translates into reduced sensitivity to the underlying event and pile-up.

Another measure of a jet resilience to soft backgrounds is the back-reaction. Let us suppose to have a hard scattering event that produces a set of jets, with given properties. If we then add soft radiation to this event and we rerun the same jet algorithm, we will obtain a different set of jets. In particular, not only jets can acquire additional soft constituents, but we are also not guaranteed that a given jet will contain the same hard particles

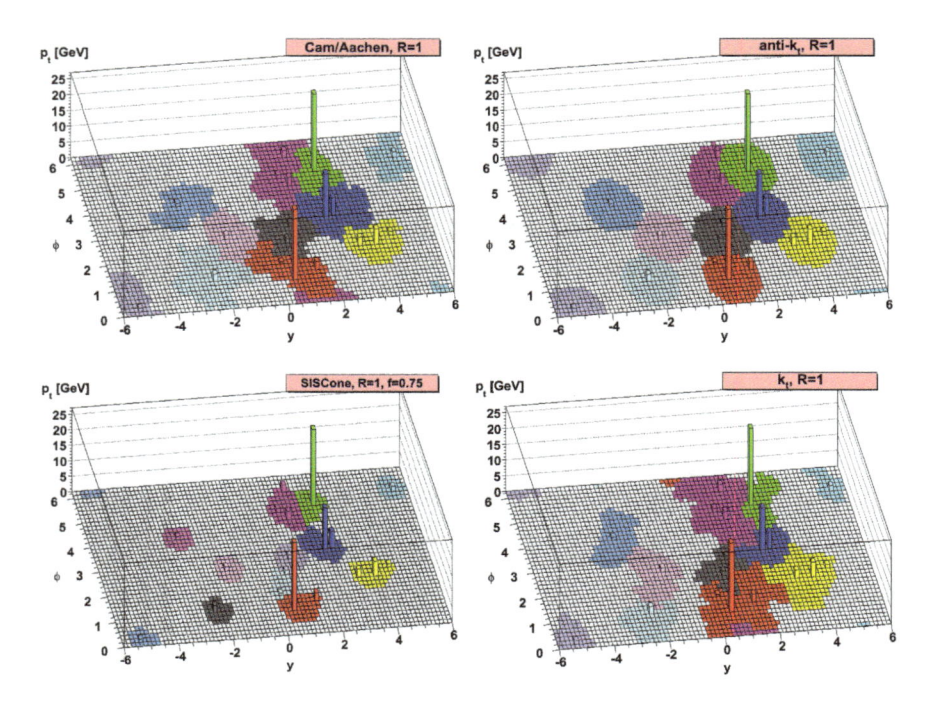

196   *Instrumentation and Techniques in High Energy Physics*

Figure 6.4. Representation of jets in the azimuth (ϕ) and rapidity (y) plane obtained with SISCone and with the three members of the generalized k_t family discussed here. All algorithms have $R = 1$, while $f = 0.75$ is the overlap parameter for the SISCone algorithm. While the jets obtained with the Cambridge/Aachen and k_t algorithm have irregular boundaries, the hard jets obtained with anti-k_t clustering are almost perfectly circular. SIScone produces smaller jets, which become more irregular as the number of constituents increases.

Source: Figure taken from Ref. [35].

of the original hard event. The back-reaction is precisely the deformation of the original jets because of the presence of the soft background. This is illustrated by the cartoon on the left-hand side of Fig. 6.5. The black dots represent the particles from the hard scattering, while the gray ones the (almost uniform) soft radiation, e.g., pile-up. The original jet, which is represented by the light gray area, is modified because of its interaction with the soft background (dark gray area).

The impact of the back-reaction on the transverse momentum of a jet is illustrated in Fig. 6.5, on the right, for different jet definitions. Positive values of $\Delta p_t^{(B)}$ correspond to transverse momentum gain, while negative ones to loss of p_t. We clearly see that back-reaction effects are strongly

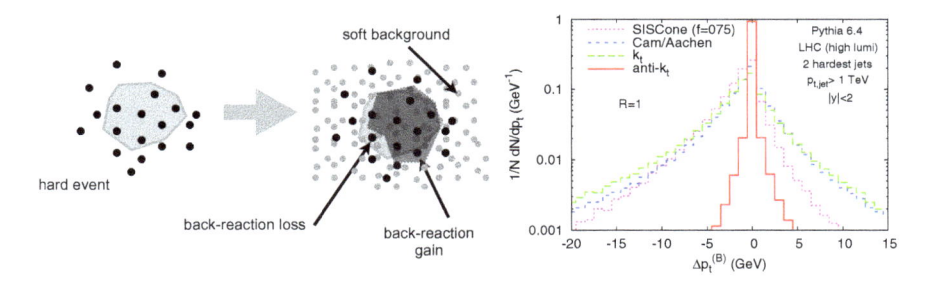

Figure 6.5. On the left, we show a cartoon describing the back-reaction effect, i.e., the modification of a hard jet due to its interactions with a soft background. On the right, we show the distribution of the transverse momentum change due to back-reaction for the anti-k_t algorithm as compared to k_t, Cambridge/Aachen, and SISCone.
Source: Figure taken from Ref. [35].

suppressed for the anti-k_t algorithm relative to the others, a feature that can help reduce the smearing of jets' momenta due to the underlying event and pile-up.

6.3 Jets as Tools

Jets are ubiquitous objects in collider phenomenology. They are employed in dedicated measurements that aim to stress-test our understanding of the Standard Model to the highest accuracy. In this context, we mention, for instance, measurements of electroweak bosons in association with many jets. Jets also appear in numerous searches for new physics, e.g., cascades of supersymmetric particles, events with one jet produced in association with missing energy in searches for dark matter, and, generically, searches for heavy states decaying into hadrons. Let us consider, for instance, a search for a new resonance X, which decays into quarks. If the mass of this new resonance is very large, it is most likely produced with a small velocity in the laboratory frame or, equivalently, with small transverse momentum. Then, its decay products move in opposite direction, fragmenting into well-separated jets, as depicted in the left-hand cartoon of Fig. 6.6. The most basic search strategy in this scenario is then to look for resonance peaks (the so-called "bump hunt") in the invariant mass distributions of the two jet with the highest transverse momenta.

We might also be interested in studying the hadronic decays of particles with mass around the electroweak scale. These can be Standard Model particles like electroweak and Higgs bosons or top quarks but also any

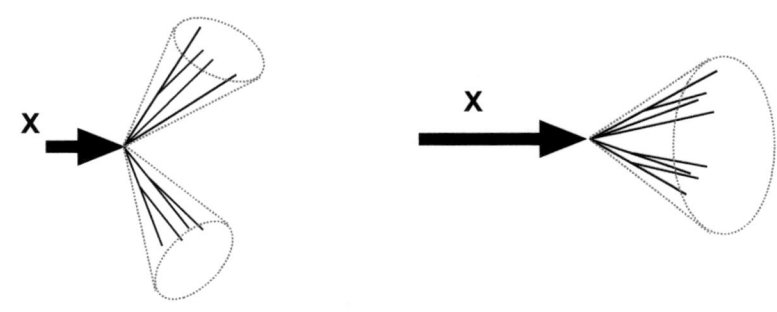

X is at rest and its decay products are reconstructed in two jets

X is boosted and its decay products are reconstructed in one jet

Figure 6.6. If a heavy state X is produced at rest, in the laboratory frame, its hadronic decay products are reconstructed as two (or more) well-separated jets, as depicted on the left. However, if its transverse momentum is large, $p_t \gtrsim 2m/R$, its decay products are collected in a single jets of radius R.

new particle with a mass of the order of the electroweak scale. Due to its unprecedentedly high colliding energy, the LHC is reaching energies far above the electroweak scale. Therefore, analyzes and searching strategies developed for earlier colliders, in which electroweak scale particles were produced with small velocities, had to be fundamentally reconsidered. In particular, as the transverse momentum of the decaying particle grows larger, its decay products become more collimated. If $p_t \gtrsim \frac{2m}{R}$, the decay products are reconstructed into a jet of radius R, as depicted in the right-hand cartoon of Fig. 6.6.

At the LHC, this scenario is particularly relevant for Higgs physics and, in particular, in the context of measurements of the couplings of the Higgs boson to the fermions. This is a crucial test for the Higgs mechanism of electroweak symmetry breaking, which predicts that the couplings to the fermions should be proportional to their masses. Despite the fact that the branching ratios into heavy (beauty b and charm c) flavors are not small, these measurements are challenging because of the large QCD background. However, when the Higgs boson is produced with a large transverse momentum, its decay products are likely to be reconstructed in a single jet. The presence of the Higgs boson can be then inferred by studying the substructure of this jet [37–39]. Consequently, jet substructure has emerged as an important tool for searches at the LHC, and a vibrant field of theoretical and experimental research has developed in the past decade, producing a variety of studies and techniques [4, 40–46].

We have already said that, in the context of resolved analyzes, the key observable to look at is the invariant mass distribution of the two jets. We can try and play the same strategy in the case of analyzes in the boosted regime and look at the jet invariant mass:

$$m_{\text{jet}}^2 = \left(\sum_{i \in \text{jet}} p_i\right)^2, \qquad (6.16)$$

where p_i are the four momenta of the jet's constituents. If the jet comprises all the debris of the decay, then its invariant mass distribution should peak around the decaying particle mass. On the other hand, background, i.e., QCD, jets have no intrinsic mass scale[6] and therefore their invariant mass must be proportional to the jet transverse momentum. Thus, one may hope that a cut on the jet invariant mass distribution will do the trick. It turns out that, despite being an important discriminant, the jet mass distribution is not enough. For instance, the jet mass turns out to be very sensitive to soft contamination, such as the underlying event and pile-up, resulting in degradation of its performance. We can see a striking example of this in Fig. 6.7, on the left. The invariant mass distribution of the leading QCD jet is shown, as measured by the ATLAS collaboration during the first run of the LHC. The different curves correspond to different pile-up situations, as measured by the number of reconstructed interaction vertices. Despite the transverse momentum of the jet being rather high, $p_t \in [600, 800]$ GeV, we can see that pile-up has a huge effect on the distribution, causing a shift of several tens of GeV. Thus, if we want to develop tools that can successfully discriminate signal and background jets in the boosted regime, we must move beyond the standard jet invariant mass and find new strategies to scrutinize the substructure of jets.

6.3.1 *Grooming and tagging*

The two key concepts in jet substructure go under the names of *grooming* and *tagging*. Broadly speaking, a grooming procedure takes a jet as an input and tries to clean it up by removing constituents which, being at wide

---

[6]The hadron-formation scale Λ is always present, but it is much lower than the energy scales considered here.

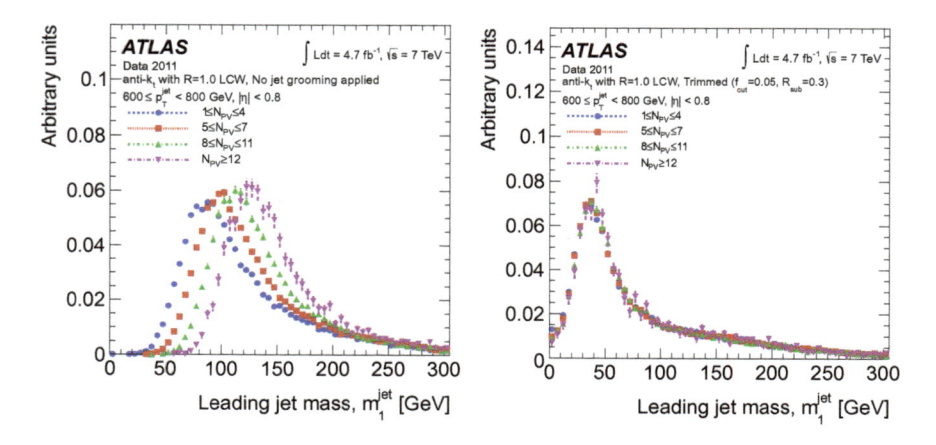

Figure 6.7. The leading jet mass distribution as measured by the ATLAS collaboration during LHC Run 1. The curves correspond to different numbers of primary vertices, a measure of the pile-up environment. The plot on the left is for standard jets, and the plot on the right for jets groomed with trimming [47].

Source: Figure taken from Ref. [48].

angle and relatively soft, are likely to come from contamination, such as the underlying event or pile-up. After this contamination has been removed, we are left with groomed jets that should be closer to our partonic picture. At this stage, we can perform a tagging step, namely, a cut on some kinematical variable that is able to distinguish signal from background. For instance, in electroweak boson decays, the energy sharing between the two daughters is symmetric. This is in contrast to QCD splittings $q \to qg$, for which the gluon tends to be soft. Thus, the energy sharing between subjets in the jets can be used as a tagging variable. We can build on this idea by noticing that high-p_t QCD jets are likely to appear as containing one prong, i.e., a hard core surrounded by a cloud of soft radiation. Electroweak (and Higgs) jets are instead two-pronged because they are initiated by a two-body decay into quarks. Jets that contain boosted top quarks feature three prongs because the top is so massive that goes through an electroweak decay before hadronizing, $t \to Wb$. If the W decays hadronically, then the top jet will contain three main subjets: one originated by the b quark and two from $W \to q\bar{q}'$. Thus, we can build tagging algorithms that distinguish jets according to the number of prongs they feature. The most famous example of such a tagger is called N-subjettiness [49, 50].

Many grooming algorithms have been developed, successfully tested, and are currently used in experimental analyzes, e.g., the mass-drop tagger [39], trimming [47], and pruning [51, 52]. A successful application of jet trimming by the ATLAS collaboration is shown in the right-hand plot of Fig. 6.7. The invariant mass distribution of the leading QCD jet is shown, but this time, jets are trimmed. We see that, in contrast to standard jets (on the left), no sensitivity to pile-up is found.[7]

By staring at the two plots in Fig. 6.7, we note a second interesting feature. The trimmed jet mass distribution is insensitive to pile-up, but it is not the same as the standard jet mass distribution, in the absence of pile-up. Thus, trimming is modifying standard jets, possibly carving away perturbative radiation too. This is something we should investigate because we do not want to undermine our perturbative understanding of jets. Regardless of their nature, substructure algorithms try to resolve jets on smaller angular and energy scales, thereby introducing new parameters. This challenges our ability of computing predictions and indeed most of the early theoretical studies of substructure tools were performed using Monte Carlo event generators. While these are powerful general-purpose tools, their essentially numerical nature offers little insight into the results produced or their detailed and precise dependence on the algorithms' parameters. A deeper, first-principle, understanding of the most used grooming and tagging techniques, both in the presence of background [53, 54] and signal jets [55, 56], was achieved when perturbative (all-order) techniques were employed to describe jet substructure. When this understanding was put at work, a second generation of substructure algorithms, which combined efficient signal-from-background discrimination together with robust theoretical understanding, was devised. One of them is SoftDrop [19], which we discuss in some detail.

The SoftDrop procedure starts with a standard jet, typically an anti-k_t jet in LHC studies. However, if we want to understand the substructure of this jet, the first thing we should do is to order the constituents in a way that reflects the jet formation history. Since the anti-k_t history does not have this feature, we recluster the jet with a more physical algorithm, namely, Cambridge-Aachen. After this procedure, we have at our disposal

---

[7]We should mention that in the more challenging pile-up environments of LHC Run 2 and 3, grooming algorithms are not enough to remove pile-up and dedicated pile-up subtraction techniques are applied.

a physically meaningful clustering tree, in which the clustering steps are ordered in angle, e.g., the final node, which corresponds to the first splitting, clusters together two prongs that are far away in the azimuth-rapidity plane. The SoftDrop procedure then performs the following steps:

(1) Break the jet j into two subjets by undoing the last stage of Cambridge-Aachen clustering. Label the resulting two subjets as j_1 and j_2.
(2) If the subjets pass the SoftDrop condition $\frac{\min(p_{t1},p_{t2})}{p_{t1}+p_{t2}} > z_{\text{cut}} \left(\frac{\Delta R_{12}}{R}\right)^\beta$, then deem j to be the final SoftDrop jet.
(3) Otherwise, redefine j to be equal to subjet with larger p_t and iterate the procedure.
(4) If j is a singleton and can no longer be declustered, then one can either remove j from consideration ("tagging mode") or leave j as the final SoftDrop jet ("grooming mode").

The difficulty posed by substructure algorithms in general, and SoftDrop in particular, is the presence of new parameters (here the angular exponent β and the energy fraction z_{cut}) that slice the phase space in a non-trivial way, resulting in potentially complicated all-order behavior of the observable at hand. This is exemplified in Fig. 6.8, where we show the

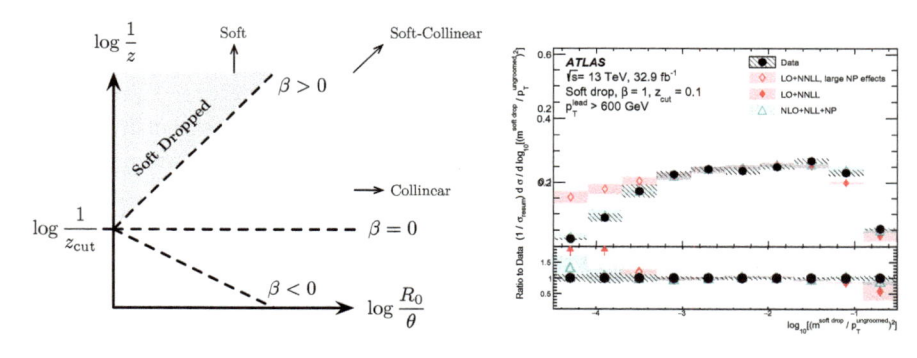

Figure 6.8. On the left, we show the SoftDrop phase space for emissions on the $(\ln \frac{1}{z}, \ln \frac{R}{\theta})$ Lund plane. For $\beta > 0$, soft emissions are vetoed while much of the soft-collinear region is maintained. For $\beta = 0$, both soft and soft-collinear emissions are vetoed. For $\beta < 0$, all (two-prong) singularities are regulated by the SoftDrop procedure. Figure taken from Ref. [19]. On the right, we show a measurement of the normalized SoftDrop jet mass distribution by the ATLAS collaboration. The data are compared to two different high-precision perturbative calculations, showing excellent agreement, across a wide range of the observable.

Source: Figure taken from Ref. [57].

phase space for soft and collinear gluon emission, from a hard parton, in the $(\ln\frac{1}{z}, \ln\frac{R}{\theta})$ plane, where $0 \leq z \leq 1$ is the energy fraction of the emitted gluon with respect to the hard parton initiating the jet, and $0 \leq \theta \leq R$ is the angle of the emission, measured from the hard parton. This representation of the soft and collinear phase space is often called the Lund plane. In the soft and collinear limit, the SoftDrop condition can be written as

$$z > z_{\text{cut}} \left(\frac{\theta}{R}\right)^\beta \quad \Rightarrow \quad \ln\frac{1}{z} < \ln\frac{1}{z_{\text{cut}}} + \beta \ln\frac{R}{\theta} \qquad (6.17)$$

Thus, vetoed emissions lie above a straight line of slope β on the $(\ln\frac{1}{z}, \ln\frac{R}{\theta})$ plane, as shown in Fig. 6.8. For $\beta > 0$, collinear splittings always satisfy the SoftDrop condition, so a SoftDrop jet still contains all of its collinear radiation. The amount of soft-collinear radiation that satisfies the SoftDrop condition depends on the relative scaling of the energy fraction z to the angle θ. As $\beta \to 0$, more of the soft-collinear radiation of the jet is removed, and in the $\beta = 0$ limit, all soft-collinear radiation is removed. In this limit, SoftDrop essentially coincides with the modified Mass Drop Tagger [53, 54]. In the strict $\beta = 0$ limit, collinear radiation is only maintained if $z > z_{\text{cut}}$. Finally, for $\beta < 0$, the soft-collinear region is removed and a hard splitting is imposed. For example, $\beta = -1$ roughly corresponds to a cut on the relative transverse momentum of the two prongs under scrutiny.

The above understanding can be formalized and precision calculations of observables measured on SoftDrop jets have been performed [58, 59]. Furthermore, while by design SoftDrop reduces the sensitivity to the underlying event and pile-up, it has been shown that this algorithm can also reduce the size of hadronization corrections, although they acquire a more complicated structure [53, 60–62].

Thus, because of their theoretical properties, i.e., good perturbative behavior and reduced sensitivity to non-perturbative physics, SoftDrop jets have emerged as an excellent playground for QCD studies at the LHC. As an example of this, we show on the right-hand side of Fig. 6.8 the comparison between a measurement of the SoftDrop jet mass performed by the ATLAS collaboration [57] (CMS also performed similar measurements, see, for instance, Ref. [63]) to high-precision perturbative calculations by two different groups: LO+NNLL [58] and NLO+NLL+NP [60], where the acronyms denote the accuracy of the calculations is apparent. The agreement is excellent and only in the three lower bins there is need for non-perturbative corrections, which are included in the NLO+NLL+NP calculation. The remarkable theoretical understanding reached for SoftDrop jets, together with the fine measurements performed by the experiments, has led to studies assessing the use of jet substructure techniques to extract Standard Model parameters, such as the strong coupling [64–66] or the top quark mass [67]. Furthermore, these observables can also be used to stress-test and improve event-simulation tools, such as parton showers and hadronization models.

6.3.2 *Jets in the era of artificial intelligence*

Our journey through jet physics would not be complete without a discussion about new approaches based on artificial intelligence. The rapid development, within and outside academia, of machine-learning techniques is having a profound impact on many aspects of society and fundamental research is not immune to this. In the context of jet physics, this revolution has brought to life a third generation of jet substructure techniques, which are now the gold standard for LHC Run 3 analyzes. However, because of its novelty and ongoing rapid progress, machine learning can still be considered an *ad hoc* field: a multitude of problems can be solved and addressed with different techniques, but some of the basic principles, the underlying structure, and a unified picture are still missing. Thus, we believe that times are not mature yet for a complete and exhaustive description of these techniques in a book.[8] Therefore, in this final section, we limit ourselves to raise a few points about the relation between deep-learning tools and expert-knowledge developed in more than ten years of jet substructure studies.

A bread and butter application of machine learning to particle physics are classification problems, including jet tagging. In this context, classification algorithms are typically trained on a control sample, which could be either Monte Carlo pseudo-data or a high-purity dataset, and then applied to an unknown sample to classify its properties. This is an example of so-called supervised learning. These ideas have been exploited in particle physics for a long time. However, because of limitations on efficiency and computing power, algorithms used to be applied to relatively low-dimensional projections of the full radiation pattern that one wished to classify. Even so, such projections usually corresponded to physically motivated observables, such as the jet mass, and therefore limitation in

---

[8]We refer the interested readers to Ref. [70].

performance was mitigated with physics understanding. Current developments in machine learning allows us to move away from low-dimensional projections and exploit deep neural networks to perform classification. This opens up the door to almost limitless possibilities that go far beyond supervised learning. Just to mention a few examples, unsupervised learning has led to the design of algorithms, which can be applied, for instance, to anomaly detection in new physics searches. Furthermore, neural network can be used not only for classification but also for simulations (e.g., parton showers, hadron formation, and detector responses) in a fast and faithful way — the particle physics equivalent of deepfake.

The most successful innovations in machine learning are coming from outside high energy physics (and chiefly from the industry giants). However, particle physics provides us with one of the few examples of a big-data system with a deep scientific understanding of the underlying model, potentially allowing us to get more insight into the broader machine-learning field. In this context, an interesting debate to mention has to do with the choice of inputs and architecture to use when building a neural network for a specific physics case. Should we be as agnostic as possible and provide a complex network with raw data from the experiments? Or should we build on our understanding of the physical processes and use physically motivated observables as input to (possibly simpler) machine learning algorithms? The former approach has the advantage of being unbiased, while following the second one we may hope, for instance, to understand what kind of information the network is learning from the data.

We close this discussion with a comparison between these two philosophies. In order to do that, we go back to our electroweak boson tagging problem. We can view a particle detector, and in particular the hadronic calorimeter, as a huge camera, taking pictures of particle collisions and, using the information from the calorimeter cells, we can build jet images [68, 71]. After appropriate averaging and pre-processing, the jet images can be input to machine-learning algorithms that are appropriate for pattern recognition, such as convolutional neural networks. Alternatively, we can build a picture of the jets based on our understanding of QCD. This is provided by the (primary) Lund jet plane [69]. The Lund jet plane is constructed by parsing backward the clustering history of a jet's Cambridge-Aachen tree, similar to the SoftDrop procedure previously described. At each step, the kinematics of the splitting, e.g., the distance between the two branches in the azimuth-rapidity plane Δ and the relative transverse momentum k_t, is recorded. The set of values that we obtain always following

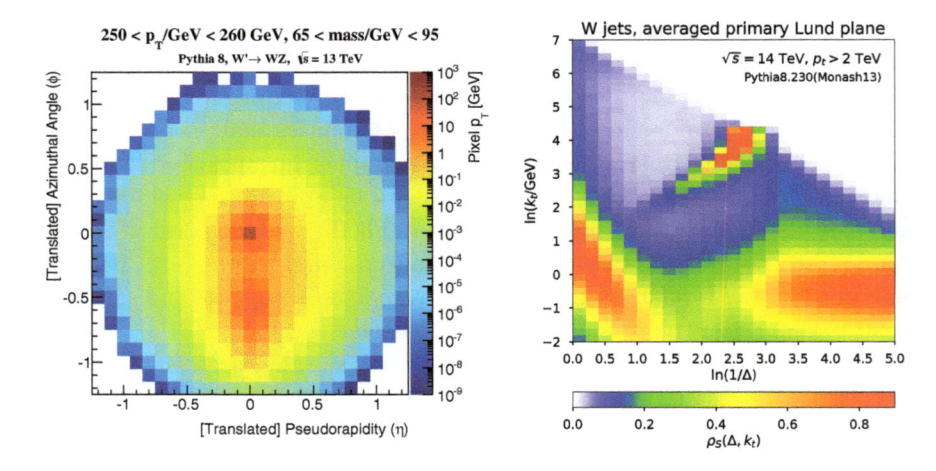

Figure 6.9. Two pictures of W-initiated jets. On the left, the average calorimetric image, after pre-processing, where pixel colors represent the energy deposited in a calorimeter cell (figure taken from Ref. [68]). On the right, the average primary Lund plane density, where colors represent the density of recorded splittings in a given $(-\log \Delta, \log k_t)$ cell (figure taken from Ref. [69]). Note that the presence of the initiating W boson appears as a two-pronged structure, on the left, and as a hot spot at $\log \frac{1}{\Delta} \simeq 0.4$ and $\log k_t \simeq 4$, on the right. Detailed comparisons between the two plots should be taken with a grain of salt because of the rather different transverse momentum selections.

the harder branch constitutes the primary Lund jet plane. Considering many jets, we can construct the density of the primary plane. Examples of a jet image and a primary Lund plane image for W jets are shown in Fig. 6.9.

6.4 Closing Remarks

We conclude this chapter by stressing once again that the key aspect that repeatedly appears in the context of jet physics is the design of algorithms that can be meaningfully used by both theory and experimental communities. Very often this implies the necessity of a tradeoff between performance and robustness. In the 1990s, one of the reason for preferring cone algorithms over sequential recombination ones was the issue of speed, an example of performance. However, as it turned out, the algorithms used at the Tevatron were not robust because they lacked IRC safety.

In the context of jet substructure studies, by performance, we usually mean the discriminating power of a tool when extracting a given signal from the QCD background, and by robustness we mean the ability to describe the tool using perturbative QCD, i.e., being as little sensitive as possible to model-dependent effects such as hadronization, the underlying event, pile-up, or detector effects, all of which likely translate into systematic uncertainties in an experimental analysis.

We can apply similar considerations to the latest-generation machine-learning tools. On the one hand, these algorithms augment performance so much that they have become standard tools for collider physics. On the other hand, they are sometimes treated as black-boxes and, more often than not, their robustness is difficult to assess with standard technologies. It is an exciting challenge for particle theorists and experimentalists to find new ways to study these tools, assess their systematics, and, ultimately, find the best metric to measure their robustness.

Acknowledgments

I am indebted to Matteo Cacciari, Gavin Salam, and Gregory Soyez for many useful discussions about jet physics and for suggesting ideas and material for this chapter. I would also like to thank Andrea Coccaro, Marc Leblanc, and Jennifer Roloff for a critical reading of this manuscript. Finally, I am grateful to four graduate students at the University of Genova, Simone Caletti, Andrea Ghira, Anna Rinaudo, and Martino Tanasini, for reading this chapter and for providing me with very valuable feedback.

References

[1] G. Soyez. Pileup mitigation at the LHC: A theorist's view. *Physics Reports*, 803:1–158, 2019. DOI: 10.1016/j.physrep.2019.01.007.
[2] A. J. Larkoski, D. Neill, and J. Thaler. Jet shapes with the broadening axis. *JHEP*, 04:017, 2014. DOI: 10.1007/JHEP04(2014)017.
[3] G. P. Salam. Towards jetography. *European Physical Journal C*, 67:637–686, 2010. DOI: 10.1140/epjc/s10052-010-1314-6.
[4] S. Marzani, G. Soyez, and M. Spannowsky. *Looking Inside Jets: An Introduction to Jet Substructure and Boosted-Object Phenomenology*, Vol. 958. Springer, 2019. DOI: 10.1007/978-3-030-15709-8.
[5] J. E. Huth, *et al.* Toward a standardization of jet definitions. In: *1990 DPF Summer Study on High-energy Physics: Research Directions for the Decade (Snowmass 90)*, December 1990, pp. 0134–136.
[6] G. Aad, *et al.* Optimisation of large-radius jet reconstruction for the ATLAS detector in 13 TeV proton–proton collisions. *European Physical Journal C*, 81(4):334, 2021. DOI: 10.1140/epjc/s10052-021-09054-3.

[7] M. Cacciari and G. P. Salam. Dispelling the N^3 myth for the k_t jet-finder. *Physics Letters B*, 641:57–61, 2006. DOI: 10.1016/j.physletb.2006.08.037.

[8] M. Cacciari, G. P. Salam, and G. Soyez. FastJet user manual. *European Physical Journal C*, 72:1896, 2012. DOI: 10.1140/epjc/s10052-012-1896-2.

[9] J. M. Butterworth, J. P. Couchman, B. E. Cox, and B. M. Waugh. KtJet: A C++ implementation of the K-perpendicular clustering algorithm. *Computer Physics Communications*, 153:85–96, 2003. DOI: 10.1016/S0010-4655(03)00156-5.

[10] F. Bloch and A. Nordsieck. Note on the radiation field of the electron. *Physical Review*, 52:54–59, 1937. DOI: 10.1103/PhysRev.52.54.

[11] T. Kinoshita. Mass singularities of Feynman amplitudes. *Journal of Mathematical Physics*, 3:650–677, 1962. DOI: 10.1063/1.1724268.

[12] T. D. Lee and M. Nauenberg. Degenerate systems and mass singularities. *Physical Review*, 133:B1549–B1562, 1964. DOI: 10.1103/PhysRev.133.B1549.

[13] G. F. Sterman and S. Weinberg. Jets from quantum chromodynamics. *Physical Review Letters*, 39:1436, 1977. DOI: 10.1103/PhysRevLett.39.1436.

[14] G. F. Sterman. Mass divergences in annihilation processes. 1. Origin and nature of divergences in cut vacuum polarization diagrams. *Physical Review D*, 17:2773, 1978. DOI: 10.1103/PhysRevD.17.2773.

[15] G. F. Sterman. Mass divergences in annihilation processes. 2. Cancellation of divergences in cut vacuum polarization diagrams. *Physical Review D*, 17:2789, 1978. DOI: 10.1103/PhysRevD.17.2789.

[16] G. F. Sterman. Zero mass limit for a class of jet related cross-sections. *Physical Review D*, 19:3135, 1979. DOI: 10.1103/PhysRevD.19.3135.

[17] A. Banfi, G. P. Salam, and G. Zanderighi. Principles of general final-state resummation and automated implementation. *JHEP*, 03:073, 2005. DOI: 10.1088/1126-6708/2005/03/073.

[18] A. J. Larkoski and J. Thaler. Unsafe but calculable: Ratios of angularities in perturbative QCD. *JHEP*, 09:137, 2013. DOI: 10.1007/JHEP09(2013)137.

[19] A. J. Larkoski, S. Marzani, G. Soyez, and J. Thaler. Soft drop. *JHEP*, 05:146, 2014. DOI: 10.1007/JHEP05(2014)146.

[20] A. J. Larkoski, S. Marzani, and J. Thaler. Sudakov safety in perturbative QCD. *Physical Review D*, 91(11):111501, 2015. DOI: 10.1103/PhysRevD.91.111501.

[21] P. T. Komiske, E. M. Metodiev, and J. Thaler. The hidden geometry of particle collisions. *JHEP*, 07:006, 2020. DOI: 10.1007/JHEP07(2020)006.

[22] F. Abe, et al. The topology of three jet events in $\bar{p}p$ collisions at $\sqrt{s} = 1.8$ TeV. *Physical Review D*, 45:1448–1458, 1992. DOI: 10.1103/PhysRevD.45.1448.

[23] G. C. Blazey, et al. Run II jet physics. In: *Physics at Run II: QCD and Weak Boson Physics Workshop: Final General Meeting*, May 2000, pp. 47–77.

[24] V. M. Abazov, et al. Measurement of the inclusive jet cross section in $p\bar{p}$ collisions at $\sqrt{s} = 1.96$ TeV. *Physical Review D*, 85:052006, 2012. DOI: 10.1103/PhysRevD.85.052006.

[25] G. P. Salam and G. Soyez. A practical seedless infrared-safe cone jet algorithm. *JHEP*, 05:086, 2007. DOI: 10.1088/1126-6708/2007/05/086.
[26] W. Bartel, *et al.* Experimental studies on multi-jet production in e+ e- annihilation at PETRA energies. *Zeitschrift für Physik*, 33:23, 1986. DOI: 10.1007/BF01410449.
[27] S. Bethke, *et al.* Experimental investigation of the energy dependence of the strong coupling strength. *Physics Letters B*, 213:235–241, 1988. DOI: 10.1016/0370-2693(88)91032-5.
[28] N. Brown and W. J. Stirling. Jet cross-sections at leading double logarithm in e+ e- annihilation. *Physics Letters B*, 252:657–662, 1990. DOI: 10.1016/0370-2693(90)90502-W.
[29] S. Catani. Jet topology and new jet counting algorithms. *Ettore Majorana International Science Series Physical Sciences*, 60:21–41, 1992. DOI: 10.1007/978-1-4615-3440-2_2.
[30] G. Leder. Jet fractions in e+ e- annihilation. *Nuclear Physics B*, 497:334–344, 1997. DOI: 10.1016/S0550-3213(97)00240-X.
[31] S. Catani, Y. L. Dokshitzer, M. H. Seymour, and B. R. Webber. Longitudinally invariant K_t clustering algorithms for hadron hadron collisions. *Nuclear Physics B*, 406:187–224, 1993. DOI: 10.1016/0550-3213(93)90166-M.
[32] S. D. Ellis and D. E. Soper. Successive combination jet algorithm for hadron collisions. *Physical Review D*, 48:3160–3166, 1993. DOI: 10.1103/PhysRevD.48.3160.
[33] Y. L. Dokshitzer, G. D. Leder, S. Moretti, and B. R. Webber. Better jet clustering algorithms. *JHEP*, 08:001, 1997. DOI: 10.1088/1126-6708/1997/08/001.
[34] M. Wobisch and T. Wengler. Hadronization corrections to jet cross-sections in deep inelastic scattering. In: *Workshop on Monte Carlo Generators for HERA Physics (Plenary Starting Meeting)*, April 1998, pp. 270–279.
[35] M. Cacciari, G. P. Salam, and G. Soyez. The anti-k_t jet clustering algorithm. *JHEP*, 04:063, 2008. DOI: 10.1088/1126-6708/2008/04/063.
[36] M. Cacciari, G. P. Salam, and G. Soyez. The catchment area of jets. *JHEP*, 04:005, 2008. DOI: 10.1088/1126-6708/2008/04/005.
[37] M. H. Seymour. Searches for new particles using cone and cluster jet algorithms: A comparative study. *Zeitschrift für Physik*, 62:127–138, 1994. DOI: 10.1007/BF01559532.
[38] J. M. Butterworth, B. E. Cox, and J. R. Forshaw. WW scattering at the CERN LHC. *Physical Review D*, 65:096014, 2002. DOI: 10.1103/PhysRevD.65.096014.
[39] J. M. Butterworth, A. R. Davison, M. Rubin, and G. P. Salam. Jet substructure as a new Higgs search channel at the LHC. *Physical Review Letters*, 100:242001, 2008. DOI: 10.1103/PhysRevLett.100.242001.
[40] A. Abdesselam, *et al.* Boosted objects: A probe of beyond the standard model physics. *European Physical Journal C*, 71:1661, 2011. DOI: 10.1140/epjc/s10052-011-1661-y.

[41] A. Altheimer, et al. Jet substructure at the tevatron and LHC: New results, new tools, new benchmarks. *Journal of Physics G*, 39:063001, 2012. DOI: 10.1088/0954-3899/39/6/063001.

[42] A. Altheimer, et al. Boosted objects and jet substructure at the LHC. Report of BOOST2012, held at IFIC Valencia, 23rd-27th of July 2012. *The European Physical Journal C*, 74(3):2792, 2014. DOI: 10.1140/epjc/s10052-014-2792-8.

[43] D. Adams, et al. Towards an understanding of the correlations in jet substructure. *European Physical Journal C*, 75(9):409, 2015. DOI: 10.1140/epjc/s10052-015-3587-2.

[44] A. J. Larkoski, I. Moult, and B. Nachman. Jet substructure at the Large Hadron Collider: A review of recent advances in theory and machine learning. *Physics Reports*, 841:1–63, 2020. DOI: 10.1016/j.physrep.2019.11.001.

[45] R. Kogler, et al. Jet substructure at the Large Hadron Collider: Experimental review. *Reviews of Modern Physics*, 91(4):045003, 2019. DOI: 10.1103/RevModPhys.91.045003.

[46] B. Nachman, et al. Jets and jet substructure at future colliders. *Frontiers in Physics*, 10:897719, 2022. DOI: 10.3389/fphy.2022.897719.

[47] D. Krohn, J. Thaler, and L.-T. Wang. Jet trimming. *JHEP*, 02:084, 2010. DOI: 10.1007/JHEP02(2010)084.

[48] G. Aad, et al. Performance of jet substructure techniques for large-R jets in proton-proton collisions at $\sqrt{s} = 7$ TeV using the ATLAS detector. *JHEP*, 09:076, 2013. DOI: 10.1007/JHEP09(2013)076.

[49] J. Thaler and K. Van Tilburg. Identifying boosted objects with N-subjettiness. *JHEP*, 03:015, 2011. DOI: 10.1007/JHEP03(2011)015.

[50] J. Thaler and K. Van Tilburg. Maximizing boosted top identification by minimizing N-subjettiness. *JHEP*, 02:093, 2012. DOI: 10.1007/JHEP02(2012)093.

[51] S. D. Ellis, C. K. Vermilion, and J. R. Walsh. Techniques for improved heavy particle searches with jet substructure. *Physical Review D*, 80:051501, 2009. DOI: 10.1103/PhysRevD.80.051501.

[52] S. D. Ellis, C. K. Vermilion, and J. R. Walsh. Recombination algorithms and jet substructure: Pruning as a tool for heavy particle searches. *Physical Review D*, 81:094023, 2010. DOI: 10.1103/PhysRevD.81.094023.

[53] M. Dasgupta, A. Fregoso, S. Marzani, and G. P. Salam. Towards an understanding of jet substructure. *JHEP*, 09:029, 2013. DOI: 10.1007/JHEP09(2013)029.

[54] M. Dasgupta, A. Fregoso, S. Marzani, and A. Powling. Jet substructure with analytical methods. *European Physical Journal C*, 73(11):2623, 2013. DOI: 10.1140/epjc/s10052-013-2623-3.

[55] I. Feige, M. D. Schwartz, I. W. Stewart, and J. Thaler. Precision jet substructure from boosted event shapes. *Physical Review Letters*, 109:092001, 2012. DOI: 10.1103/PhysRevLett.109.092001.

[56] M. Dasgupta, A. Powling, and A. Siodmok. On jet substructure methods for signal jets. *JHEP*, 08:079, 2015. DOI: 10.1007/JHEP08(2015)079.

[57] M. Aaboud, *et al.* Measurement of the soft-drop jet mass in pp collisions at $\sqrt{s} = 13$ TeV with the ATLAS detector. *Physical Review Letters*, 121(9): 092001, 2018. DOI: 10.1103/PhysRevLett.121.092001.

[58] C. Frye, A. J. Larkoski, M. D. Schwartz, and K. Yan. Factorization for groomed jet substructure beyond the next-to-leading logarithm. *JHEP*, 07: 064, 2016. DOI: 10.1007/JHEP07(2016)064.

[59] A. Kardos, A. J. Larkoski, and Z. Trócsányi. Groomed jet mass at high precision. *Physics Letters B*, 809:135704, 2020. DOI: 10.1016/j.physletb.2020.135704.

[60] S. Marzani, L. Schunk, and G. Soyez. The jet mass distribution after soft drop. *European Physical Journal C*, 78(2):96, 2018. DOI: 10.1140/epjc/s10052-018-5579-5.

[61] A. H. Hoang, S. Mantry, A. Pathak, and I. W. Stewart. Nonperturbative corrections to soft drop jet mass. *JHEP*, 12:002, 2019. DOI: 10.1007/JHEP12(2019)002.

[62] A. Pathak, I. W. Stewart, V. Vaidya, and L. Zoppi. EFT for soft drop double differential cross section. *JHEP*, 04:032, 2021. DOI: 10.1007/JHEP04(2021)032.

[63] A. M. Sirunyan, *et al.* Measurements of the differential jet cross section as a function of the jet mass in dijet events from proton-proton collisions at $\sqrt{s} = 13$ TeV. *JHEP*, 11:113, 2018. DOI: 10.1007/JHEP11(2018)113.

[64] J. Baron, S. Marzani, and V. Theeuwes. Soft-drop thrust. *JHEP*, 08:105, 2018. DOI: 10.1007/JHEP08(2018)105. [Erratum: *JHEP*, 05:056, 2019].

[65] S. Marzani, D. Reichelt, S. Schumann, G. Soyez, and V. Theeuwes. Fitting the strong coupling constant with soft-drop thrust. *JHEP*, 11:179, 2019. DOI: 10.1007/JHEP11(2019)179.

[66] H. S. Hannesdottir, A. Pathak, M. D. Schwartz, and I. W. Stewart. Prospects for strong coupling measurement at hadron colliders using soft-drop jet mass, October 2022.

[67] A. H. Hoang, S. Mantry, A. Pathak, and I. W. Stewart. Extracting a short distance top mass with light grooming. *Physical Review D*, 100(7), 074021, 2019. DOI: 10.1103/PhysRevD.100.074021.

[68] L. de Oliveira, M. Kagan, L. Mackey, B. Nachman, and A. Schwartzman, Jet-images — deep learning edition. *JHEP*, 07:069, 2016. DOI: 10.1007/JHEP07(2016)069.

[69] F. A. Dreyer, G. P. Salam, and G. Soyez. The Lund jet plane. *JHEP*. 12: 064, 2018. DOI: 10.1007/JHEP12(2018)064.

[70] M. Feickert and B. Nachman. A living review of machine learning for particle physics, February 2021.

[71] J. Cogan, M. Kagan, E. Strauss, and A. Schwarztman. Jet-images: Computer vision inspired techniques for jet tagging. *JHEP*, 02:118, 2015. DOI: 10.1007/JHEP02(2015)118.

Chapter 7

Instrumentation and Techniques in High Energy Physics: Liquid Argon Neutrino Detectors

David Caratelli

Department of Physics, Broida Hall
University of California, Santa Barbara, CA 93106-9530
dcaratelli@ubcs.edu

Liquid Argon Time Projection Chambers (LArTPCs) have become a prominent detector technology employed in experimental neutrino physics. This chapter presents a review of the LArTPC technology, with an emphasis on its use in large-scale neutrino experiments which study MeV to GeV neutrino interactions from accelerator or astrophysical sources. First, we provide a description of how these detectors came about and what motivates their construction in Section 7.1. Next, we describe the operational principle of LArTPC detectors, covering how signals are formed (Section 7.2), propagated (Section 7.3), and detected (Section 7.4). We then conclude with a description of how signals from LArTPCs are reconstructed and analyzed, and provide a brief overview of how such detectors are calibrated in Section 7.5.

7.1 Introduction: Why and How LArTPC Neutrino Detectors Came About

In the literature, several people are often highlighted for their contributions to the development of the LArTPC technology. The potential for liquid argon detectors as fully active detectors for particle physics was recognized by Willis and Radeika [1] and Nygren [2] in the early 1970s. Their

---

2024 © The Author(s). This is an Open Access chapter published by World Scientific Publishing Company, licensed under the terms of the Creative Commons Attribution 4.0 International License (CC BY 4.0). https://doi.org/10.1142/9789819801107_0007

applicability to neutrino detection was put forward by Rubbia [3] and others [4] shortly after. Since these early proposals, the successful development of the technology came about thanks to the effort of entire communities tied to research in detector instrumentation, electronics, engineering, software, and offline tool development. Before delving into a description of the technology itself, it is worth understanding why and how this detector technology came to be such a significant player in experimental neutrino physics.

Detectors are developed because they solve a given technological challenge. Liquid argon neutrino detectors do so in two fundamental ways: they provide fully instrumented large-mass targets for neutrino interactions and do so while faithfully tracking with millimeter accuracy the particles produced in such interactions. These features address the need for large-scale detectors in neutrino experiments, where event statistics are a precious commodity, and satisfy an important requirement to accurately measure the leptons and hadronic system needed to reconstruct the neutrino flavor and energy which are input to neutrino oscillation measurements. LArTPCs achieve this by leveraging argon's large yield of ionization charge and scintillation light. In this way, the detector acts both as the active target for incoming neutrinos and the source of ionization charge and scintillation light signals. Transport of the charge signals through a uniform electric field allows us to efficiently instrument a single 2D detector wall while preserving the 3D pattern of charge deposition, providing accurate 3D imaging capabilities. Finally, the ability to instrument such a detector with modern readout electronics elevates the bubble chamber like imaging capabilities of a LArTPC providing detailed calorimetric information necessary for quantitative measurements with modern computing tools. These features combined make the LArTPC an ideal technology for precision measurements of neutrino interactions and oscillations. The exquisite imaging capabilities of the LArTPC further enable searches for rare interaction modes which are the signature of Beyond the Standard Model (BSM) physics. These searches often complement the experimental program at intense neutrino beam facilities.

Making the simple detection principle of the LArTPC a reality has required significant technological development. Advances in cryogenic engineering as well as argon purification have made possible constructing large-scale detectors which allow for the efficient propagation of electron and photon signatures over meter-scale distances. Low-power electronics that operate in liquid argon with high channel count have enabled high-resolution imaging with low noise. Advances in computing and analysis methods have made the study of dataset comprising $\mathcal{O}(10^6)$ neutrino interactions a reality. These are only some of the developments that have contributed to making the LArTPC technology successful today. While this chapter provides a comprehensive overview of the operation and performance of liquid argon neutrino detectors, further references can provide more detailed information. Of note are references that discuss aspects of instrumentation and detector performance tied to this technology. Reference [5] detailing the performance of the 3-ton prototype for the ICARUS experiment, a launching pad for the development of this detector technology, provides a comprehensive overview of many detector effects and operation parameters. The "Detector Papers" from ICARUS [6, 7], MicroBooNE [8], LArIAT [9], and protoDUNE [10, 11] are useful references that discuss in detail the design of important LArTPC detectors while also providing insight into detector design, construction, and operation more broadly. More references from the broad literature of liquid argon neutrino experiments tied to specific subsystems, detector effects, or calibration methods are provided throughout this chapter when relevant.

7.1.1 *Operational principle of a LArTPC*

This section gives a practical overview of how a LArTPC is used to image neutrino interactions. The many concepts and detector components introduced here are expanded on in subsequent sections. A neutrino LArTPC aims to detect the visible signature of neutrino interactions comprised of the ionization charge and scintillation light produced as charged particles propagate through the detector. Figure 7.1 shows a schematic of the main components of a TPC and how ionization electrons are used to record signals on the electronics. A uniform electric field is applied across the TPC through a negative potential on the cathode plane. This field causes ionization electrons to drift toward the anode, where charge sensors (wires or pixels) record the current induced by the drifting electrons, or collect their charge directly. Each sensor records an analog pulse associated with the drifting electron cloud. The uniform drift allows ionization electrons recorded at the anode to faithfully map out the 3D pattern of energy deposition produced by charged particles resulting from the neutrino interaction. Multiple wire planes, oriented at different angles, are used to record multiple complementary signatures of the same ionization electron cloud. Time-coincident signals on different wire planes can then be used to "triangulate" the charge and obtain the exact 2D coordinate for the ionization cloud on the 2D wire plane. Finally, the depth in the detector

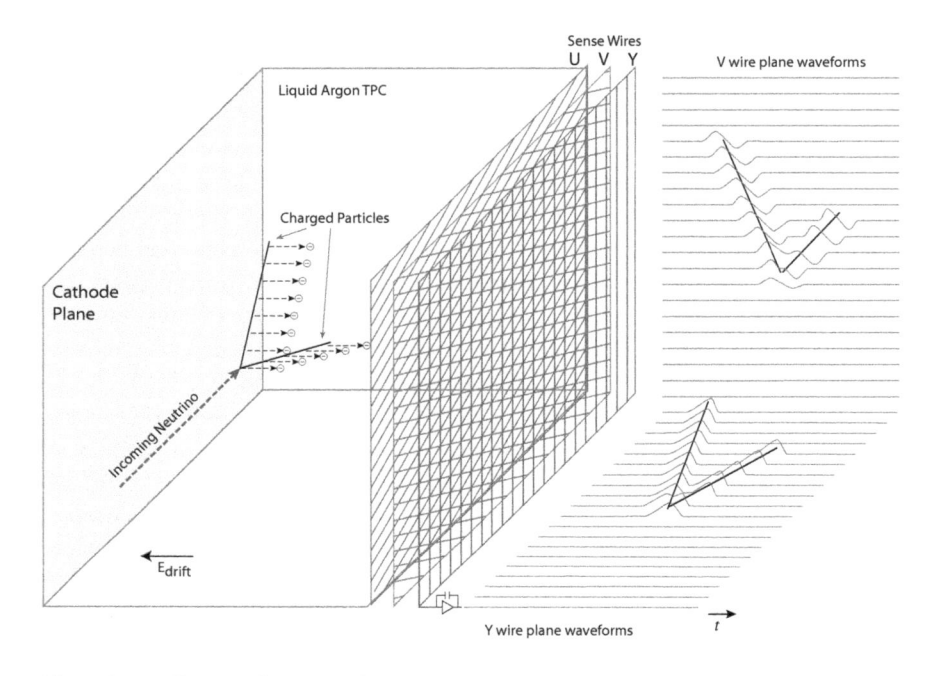

Figure 7.1. Cartoon depicting the operational principle of a LArTPC. Ionization charge drifts uniformly across the detector volume, producing signatures on different sense wire planes which when combined in offline reconstruction are used to recover the 3D charge deposition pattern.
Source: Figure from Ref. [8].

along the direction of the electric field can be calculated leveraging the fact that electrons drift at a constant speed of $\mathcal{O}(1)$ mm/μs. The TPC provides full 3D information up to a degeneracy in the absolute distance of energy deposition along the direction of the electric field associated with the drift velocity. Breaking this ambiguity requires an independent measurement of the interaction time which is provided by the scintillation light. Scintillation photons travel across the detector in a few ns but unlike ionization electrons propagate isotropically. Collecting them while maintaining the uniform and fully active nature of the TPC requires placing light sensors — typically Photo Multiplier Tubes (PMTs) or Silicon Photo-Multipliers (SiPMs) — on the edge of the TPC, often right behind the anode plane. Scintillation light provides accurate timing which is essential for absolute 3D position reconstruction as well as background rejection in surface-based LArTPC detectors studying GeV–scale neutrino interactions.

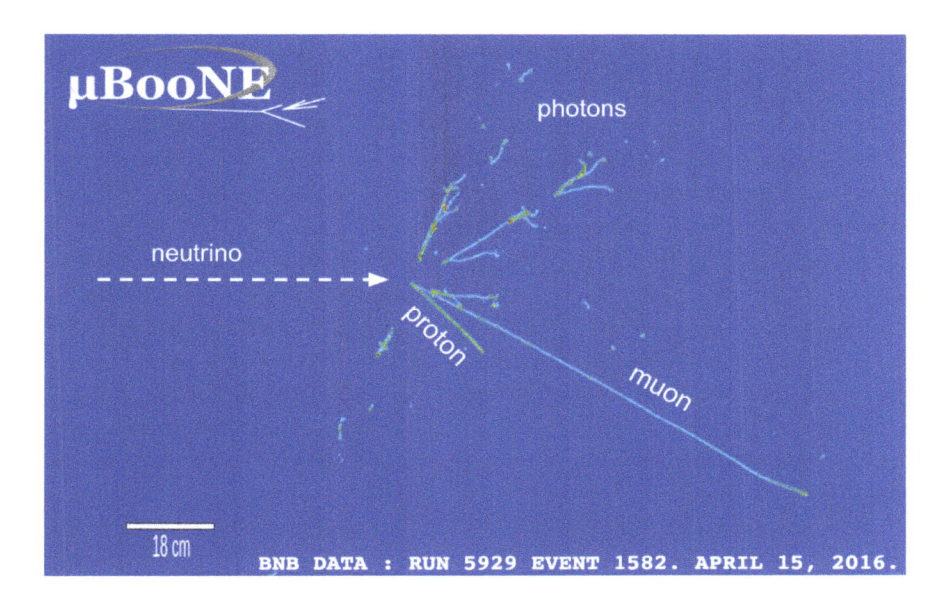

Figure 7.2. Event display showing a candidate neutrino interaction recorded with the MicroBooNE LArTPC. The 2D image shows the signals recorded on the wires of one of the three wire planes with which the detector is instrumented. Data from each wire is represented as a vertical strip in the image, with the vertical axis denoting the recorded time of the signal on the wire. The color scale in the image denotes the amount of charge collected. The large portions of blue in the image represent regions with no collected signals. The neutrino (which originates from the beam and enters from the left in the image) is not directly visible in the detector but produces charged particles which ionize the argon and leave behind trails of ionization electrons. The activity in green and red is the signature of these ionization electrons as recorded on the sense wires. The topology and charge profile of the ionization pattern can be leveraged to reconstruct each particle's species and its kinematics.

Source: Figure reproduced from microboone-exp.fnal.gov.

In addition to providing millimeter-scale position resolution on final-state particle trajectories, the large signal-to-noise of modern LArTPC detectors provides up to percent-level charge resolution for individual energy deposits on the collection wires and thresholds of $\mathcal{O}(100 \text{ keV})$. Figure 7.2 shows a neutrino interaction candidate collected with the Micro-BooNE LArTPC with several final-state particles coming out of the interaction vertex. The charge collected on the wires, represented by the color scale on the figure, measures the energy loss profile which is used for calorimetric energy measurements as well as particle identification (PID).

7.1.2 Experimental landscape in the early 2020s

Liquid argon neutrino detectors are now at the forefront of the experimental neutrino physics program aiming to perform precision measurements of neutrino oscillations and search for new physics. This experimental program, international in nature, is centered around Fermilab's intense neutrino beams which provide a focused source of 0.1–10 GeV neutrinos delivering $1\text{–}10 \times 10^{20}$ protons on target (POT) per year. Two experimental projects currently make use of the LArTPC technology in Fermilab's neutrino beamline: the Short Baseline Neutrino (SBN) program [12] and the Deep Underground Neutrino Experiment (DUNE). Available references in the literature provide a useful overview of the SBN [13] and DUNE [14–16] experimental programs. Here, we provide a brief summary of both to help better contextualize the LAr neutrino detector description in relation to the physics being pursued.

The physics reach of this program is very broad and encompasses several areas. Driving the detector and beam design is the precision measurement of neutrino oscillations, with emphasis on the measurement of the δ_{CP} violating phase in the lepton sector, and the neutrino mass ordering with DUNE. The broader DUNE and SBN programs have a varied BSM physics program which leverages the intense neutrino beams as a possible source of new particles produced through feeble couplings to the standard model. The DUNE far detectors, thanks to their underground location, serve as a unique astrophysics observatory for neutrinos from a possible galactic supernova burst and from solar neutrinos. The quiet far detector environment further makes the experiment well suited for searches for proton decay or neutron-antineutron oscillation which test fundamental symmetries. Finally, the SBN program provides a venue for key measurements of neutrino scattering on argon, an important source of systematic uncertainty for oscillation measurements and BSM searches alike. The breadth of this program is possible in part thanks to the versatile capabilities of the LArTPC detector's ability to image interactions with exceptional resolution across the MeV to several GeV energy regime.

The SBN program is comprised of three LArTPC detectors placed at different distances from the neutrino beam target. The three detectors are the Short Baseline Neutrino Detector (SBND), the MicroBooNE detector, and ICARUS. In addition to multi-detector searches for eV-scale sterile neutrinos, each detector carries out an independent physics program. The MicroBooNE experiment [8] collected data from 2015 to 2021, making it the longest-running LArTPC experiment to date. Throughout this time, MicroBooNE has played a key role in transforming the LArTPC technology into one capable of delivering high-precision measurements of neutrino interactions with advanced analysis techniques with large-scale datasets. This includes MicroBooNE's first results investigating the nature of the MiniBooNE excess [17–21]. The ICARUS experiment [22] serves as the primary far-detector for the SBN program thanks to its 600 tons of active argon mass. The detector began data-taking in 2020. SBND, the near detector for the program, sits only ~100 m from the beam target and is therefore exposed to a very large flux of neutrino interactions. This makes the SBND detector uniquely positioned for high-statistics measurements of neutrino–argon interactions and searches for rare processes tied to BSM signatures. The SBND detector is expected to begin data-taking in 2023.

DUNE will represent the culmination of the long-baseline oscillation program and leverages a powerful beam, longer baseline from the neutrino source, and large LArTPC active mass to achieve the statistics and precision needed for its oscillation physics program. The experiment is comprised of near and far detectors located in the Homestake mine in South Dakota and on the Fermilab campus, respectively. The detector is expected to start data-taking in the late 2020s, with a staging of the several modules and detectors which make up the experimental facility. The far detector, situated one mile underground at the Sanford Underground Research Facility in Lead, South Dakota, will additionally serve as a unique facility for the observation of astrophysical neutrinos from a galactic supernova burst and the Sun [15]. The near detector provides a rich physics program itself, centered on searches for BSM physics [16].

The development of the LArTPC technology benefits from a continuous stream of small-scale demonstrators and R&D test stands. Those that have documented their operations in the literature provide an invaluable source for readers interested in learning the details of hardware components, engineering requirements, cryogenic, DAQ, and many other topics related to the design, construction, and operation of a LArTPC. Details can be found in references from the LArPD [23] and LongBo [24] test stands at Fermilab, the ICARUS 3-ton demonstrator [5], and BNL's 20-liter test stand [25]. Large-scale prototype detectors such as the protoDUNE single [10, 11] and double-phase TPC complement this list and are providing insight for the construction of multi-kiloton LArTPC detectors in DUNE.

7.2 Particle Propagation and Signal Formation in Liquid Argon Neutrino Detectors

Liquid argon neutrino detectors take advantage of their ability to precisely track the path of charged particles propagating through the detector volume. This section describes how particles propagate in liquid argon and how signals of ionization charge and scintillation light are formed.

7.2.1 What does a neutrino LArTPC see?

Neutrino experiments rely on the detection of final-state particles produced in neutrino interactions. Of particular relevance to the study of neutrino oscillations is the signature of the charged lepton resulting from charged current neutrino interactions. This primarily means electrons and muons from ν_e and ν_μ charged current (CC) interactions, respectively. The hadronic response in neutrino interactions from the MeV to the few-GeV scale is vastly complex and tied to the details of the neutrino interaction mode and effects that impact the propagation of particles through the argon nucleus. A review of such processes is outside of the scope of this text. Here we focus on describing how the most common particles produced in neutrino interactions propagate through and manifest themselves in LArTPC detectors. One of the features that make LArTPC detectors so powerful in the study of neutrinos is their ability to record with low threshold particles produced in the interaction's hadronic recoil. Photons, charged and neutral pions, protons, and neutrons are the most commonly produced particles for interactions of up to several GeV of energy. The description of particle propagation is subdivided between particles that are often referred to as "shower-like", such as electrons and photons, and those that are "track-like", such as muons, protons, and charged pions. A final section on the propagation of neutrons concludes this presentation. While the description presented centers on propagation in LAr, relevant background can be found in the PDG's "Passage of Particles through Matter" chapter of the Particle Data Group's "Review of Particle Physics" [26].

7.2.1.1 Propagation of electrons and photons

The propagation of electrons and photons in argon is governed by Compton scattering and pair production for photons and ionization and Bremsstrahlung for electrons. Figure 7.3 shows the energy loss profile for

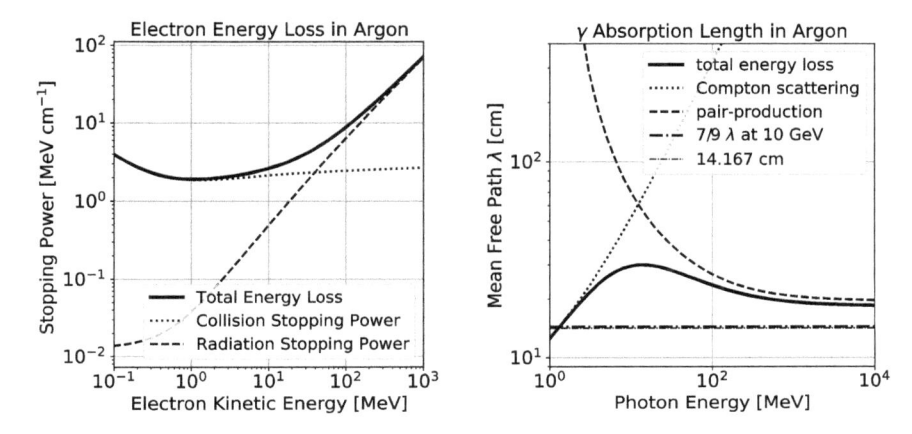

Figure 7.3. Left: Energy loss for electrons as a function of energy divided in the two primary components: Collision stopping power (or ionization) and radiative losses (Bremsstrahlung photon production). Right: Mean-free path for photons propagating in argon contributed by two main processes: Compton scattering and pair-production, dominating below and above 10 MeV, respectively.

electrons (left) and the interaction length for photons (right) as a function of energy. Pair production ($\gamma \to e^+e^-$) for photons and Bremsstrahlung ($e + N \to e + \gamma$) for electrons cause the propagation of these two particles to be interconnected and leads to the formation of electromagnetic (EM) cascades which manifest as "showers" in the detector. Energy has a significant impact on the topology of EM showers. Electrons and photons of several hundred MeV or more lead to fully developed EM showers that are easily recognized as such in the detector. As one approaches the $\mathcal{O}(100)$ MeV regime, the stochastic nature of pair production and Bremsstrahlung photon production lead to "fragmented" EM showers. Below $\mathcal{O}(100)$ MeV, EM showers appear largely as a single track-like ionization segment caused by an electron, followed by several isolated Compton scatters contributed by low-energy Bremsstrahlung photons. Figure 7.4 shows examples of EM showers in LAr in all three energy regimes. Shower length scales logarithmically with energy and generally propagates for a distance $\mathcal{O}(1)$ m for showers in the 10s of MeV to GeV range, while they are contained transversely by a Moliere radius of ~ 9 cm.

Having highlighted the similarity between electron and photon propagation in LAr, it is worth mentioning the important differences that allow the LArTPC technology to distinguish between these two particles. Since photons do not carry electric charge, they do not deposit energy in LAr

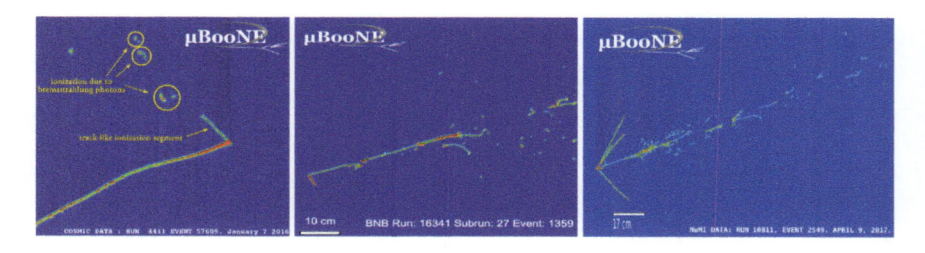

Figure 7.4. Examples of EM interactions in LAr. Left: Michel electron of several tens of MeV (reproduced from Ref. [27]). Center: Few hundred MeV EM showers from a ν_e interaction candidate (reproduced from Ref. [19]). Right: Higher energy EM shower from a ν_e interaction candidate (reproduced from microboone-exp.fnal.gov).

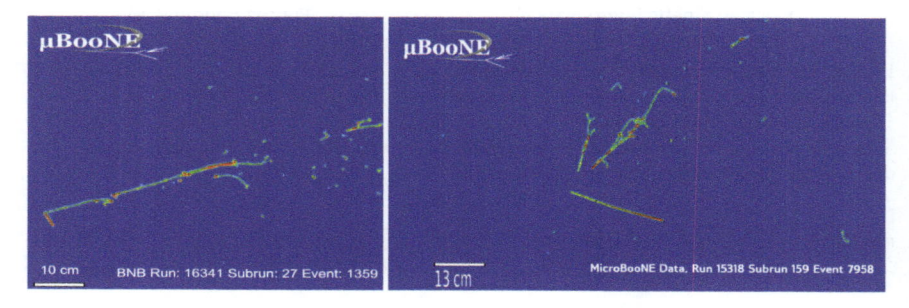

Figure 7.5. Left: Electron shower candidate from a ν_e interaction. Right: Two photon candidates from the decay of a π^0 produced in a neutrino interaction. In the event with photons, one can easily make out the gap from the shower start point to the interaction vertex (reproduced from microboone-exp.fnal.gov).

until they have interacted. The initial propagation distance covered by the photon before pair-converting leads to a visible "gap" from the neutrino interaction vertex. This gap is characteristic of photons' $\sim$20 cm conversion distance in LAr. Electrons, which ionize the argon as soon as they are produced, leave no such gap from their production point. Given the large conversion distance for photons, this gap is clearly visible in the detector but requires the presence of coincident activity at the interaction point such as visible tracks from the hadronic recoil. Figure 7.5 illustrates how the presence of a gap helps distinguish photons vs. electrons. The second distinguishing feature is associated with the energy loss profile at the start of the shower. The electron–positron pair produced by a photon will lead to double the ionization stopping power compared to a single electron. This difference will remain until the EM shower has a chance to develop, typically after several centimeters. The segment before the EM shower starts to fully

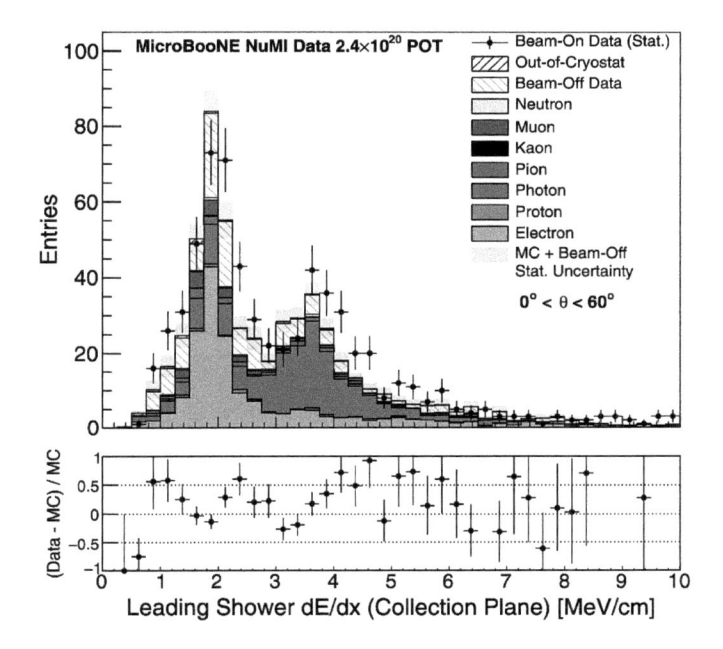

Figure 7.6. Distribution of shower trunk dE/dx for electrons and photons observed in the MicroBooNE detector from the NuMI neutrino beamline. The distribution is dominated by electrons, in light gray, contributing at 2 MeV/cm, and photons, in dark gray, peaking at 4 MeV/cm.
Source: Figure reproduced from Ref. [28].

develop is referred to as the shower "trunk". Figure 7.6 shows one of the first demonstrations of this e/γ separation technique, measuring the dE/dx energy loss in the first four cm of the shower trunk and clearly distinguishing electrons, peaking at 2 MeV/cm from photons at twice that value.

7.2.1.2 Propagation of muons, charged pions, kaons, and protons

Muons, charged pions, and protons do not radiate much of their energy at GeV energies. Instead, they primarily lose energy through direct ionization of the argon. This leads to "track-like" signatures: fairly linear segments of uniform ionization. Unless the particle re-interacts, it will come to a stop after having lost all its kinetic energy. The energy loss profile follows the Bethe–Bloch distribution, which describes the energy-dependent energy loss per unit distance (dE/dx) for a particle of a given mass. This

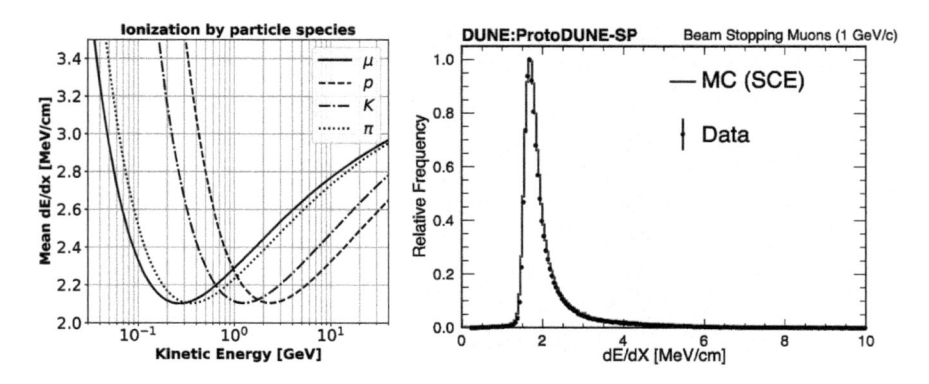

Figure 7.7. Left: Ionization energy loss for different particle species as a function of kinetic energy. Right: Measured dE/dx distribution from the protoDUNE detector for minimally ionizing muons.
Source: Figure from Ref. [10].

distribution is characterized by a minimum (occurring roughly when the particle's kinetic energy is equivalent to its mass) and a "Bragg peak": a sharp increase in the rate of ionization as the particle's energy decreases. Figure 7.7 shows the mean energy loss to ionization for muons, protons, charged pions, and kaons as a function of the particle's kinetic energy. While the Bethe–Bloch function describes the mean energy loss, the probability distribution function of the energy lost by a particle traversing a layer of argon follows a Landau–Vavilov distribution [29]. A useful metric by which ionization energy loss is quantified is the most probable value (MPV) of this distribution. Unlike the mean energy loss, the MPV depends on the thickness of the layer of material traversed, dx. Figure 7.7 (right) shows the distribution of dE/dx measured with the protoDUNE detector with a sample of minimally ionizing muons, which have an MPV energy loss of roughly 1.7 MeV/cm.

While ionization is the primary means of energy loss for track-like particles, other interaction modes are relevant to the overall topology and energy loss profile in the TPC. Muons will occasionally produce δ-rays: energetic electrons which branch out from the primary muon track. Protons and pions often reinteract with an argon nucleus before coming to rest, leading to more complicated topologies which can include a number of final state hadrons. These reinteractions are energy dependent, becoming more likely as the proton or pion's energy increases. Finally, muons, charged pions, and charged kaons will decay via the weak force once stopped, each producing a characteristic signature in the detector. The decay chain differs depending on the charge of the particle. Positive muons will decay promptly to a Michel electron, while negatively charged muons will capture on argon and decay to an electron only ∼3/4 of the time. The resulting decay electron, referred to as a "Michel" electron, has an energy of up to half the muon mass and appears in the TPC as a secondary particle track branching from the muon end point (as can be seen in the left-most image of Fig. 7.4). When charged pions come to rest, they will decay promptly to a muon which subsequently follows the decay chain described above. The decay muon is produced with a small kinetic energy of $m_\pi - m_\mu \sim 35$ MeV, making it hard to detect and causing the pion-decay and muon-decay topologies to resemble each other. Finally, charged kaons can follow several possible decay modes. The most common (64% branching ratio) is to a muon and ν_μ. This decay path leads to a characteristic topology in the detector with a three-track segment of $K \to \mu \to e$ where the muon has a fixed energy of $m_K - m_\mu$ which can be leveraged for kaon identification.

7.2.1.3 *Propagation of neutrons*

Neutrons are an important final-state particle in neutrino interactions yet are largely invisible in liquid argon. Their large interaction length often causes them to escape the detector entirely. This makes neutrons one of the main contributors to bias and uncertainty in neutrino energy reconstruction. When neutrons interact in the argon, they do so through neutron capture or inelastic scattering. Both processes lead to final-state low-energy photons, producing isolated point-like activity in the detector which rarely spans more than one wire on the TPC. These signatures can help determine the presence of neutrons but are not adequate for measuring their energy or direction. Higher energy inelastic scatters on argon can lead to final-state particles from the struck nucleus exiting with considerable energy, often producing visible protons. For the study of $\mathcal{O}(0.1 - 10)$ GeV neutrinos, the primary efforts tied to the measurement of neutrons are focused on identifying their presence in a neutrino interaction in order to improve neutrino energy reconstruction.

7.2.2 *Signal formation in LArTPCs*

As charged particles traverse the argon, their passage induces local perturbations to the electric field which transfer energy from the particle to electrons orbiting the argon nucleus. These electrons will either completely

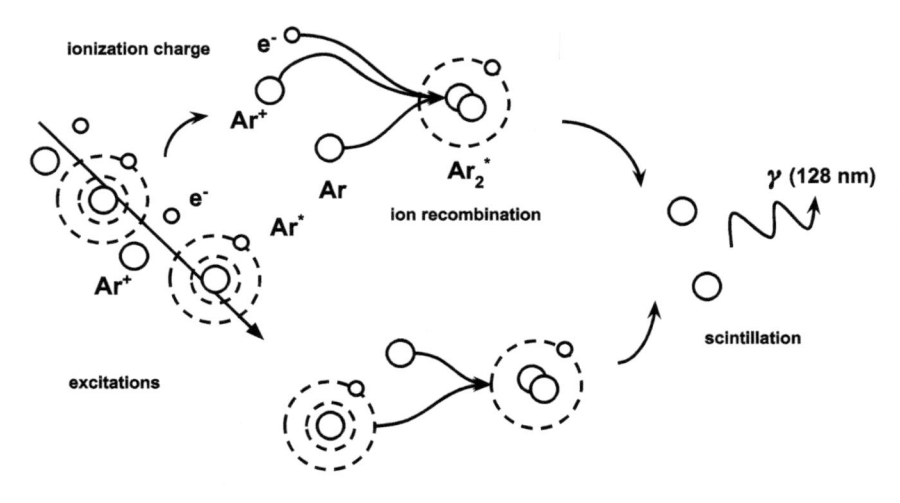

Figure 7.8. Ionization and excitation signal formation by the passage of a charged particle in liquid argon.

disassociate from the atom, leading to free electrons, or will reach a higher energy orbital, leading to an excitation which de-excites producing scintillation photons. One of the features which makes argon an ideal detector medium is the high yield for both signals, which are produced in roughly equal amounts. Roughly 40,000 scintillation photons and ionization electrons are liberated for every MeV of energy deposited. Figure 7.8 shows a schematic of the processes and ionization and excitation, and how these two paths are interconnected.

We describe the specific features of both signatures individually. Since the production of the two signatures is interconnected, we come back and discuss their interplay through the effect of ion recombination which influences particle identification and calorimetric measurements.

7.2.2.1 *Ionization charge*

Ionization electrons are freed from their orbit as a consequence of the passage of a nearby energetic charged particle. The average energy expended per ion pair, referred to as W, determines the number of free electrons produced per amount of energy deposited. For liquid argon, this quantity is $23.6^{+0.5}_{-0.3}$ eV [30]. Given the ~ 0.5 MeV of energy deposited by a particle on a single wire in a TPC, this leads to electron clouds of $\sim 10^4$ electrons drifting toward each wire. Ionization electrons produced in the detector can have $\mathcal{O}(\text{eV})$ of energy but quickly thermalize to energies of a fraction of an eV as a consequence of random collisions with nearby argon atoms [31]. How these electrons behave next is complicated by the busy environment in which they find themselves: a sea of positive (Ar^+) and negative (e) ions which together cause an effect known as ion recombination, which is discussed in Section 7.2.2.3. The propagation of free electrons across the detector is discussed in Section 7.3.1.

7.2.2.2 Scintillation light

Excited argon atoms rapidly combine with a ground-state argon atom to form excited dimer Rydberg states: an Ar_2^* core with a bound, shared electron. Two different excitation states are populated, referred to as the "singlet" and "triplet", which get their name from how the spin of the electron and argon dimer are coupled. Both states de-excite to produce photons in the Visible Ultra-Violet (VUV) spectrum in a narrow wavelength range centered at 128 nm. The singlet and triplet excimers have significantly different lifetimes of 6 ns ("fast", or "prompt") and $\sim$1.3 μs ("late-light"), respectively, leading to two components in the scintillation light response. The ratio of scintillation photons produced via the singlet vs. triplet decay is roughly 1:3, though this ratio is dependent on the particle species and local electric field. The total light yield for scintillation light in argon is $\sim$40,000 γ/MeV [32], though again this quantity is dependent on the local environment and particle species, as become apparent in the following section describing ion recombination.

7.2.2.3 Ion recombination and impact on charge and light

Ionization electrons are surrounded by their positively charged argon ion counterparts. The positive attraction between the Ar^+ and e^- ions will cause them to recombine, an effect referred to as "ion recombination". The amount of electrons which will recombine with nearby positively charged ion atoms is a function of two key parameters: the local density of ions and the amount of time the electrons and Ar^+ are in close proximity. Ion density is proportional to the energy lost by the ionizing particle, dE/dx. How quickly ionization electrons will drift away from the positive ions instead depends on the strength of the external electric field E applied in the TPC. Recombination, denoted as $\mathcal{R}$, is therefore expressed as a function of these two macroscopic observables, $\mathcal{R}(dE/dx, E)$.

Ion recombination has two key consequences for LArTPC particle detection. First, ion recombination leads to a nonlinear relation between deposited energy and observable charge. This is due to the larger amount of recombination (and therefore larger fractional charge quenching) which takes place for larger local energy deposition, where a higher density of ions leads to an increased probability of electrons recombining. The second key impact is to closely link ionization and scintillation signatures. As electrons recombine with positively charged ions, they do so forming excited states which in turn produce scintillation photons. The charge quenching caused by recombination thus leads to an increase in scintillation light. This leads to an anti-correlation between the yield of ionization electrons and scintillation photons. For a sense of scale, minimally ionizing muons ($dE/dx \sim 2.1$ MeV/cm) observe charge quenching of 30–50%, while for highly ionizing particles such as protons in the Bragg peak, the amount of charge quenching can reach 80% or more. Two detailed studies and most widely used parametrizations for ion recombination were produced by the ICARUS [33] and ArgoNeuT [34] experiments.

7.3 LArTPC Signal Propagation

Argon is transparent to its own scintillation light and ionization charge, making it an excellent detector target for signal transport over the large volumes necessary to observe neutrino interactions with high statistics. Ionization charge drifts uniformly under the influence of the TPC's electric field producing a 3D map which faithfully tracks the trajectory of charged particles in the detector, while scintillation photons propagate isotropically from their point of origin. This section focuses on describing the details of how ionization electrons and scintillation photons propagate through the argon to reach their respective sensors.

7.3.1 *Ionization charge transport*

Once ionized, electrons begin to drift under the influence of the external TPC electric field. While this causes electrons to accelerate in the direction opposite the electric field, the intrinsic thermal motion of the electrons, and their constant collision with nearby argon atoms, leads them to quickly reach terminal velocity and drift at a uniform speed. The drift velocity grows logarithmically with increasing E-field strength. The relatively slow drift velocity causes ionization charge to take $\mathcal{O}$ (ms) to drift over $\mathcal{O}$ (m)

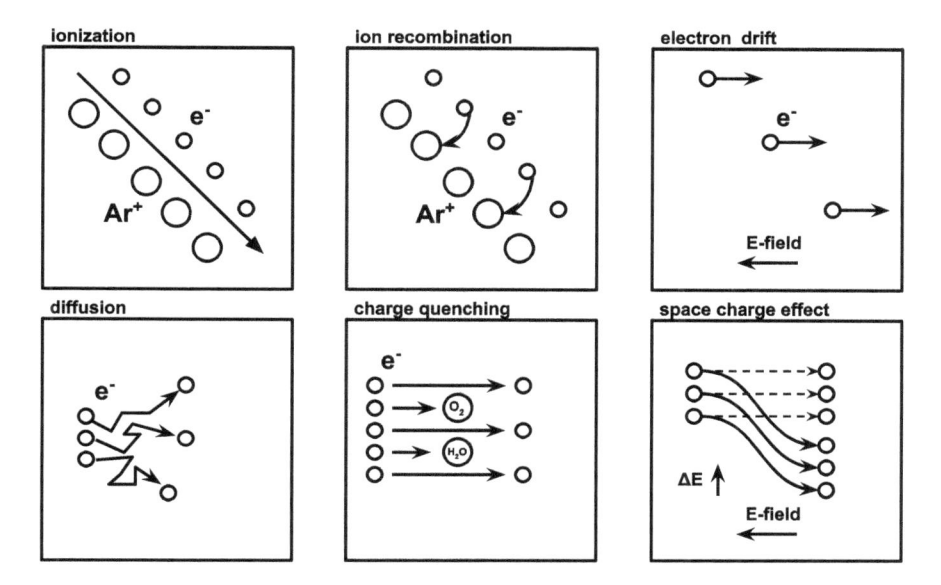

Figure 7.9. Detector effects that impact the propagation and transport of ionization charge across the TPC.

scales. This has significant impact on the design of electronics and DAQ systems which must record charge from the TPC for this timescale to fully image interactions and can lead to pile-up of interactions, such as coincident cosmic rays. During this relatively slow journey across the detector, several factors can interfere with the propagation of free electrons. Quenching due to electronegative impurities, diffusion of ionization clouds, and distortions due to "Space-Charge" effects are all described next. Figure 7.9 illustrates the detector effects that impact the propagation and transport of ionization charge across the TPC.

7.3.1.1 Charge quenching due to impurities

Electronegative impurities in the argon attract the drifting electrons, quenching the charge signal. The rate at which electrons are absorbed is directly proportional to the impurity concentration. Water and oxygen, natural contaminants in argon, are the primary contributors to charge quenching. With concentrations at the part-per-billion (ppb) level, the attenuation length of ionization electrons becomes of order meters. Limiting the amount of impurities in the argon is therefore essential in order to operate LArTPC detectors with meter-scale drift lengths. Achieving high argon purity was one of the key technological developments that has enabled the LArTPC technology to mature into a viable detection method for neutrino experiments. Charge quenching leads to a non-uniform detector response, due to the exponential suppression of ionization charge as a function of depth in the TPC (along the direction of the TPC's electric field). If impurity concentrations are too high, charge signals will fade away and not be detected on the TPC wires. Significant effort has gone into achieving high-purity through advances in engineering of the cryogenic recirculation and purification systems for LArTPC. Figure 7.10 from Ref. [35] shows the measured water and oxygen concentration in the MicroBooNE cryostat during detector commissioning as a function of time after initiating argon filtration. While water and oxygen have the largest impact due to their relatively larger concentrations, other contaminants can quench argon signals. A review and assessment of the impact on electron attenuation due to different contaminants can be found in Ref. [36].

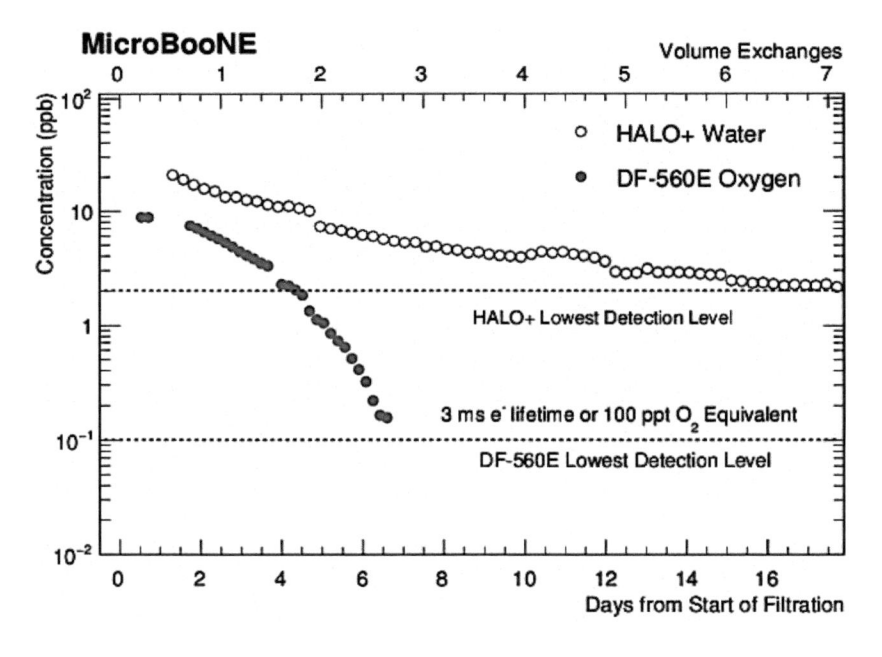

Figure 7.10. Concentration of water and oxygen in the MicroBooNE cryostat as a function of days since filtration of the argon. The detector reaches concentrations at or below one part per billion, enabling high electron lifetime and therefore high signal transparency across the TPC volume.

Source: Figure from Ref. [35].

7.3.1.2 Ion diffusion

The collisions occurring on a microscopic scale between drifting electrons and argon atoms lead to a broadening of the electron clouds produced by ionization. Diffusion affects aspects of detector performance tied to position resolution, peak amplitude (impacting detection thresholds), and the optimization of aspects of TPC readout, such as sampling frequency and electronics shaping response. Furthermore, diffusion can affect calculations of charge-sharing across TPC wires [37, 38], impacting calibrations. Macroscopically, diffusion leads to the broadening of electron clouds of one micron after a propagating over a drift distance of 1 m. The exact magnitude is different in the component along the direction of the electron drift (longitudinal, D_L) and that perpendicular to it (transverse, D_T) and is impacted by the strength of the local electric field. Diffusion is quantified as the spread in an electron cloud in cm^2/s. Measurements are typically performed for D_L, with values in the 4–6 cm^2/s range [39–41]. Values of the transverse diffusion D_T are typically slightly larger and also depend on the value of the local electric field.

7.3.1.3 Space charge effect

Positive argon ions drift as well, though much more slowly than their electron counterparts. In surface LArTPCs, a large steady-state rate of ionization from cosmic rays leads to a build-up of positively charged Ar ions in the detector, which in turn distorts the electric field in the TPC, causing what is referred to as the space charge effect (SCE). SCE in a TPC has the same effect as atmospheric aberration for telescopes: by causing local distortions in the TPC's uniform electric field, it bends the otherwise straight trajectory of drifting electrons. Uncorrected, it leads to a distortion of the reconstructed 3D charge map for ionization taking place in the TPC. Figure 7.11 shows the impact of this effect in the MicroBooNE detector: dotted lines represent the actual detector boundaries in the drift (x) and vertical (y) coordinates, while the scatter points are the start and endpoint of reconstructed cosmic-ray tracks in the detector before any SCE corrections are applied. The large differences between the two, particularly close to the cathode plane (right boundary in the figure), are a consequence of the deflection of ionization drift due to positive ions.

Positively charged ions tend to build up near the TPC cathode, where distortions are greatest. The varying electric field further impacts detector response by impacting the amount of ion recombination (see Section 7.2.2.3)

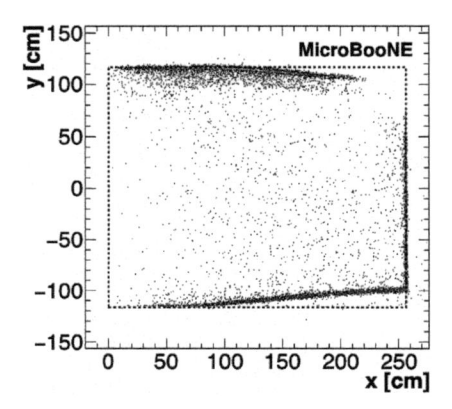

Figure 7.11. Effect of SCE as seen in MicroBooNE.
Source: Figure from Ref. [42].

and therefore the charge and light yield. The magnitudes of SCE distortions are both dependent on the position within the detector and on the rate of positive ion build-up in the detector. This second quantity depends primarily on the steady-state rate of ionizing radiation to which the detector is exposed: more ionization leads to a larger SCE effect. This means that detectors operating on the surface will be impacted by a significantly larger SCE effect than those underground. Quantitative studies of SCE effects have been carried out with the ICARUS [43], MicroBooNE [42, 44], and protoDUNE [10] detectors. What is a macroscopic offset of $\mathcal{O}(1\text{-}10)$ cm on the surface becomes a negligible effect underground, where the cosmic-ray rate is suppressed by several orders of magnitude.

7.3.2 *Scintillation light propagation*

The fact that argon is transparent to its own 128 nm scintillation photons means that to first-order photons collected as a function of distance from the production point follow a $1/r^2$ dependence. Rayleigh scattering and attenuation due to impurities impact the total amount of light collected and cause to deviate from this relationship. These two effects are described next.

7.3.2.1 *Rayleigh scattering*

Rayleigh scattering describes elastic scattering of light on particles of wavelength smaller than that of the incident photon. The Rayleigh scattering rate is proportional to $1/\lambda^4$ and therefore has a strong dependence on the photon wavelength. Recent measurements [10, 45] of the Rayleigh scattering length on argon report a value of ∼90–100 cm. These measurements are complicated by the fact that detector volumes for existing LArTPCs are of similar scale to the scattering length itself, and numerous effects can contribute to impacts that mimic that of scattering. The primary impact Rayleigh scattering has on detecting scintillation photons is to modify the simple $1/r^2$ dependence expected on the number of photons observed as a function of distance and direction from their production location. Rayleigh scattering also directly impacts the group velocity of photons traveling in argon, with a measured value from Ref. [45] of 13.4 ± 0.1 cm/ns. In this context, doping of argon with Xenon has been proposed as a way to mitigate the impact of Rayleigh scattering over large volumes due to the larger emission wavelength of ∼178 nm for xenon.

7.3.2.2 *Light attenuation*

Impurities in the argon can quench scintillation photons. Nitrogen, a contaminant found at the part-per-million level, has the largest impact, quenching scintillation light by "stealing" the shared electron from excited argon dimers. The very fast decay timescale of the singlet state leaves it almost completely unaffected by nitrogen contaminants, which instead compete with the de-excitation of the longer-lived (1.3 μs) triplet state. Quenching of the triplet excited state causes, in addition to an overall decreased light yield, a reduction of the effective late-light time constant. This in turns impacts the use of scintillation light for pulse-shape discrimination and other application of timing obtained from scintillation light.

7.4 LArTPC Signal Detection

Once drifting electrons and scintillation photons have reached the edges of the detector, they are detected by charge and light sensors, respectively. In this section, we describe the methods and hardware used to record these two signatures.

7.4.1 *Ionization charge detection*

The method employed to measure the ionization signature of an LArTPC traditionally relies on wire sensors. More recently, pixel-based detectors have been proposed and successfully operated. We discuss largely wire-based readout, which has been used in LArTPC detectors for many decades. While the core detection principle is the same, details tied to hardware and operations of pixel-based detectors can be found in Refs. [46, 47].

A wire-based TPC must be able to produce multiple signatures for the same cloud of ionization electrons on different wire planes in order to recover the 3D position of the energy deposition. To achieve this, wires are arranged in distinct planes at different orientations. Each plane is operated at a specific voltage in order to allow clouds of electrons to drift past the first wire planes, while still inducing a detectable signal, finally causing the charge to collect on the last, "collection" plane. The left panel of Fig. 7.12 shows a top-down view of the MicroBooNE detector with different wires denoted as dots arranged on three planes. The field lines followed by ionization electrons converge on the bottom (collection) plane. The image on the right shows the induced signature (top and middle) and collected charge (bottom) caused by the same cloud of ionization electrons. The time-coincidence of these signals, together with knowledge of the geometric location of the wires, allows us to recover the 3D coordinates of the energy deposited. The need for distinct wire planes comes with significant design challenges tied to achieving full transparency through the application of appropriate bias voltages on the three wire planes. This in turn motivates a

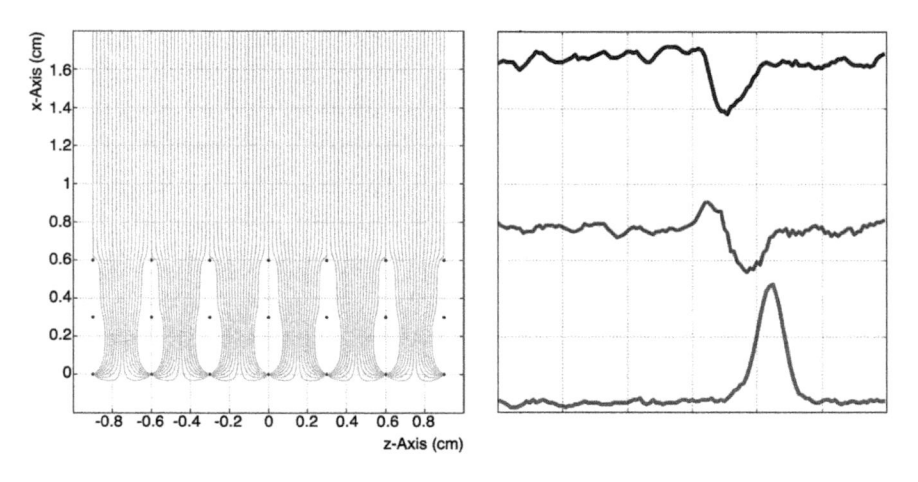

Figure 7.12. Wire geometry and simulated trajectory of drifting electrons for the MicroBooNE TPC wire plane.

Source: Figure from Ref. [48].

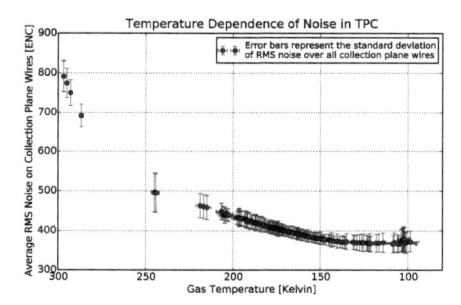

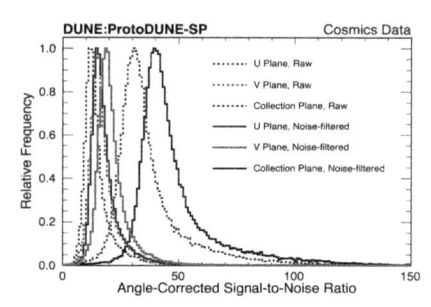

Figure 7.13. Left: Temperature dependence of TPC noise with cold electronics from MicroBooNE [51]. Right: Signal-to-noise measured in protoDUNE data [11].

detailed understanding of the field response which causes the induced and collected signals. Extensive details on this topic are provided in references on TPC signal processing from the MicroBooNE collaboration [48, 49].

The ability to effectively operate the TPC depends on achieving sufficient signal-to-noise to detect drift electrons. The development of low-power cold electronics for TPC wire readout [50] have enabled further noise suppression that has brought noise levels in TPCs to the few-hundred electrons in Equivalent Noise Charge (ENC). Figure 7.13 (left) shows the noise dependence on temperature for MicroBooNE's TPC collection plane wires, indicating the significant reduction at LAr temperatures. The right panel of Fig. 7.13 shows the signal-to-noise for ionization in protoDUNE measured on all three wire-planes. Signal-to-noise levels of $\sim$40 are achieved for minimally ionizing muons on the collection plane.

7.4.2 *Scintillation light detection*

Detecting scintillation light photons produced in argon requires being sensitive to their 128 nm wavelength. Wavelength shifters are used to convert the 128 nm photons to photons of longer wavelength which can be detected by PMTs or SiPMs. TetraPhenyl-Butadiene (TPB) is an organic scintillator molecule which efficiently absorbs VUV photons re-emitting them with a wavelength of $\sim$400–500 nm, a range that allows them to penetrate the glass surrounding a PMT and undergo photoelectric conversion on the photocathode. Other wavelength shifters are also available, and a review of LAr scintillation light wavelength shifters is available in Ref. [52]. Integrating the TPB wavelength shifter in LAr detector design has been achieved through multiple methods, including coating directly the PMT's glass surface, placing a TPB-coated plate directly in front of the PMT [8], or coating the detector surface [53]. In coincidence with development of PD systems for the DUNE experiments, further solutions for LAr scintillation detection have been designed. In particular, the ARAPUCA [54,55] concept acts as a light trap through the use of a dichroic filter and leads to two wavelength shifting stages: the first to allow photons to enter the detector and the second to trap them in the scintillator strip. Since the ARAPUCA concept was introduced, this detector technology has expanded in several directions and is being deployed in detectors such as protoDUNE [10] SBND, and the DUNE Vertical Drift technology [56]. Increasing the light yield is also achieved through the installation of wavelength-shifting and/or reflective surfaces on the inner walls of the TPC. This solution was employed by the WarP [53] detector and is being implemented in the SBND TPC, where it will provide high light yield of $\mathcal{O}(100)$ PE/MeV uniformly across the TPC volume, countering the $1/r^2$ dependence typically seen for scintillation signals due to the asymmetric positioning of light sensors in the cryostat.

7.4.3 Auxiliary detectors

LArTPC detectors are sometimes accompanied by external Cosmic Ray Taggers (CRTs) which consist of scintillator strips mounted externally to the detector's cryostat and are leveraged to identify coincident cosmic-ray particles which traverse the detector volume. CRTs such as those described in Refs. [57, 58] are primarily used to form anti-coincidences for vetoing cosmic-ray activity that may mimic neutrino interactions.

7.5 Operation, Performance, and Calibration of Liquid Argon Neutrino Detectors

Achieving stable operation of large-scale LArTPC detectors has been a major technological milestone that has enabled the technology to produce high-quality physics results and set the stage for decade-long operation of detectors, such as DUNE. A significant component of LArTPC operations is the need to operate a reliable cryogenic infrastructure to circulate, replenish, and purity the LAr inside the detector cryostat. While not discussed in this chapter, Refs. [5, 8, 23, 59] provide valuable information on the topic.

One of the advantages of the LArTPC technology is the ability to not only image particle interactions with high resolution but also provide the detailed calorimetry necessary for quantitative measurements of particle kinematics. For modern experiments, this inevitably requires having access to efficient and accurate reconstruction and calibration techniques which are able to process and interpret data in an automated way, providing reconstructed particle kinematics on an event-by-event basis to be used to develop event selections and carry out physics measurements. This is particularly true for neutrino experiments which aim to study complex particle interactions where the final-state observables vary both in particle species and kinematics. Having established the successful operation of LArTPC detectors and their ability to collect rich datasets of neutrino interactions, it is important to discuss the technological challenge of leveraging these data for precise neutrino measurements with sophisticated pattern recognition and analysis tools. Furthermore, it is worth highlighting the significant progress made in recent developments in reconstruction and analysis methods for LArTPC detectors. Due to these reasons, we dedicate this section to reviewing the status and progress in the calibration and analysis of LArTPC datasets.

7.5.1 *LArTPC reconstruction*

Converting the raw signals obtained from an LAr neutrino detector into quantitative measurements of particle kinematics for use in physics analyses is a task that falls under the name of "event reconstruction". This is a fundamental operation without which the quality of the data obtained from the detector cannot be put to use for quantitative physics measurements. This operation needs to be performed in an automated way, through computing tools that can be run at-scale over millions of events and are reproducible. Significant progress has been made to deliver high-quality reconstruction methods for LArTPC detectors. Contributions from the ICARUS, ArgoNeuT, MicroBooNE, and protoDUNE experiments have, in different phases, spearheaded the publication of physics results with LArTPC data. We briefly discuss the major components in LArTPC event reconstruction for LArTPC detectors.

7.5.1.1 *Signal processing and hit finding*

Converting signals recorded on individual wires (or pixels for pixel-based LArTPC detectors) into measurements of charge deposition is the first step in most reconstruction workflows. This operation takes several forms but often starts with an initial signal processing stage which is responsible for filtering TPC electronics noise and applying signal deconvolution to correct for the impact of electronics response on signals induced by drifting electrons. The outcome of this process is a measurement of the amount and arrival time of charge for each recorded energy deposit on a wire. While these tasks are not new to LArTPC reconstruction, they have been significantly expanded in recent years. Reference [60] discusses the noise-filtering techniques utilized both in hardware and offline analysis, which lead to achieving signal-to-noise of up to 50-to-1 on the collection plane. References [48, 49] describe the signal processing developed to reconstruct signals from raw waveforms. Of note is the innovative 2D deconvolution methods implemented to account for angle-dependence in wire response and the impact of charge-sharing across wires, which when incorporated in the deconvolution procedure provide a more uniform detector response and higher signal efficiency for tracks oriented with a large pitch with respect to the wire plane. Figure 7.14 shows the impact of 2D deconvolution on

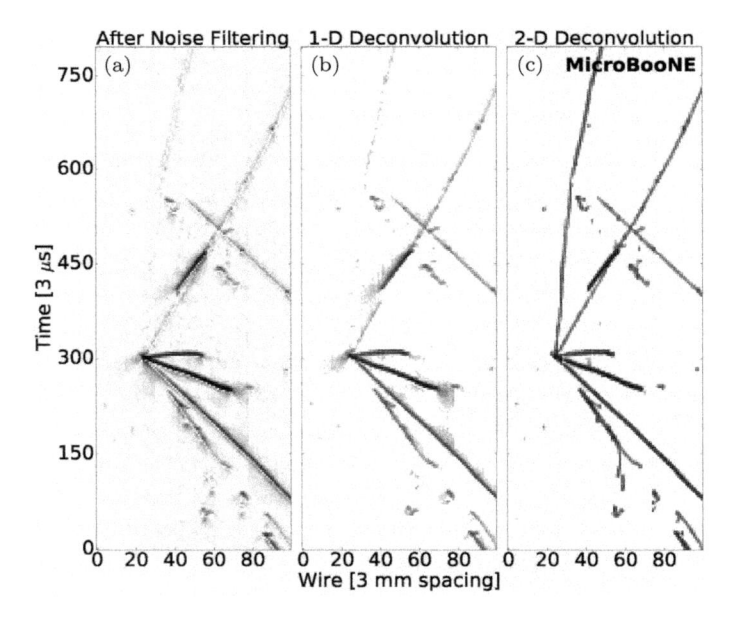

Figure 7.14. Two-dimensional deconvolution applied to a MicroBooNE neutrino interaction candidate. The sharpness of the tracks originating from the neutrino interaction vertex (at coordinates 20 mm, 300 μs) is significantly improved in the 2D deconvolution method (right panel) with respect to a 1D approach (center). This can be seen clearly in the tracks that are aligned with the vertical coordinate in the image.

Source: Figure from Ref. [48].

signal from a neutrino interaction candidate recorded with the MicroBooNE LArTPC. Charge measurements on each plane, often referred to as "hits", need to then be correlated across planes to produce 3D patterns of energy deposition. This is achieved by relying on the geometric overlap of wire planes and charge-matching deposits of equal charge on different wire planes. Different reconstruction paradigms perform the transition from 2D to 3D differently, and several workflows which employ image-like pattern recognition through machine learning techniques rely directly on wire images as input to tracking and PID algorithms.

7.5.1.2 *Tracking and particle flow reconstruction*

Given the measured energy deposits in the TPC, reconstruction algorithms are next tasked with isolating individual interactions taking place in the detector. For surface detectors, where the large rate of external cosmic-ray interactions leads to "busy" events, this can be a fairly complex task. An example event from the MicroBooNE detector is shown in Fig. 7.15, where dozens of cosmic-ray interactions pile up on top of a single neutrino interaction, shown in the zoomed in panel. The image gives a sense of the reconstruction complexity involved. The individual particles making up each interaction need to be isolated and tracked. These tasks are

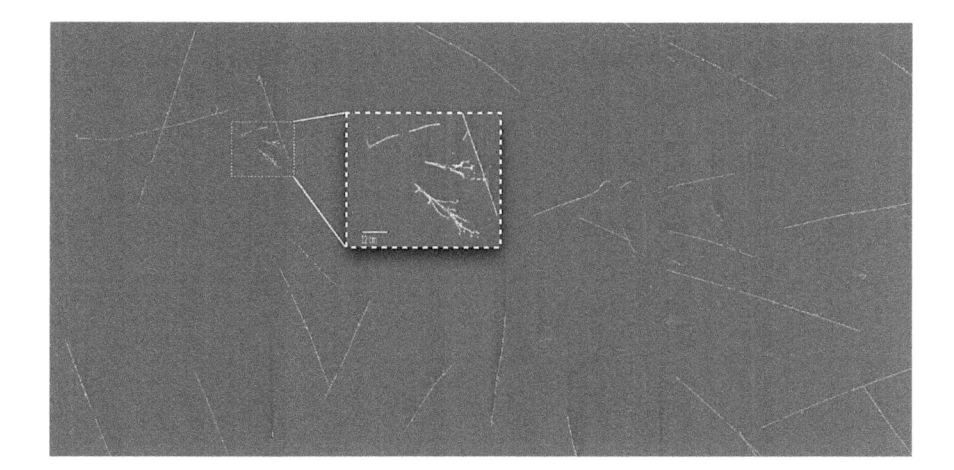

Figure 7.15. Collection-plane view of an event recorded by MicroBooNE. The event is filled with cosmic-ray interactions which reach the detector during the TPC's $\mathcal{O}(\mathrm{ms})$ drift window. The single beam neutrino interaction in the event is isolated in the zoom-in panel.

achieved through the employment of multi-algorithm chains which leverage the position and charge map of energy deposits to identify particle tracks and their parentage relationship. Comprehensive reconstruction toolkits such as Pandora [61, 62] and Wire-Cell [63, 64] offer fully automated reconstruction with tracking and particle-flow reconstruction for generic neutrino and cosmic-ray interactions. Machine learning techniques have also been developed to leverage the image-like quality of LArTPC data for the purpose of event reconstruction. Details on such methods are documented in Refs. [65–71].

7.5.1.3 Cosmic rejection and charge-to-light matching

Surface LArTPC detectors suffer from a high rate of cosmic-ray activity which swamps neutrino signals, in part due to the slow TPC readout. Even for events where a neutrino interaction occurs in the detector, $\mathcal{O}(10)$ cosmic-ray interactions pile-up on the neutrino image, requiring tools that are capable of identifying and isolating the TPC charge deposited by the neutrino interaction. This task is referred to as "flash-matching", or "charge-to-light" matching. Leveraging the prompt scintillation light-signature in-time with the beam, flash-matching identifies the interaction in the TPC that is compatible with the observed scintillation light signal. This is achieved by comparing the geometric pattern of scintillation light recorded in the detector to the light that is predicted to hit each photosensor based on the spatial charge distribution. This matching leverages the absolute charge and position-dependence collected with both the PDS and TPC systems and is capable of reducing cosmic-ray backgrounds from signal-to-background levels of 1 : 100 to 1 : 1 or better [72, 73]. Figure 7.16 shows how this operation takes place. The left panel shows a side view of the MicroBooNE TPC with the PMT positions denoted by circles. The

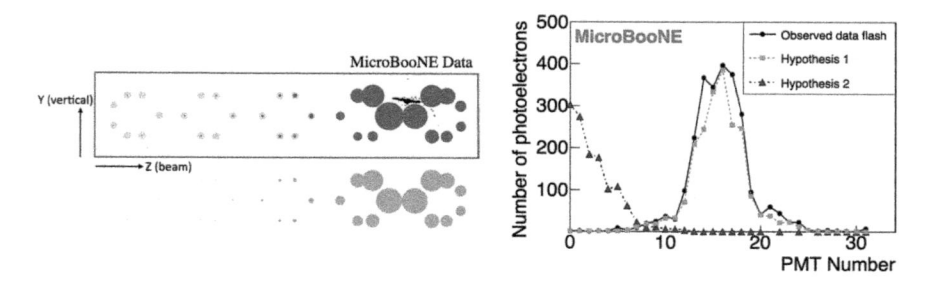

Figure 7.16. Flash matching illustrations from Refs. [72, 73].

elongated splash in the image is a candidate ν_e interaction reconstructed with TPC information. The circles within the TPC rectangle indicate the number of photoelectrons recorded by the PMT array in-time with the neutrino beam, while the circles below the TPC represent the expected light predicted based on the observed charge recorded by the TPC. The good match between them indicates that the TPC interaction is consistent with having occurred in-time with the beam. The image on the right shows an analogous quantitative study comparing the spectrum of photoelectrons recorded on the PMT array (solid line) to the predicted number of photons based on the pattern of charge from two different TPC interactions: one of them clearly matches, while the other can be rejected.

7.5.1.4 *Calorimetry, particle kinematics measurements, and particle ID*

Once individual particle tracks are reconstructed in the TPC, their kinematics (momentum and direction) and particle species are reconstructed leveraging the TPC's spatial and calorimetric information. Particle ID first aims to determine the track-like or shower-like nature, which typically separates track-like μ, p, $\pi^{+/-}$, and K from shower-like e and γ particles. Within track-like particles, calorimetric PID algorithms make use of the dE/dx vs. residual range profile characteristic of each particle to determine the particle species. Figure 7.17 shows the characteristic separation in energy-loss profile for protons and muons. For contained tracks, compatibility with the expectation for each particle species (given by the solid theoretical lines superimposed on the figure) allows us to provide accurate PID. Electron–photon separation is a task that was introduced in Section 7.2.1.1 and leverages two key features: the "gap" between an interaction vertex (when visible) and the start of the EM shower which is present for photons but not for electrons, and the energy loss at the shower trunk, which for photons which pair convert is twice as large as for electron showers. Figure 7.6 shows an example of this second, calorimetric classification method.

While LArTPCs provides excellent PID capabilities, charge sign determination is a task that is not performed on a particle-by-particle basis in LArTPCs given the lack of an external magnetic field.

Once the particle species is determined, a measurement of its kinematics follows. For track-like particles (muons, protons, and pions) which come to stop in the detector, their momentum is typically estimated by range,

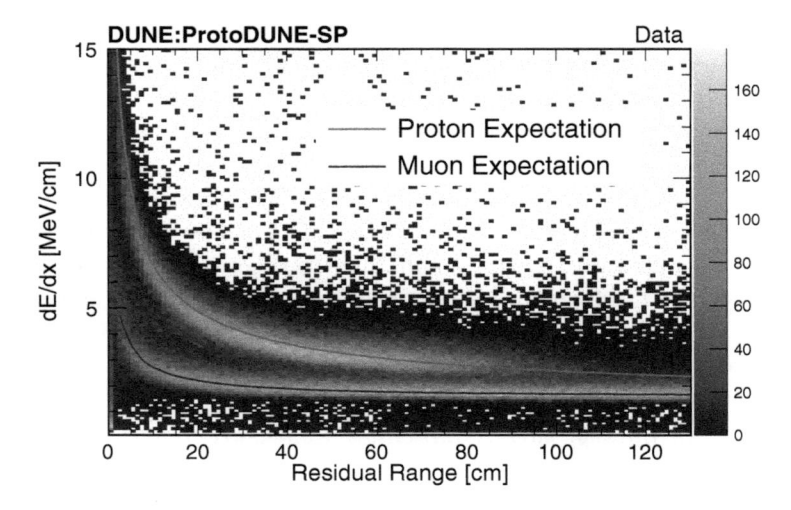

Figure 7.17. Muon and proton dE/dx profile vs. residual range from protoDUNE [10]. The grayscale histogram shows measured energy loss as a function of residual range for proton and muon candidates. The population with larger dE/dx values is associated with protons which ionize more heavily. The difference in energy loss profile between different particle species allows for precise particle ID in LArTPC detectors.

leveraging the strong correlation between track length and momentum. This method gives an energy resolution of a few percent for protons and muons. While the method works well for charged pions as well, their higher probability of re-interacting makes a momentum determination more difficult to obtain reliably. The effect of Multiple Coulomb Scattering (MCS) is used for energy estimation for uncontained muons. MCS leverages the correlation between a muon's energy and the average angular deflection due to Coulomb scattering to measure the muon energy with an accuracy of ∼20%. This method, employed through the use of a likelihood and demonstrated in both the ICARUS [74] and MicroBooNE [75] experiments, proves reliable up to $\mathcal{O}(1)$ GeV energies, at which point the scattering amplitudes become comparable with the detector's angular resolution. For EM showers from electrons and photons, energy reconstruction is calculated calorimetrically by integrating the visible charge deposited in the TPC. This method, which requires careful considerations tied to absolute energy scale calibration, including the impact of ion recombination on EM showers, achieves performance which ranges in the 10–20% resolution. Importantly, the method is often limited by reconstruction inefficiencies in fully collecting charge produced by the EM shower which either falls below hit-reconstruction thresholds or is incorrectly missed by charge-clustering algorithms. This inefficiency also leads to a need for corrections to any potential bias in determining the shower energy, which lead to an underestimation of the shower energy if left uncorrected. Extensive details on EM shower energy-scale reconstruction and calibrations can be found in Refs. [27, 76–79].

7.5.1.5 Machine learning in LArTPC reconstruction

Machine Learning (ML) and Artificial Intelligence (AI) have become integral components of data analysis in high energy physics. The development of reconstruction tools based on ML methods has in particular leveraged the image-like features of LArTPC datasets. These methods often rely on the use of deep or sparse neural networks which have found broad application in AI and computing beyond physics. Tasks such as neutrino identification [80,81], pixel-based particle ID [82,83], and reconstruction [84] are all rapidly making significant advances in reconstruction capabilities and being deployed in physics measurements [20].

7.5.2 Detector calibrations

Calibrating for the multiple effects which impact charge and light signals described earlier in this chapter requires significant effort. Several strategies and sources are leveraged to perform calibrations of different signals. Intrinsic sources of energy loss provide valuable samples for detector calibration. In particular, surface LArTPCs can leverage a uniform and steady-state rate of cosmic-ray muons for relative position-dependent and absolute energy-scale calibrations. Samples of both neutrino and cosmic-ray induced particles often leveraged include $\pi^0 \to \gamma\gamma$ decays and Michel electrons for EM shower energy-scale calibrations, as well as stopping protons and muons for absolute energy-scale and ion recombination calibrations. Figure 7.18 shows the reconstructed invariant mass from diphotons coming from π^0 candidates. The kinematics of the π^0 decay allow us to recover the π^0 mass and provide a valuable way to validate and calibrate EM shower energy-scale reconstruction.

In addition to this "free" calibration source, several detector subsystems and components are installed to provide additional tools intended to meet particular calibration challenges. Ionizing lasers such as that described in Ref. [85], which produce continuous track-like segments of ionization in the TPC, are used to map out the electric field in the TPC and calibrate

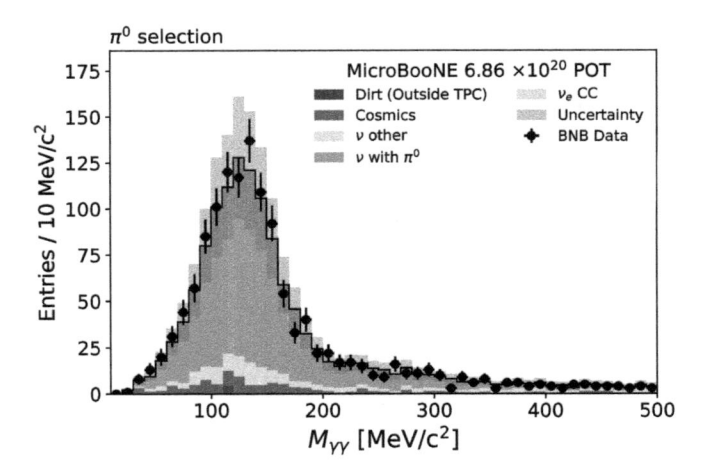

Figure 7.18. Reconstructed invariant diphoton mass from π^0 candidates. *Source*: Figure from MicroBooNE [19].

effects such as SCE. Scintillation light LED calibration systems such as those described in Refs. [10, 86] also help complement natural sources of photons.

Calibrations for scintillation light signatures have been developed by several experimental collaborations leveraging either test-beam facilities (protoDUNE [11] and LArIAT [87]) or leveraging cosmogenic muons and protons as in the case of MicroBooNE [88, 89].

Calibration of the TPC response to ionization is critical in order to obtain accurate calorimetric measurements necessary for PID and particle kinematics measurements. Calibrations that cover relative distortions in the charge response across the TPC volume need to account for multiple effects, including dependence on the local wire response, quenching due to impurities, and SCE distortions. Absolute energy-scale calibrations responsible for converting measured currents on the wires into total energy deposited are instead primarily impacted by electronics gain and ion recombination. While multiple calibration techniques are employed to measure each effect independently, they are often integrated in a comprehensive calibration strategy, described in Ref. [90] for the MicroBooNE experiment and Ref. [10] for protoDUNE. A comprehensive review of these methods is presented in Ref. [91]. The calibration of EM showers poses specific challenges tied to the topology of shower propagation in liquid argon. This is in part due to the lossy nature of EM shower energy reconstruction, which relies on calorimetric charge measurements but may suffer from charge falling below threshold or reconstruction inefficiencies. The importance of EM shower energy-scale calibrations is paramount to both the ν oscillation program searching for $\nu_\mu \to \nu_e$ signatures with electron showers in the final state as well as BSM searches in the sub-GeV regime which often include EM shower final states. The topic of EM shower calibrations in LArTPCs is discussed in several articles, including Refs. [27, 76–79].

References

[1] W. J. Willis and V. Radeka. Liquid argon ionization chambers as total absorption detectors. *Nuclear Instruments and Methods in Physics Research Section A: Accelerators, Spectrometers, Detectors and Associated Equipment*, 120:221–236, 1974.

[2] D. R. Nygren. The time projection chamber: A new 4 pi detector for charged particles. *eConf*, C740805:58, 1974.

[3] C. Rubbia. The liquid argon time projection chamber: A new concept for neutrino detectors, May 1977.

[4] H. H. Chen, P. E. Condon, B. C. Barish, and F. J. Sciulli. A neutrino detector sensitive to rare processes. I. A study of neutrino electron reactions, May 1976.

[5] P. Cennini, et al. Performance of a 3-ton liquid argon time projection chamber. *Nuclear Instruments and Methods in Physics Research Section A: Accelerators, Spectrometers, Detectors and Associated Equipment*, 345:230–243, 1994.

[6] S. Amerio, et al. Design, construction and tests of the ICARUS T600 detector. *Nuclear Instruments and Methods in Physics Research Section A: Accelerators, Spectrometers, Detectors and Associated Equipment*, 527:329–410, 2004.

[7] P. Abratenko, et al. ICARUS at the Fermilab short-baseline neutrino program: Initial operation. *The European Physical Journal C*, 83(6):467, 2023.

[8] R. Acciarri, et al. Design and construction of the MicroBooNE detector. *JINST*, 12(02):P02017, 2017.

[9] R. Acciarri, et al. The liquid argon in a Testbeam (LArIAT) experiment. *JINST*, 15(04):P04026, 2020.

[10] B. Abi, et al. First results on ProtoDUNE-SP liquid argon time projection chamber performance from a beam test at the CERN neutrino platform. *JINST*, 15(12):P12004, 2020.

[11] A. Abed Abud, et al. Design, construction and operation of the ProtoDUNE-SP liquid argon TPC. *JINST*, 17(01):P01005, 2022.

[12] M. Antonello, et al. A proposal for a three detector short-baseline neutrino oscillation program in the Fermilab booster neutrino beam, March 2015.

[13] P. A. N. Machado, O. Palamara, and D. W. Schmitz. The short-baseline neutrino program at Fermilab. *Annual Review of Nuclear and Particle Science*, 69:363–387, 2019.

[14] B. Abi, et al. Long-baseline neutrino oscillation physics potential of the DUNE experiment. *European Physical Journal C*, 80(10):978, 2020.

[15] B. Abi, et al. Supernova neutrino burst detection with the deep underground neutrino experiment. *European Physical Journal C*, 81(5):423, 2021.

[16] B. Abi, et al. Prospects for beyond the standard model physics searches at the deep underground neutrino experiment. *European Physical Journal C*, 81(4):322, 2021.

[17] P. Abratenko, et al. Search for an excess of electron neutrino interactions in MicroBooNE using multiple final-state topologies. *Physical Review Letters*, 128(24):241801, 2022.

[18] P. Abratenko, et al. Search for neutrino-induced neutral-current Δ radiative decay in MicroBooNE and a first test of the MiniBooNE low energy excess under a single-photon hypothesis. *Physical Review Letters*, 128:111801, 2022.

[19] P. Abratenko, et al. Search for an anomalous excess of charged-current νe interactions without pions in the final state with the MicroBooNE experiment. *Physical Review D*, 105(11):112004, 2022.

[20] P. Abratenko, et al. Search for an anomalous excess of charged-current quasielastic νe interactions with the MicroBooNE experiment using deep-learning-based reconstruction. *Physical Review D*, 105(11):112003, 2022.

[21] P. Abratenko, et al. Search for an anomalous excess of inclusive charged-current ν_e interactions in the MicroBooNE experiment using wire-cell reconstruction. *Physical Review D*, 105(11):112005, 2022.

[22] L. Bagby, et al. Overhaul and installation of the ICARUS-T600 liquid argon TPC electronics for the FNAL short baseline neutrino program. *JINST*, 16(01):P01037, 2021.

[23] M. Adamowski, et al. The liquid argon purity demonstrator. *JINST*, 9:P07005, 2014.

[24] C. Bromberg, et al. Design and operation of LongBo: A 2 m long drift liquid argon TPC. *JINST*, 10(07):P07015, 2015.

[25] Yichen Li, et al. A 20-liter test stand with gas purification for liquid argon research. *JINST*, 11(06):T06001, 2016.

[26] R. L. Workman, et al. Review of particle physics. *PTEP*, 2022:083C01, 2022.

[27] R. Acciarri, et al. Michel electron reconstruction using cosmic-ray data from the MicroBooNE LArTPC. *JINST*, 12(09):P09014, 2017.

[28] P. Abratenko, et al. Measurement of the flux-averaged inclusive charged-current electron neutrino and antineutrino cross section on argon using the NuMI beam and the MicroBooNE detector. *Physical Review D*, 104(5):052002, 2021.

[29] P. V. Vavilov. Ionization losses of high-energy heavy particles. *Soviet Physics JETP*, 5:749–751, 1957.

[30] M. Miyajima, T. Takahashi, S. Konno, T. Hamada, S. Kubota, H. Shibamura, and T. Doke. Average energy expended per ion pair in liquid argon. *Physical Review A*, 9:1438–1443, 1974.

[31] Z. Beever, D. Caratelli, A. Fava, F. Pietropaolo, F. Stocker, and J. Zettlemoyer. TRANSLATE — A Monte Carlo simulation of electron transport in liquid argon, November 2022.

[32] T. Doke, A. Hitachi, J. Kikuchi, K. Masuda, H. Okada, and E. Shibamura. Absolute scintillation yields in liquid argon and xenon for various particles. *Japanese Journal of Applied Physics*, 41(3A, Part 1):1538–1545, 2002.

[33] S. Amoruso, et al. Study of electron recombination in liquid argon with the ICARUS TPC. *Nuclear Instruments and Methods in Physics Research Section A: Accelerators, Spectrometers, Detectors and Associated Equipment*, 523:275–286, 2004.

[34] R. Acciarri, et al. A study of electron recombination using highly ionizing particles in the ArgoNeuT liquid argon TPC. *JINST*, 8:P08005, 2013.

[35] Measurement of the electronegative contaminants and drift electron lifetime in the MicroBooNE experiment. May 2016.

[36] Y. Li, C. Bromberg, M. Diwan, S. Kettell, S. Martynenko, X. Qian, V. Paolone, J. Stewart, C. Thorn, and C. Zhang. Parameterization of electron attachment rate constants for common impurities in LArTPC detectors May 2022.

[37] A. Lister and M. Stancari. Investigations on a fuzzy process: Effect of diffusion on calibration and particle identification in liquid argon time projection chambers. *JINST*, 17(07):P07016, 2022.

[38] G. Putnam and D. W. Schmitz. Diffusion changes the peak value of energy loss observed in a LArTPC, May 2022.

[39] P. Abratenko, et al. Measurement of the longitudinal diffusion of ionization electrons in the MicroBooNE detector. *JINST*, 16(09):P09025, 2021.

[40] Y. Li, et al. Measurement of longitudinal electron diffusion in liquid argon. *Nuclear Instruments and Methods in Physics Research Section A: Accelerators, Spectrometers, Detectors and Associated Equipment*, 816:160–170, 2016.

[41] M. Torti. Electron diffusion measurements in the ICARUS T600 detector. *Journal of Physics: Conference Series*, 888(1):012060, 2017.

[42] P. Abratenko, et al. Measurement of space charge effects in the MicroBooNE LArTPC using cosmic muons. *JINST*, 15(12):P12037, 2020.

[43] M. Antonello, et al. Study of space charge in the ICARUS T600 detector. *JINST*, 15(07):P07001, 2020.

[44] C. Adams, et al. A method to determine the electric field of liquid argon time projection chambers using a UV laser system and its application in MicroBooNE. *JINST*, 15(07):P07010, 2020.

[45] M. Babicz, S. Bordoni, A. Fava, U. Kose, M. Nessi, F. Pietropaolo, G.L. Raselli, F. Resnati, M. Rossella, P. Sala, F. Stocker, and A. Zani. A measurement of the group velocity of scintillation light in liquid argon. *Journal of Instrumentation*, 15(09):P09009–P09009, September 2020.

[46] D. A. Dwyer, et al. LArPix: Demonstration of low-power 3D pixelated charge readout for liquid argon time projection chambers. *JINST*, 13(10):P10007, 2018.

[47] D. Nygren and Y. Mei. Q-Pix: Pixel-scale signal capture for kiloton liquid argon TPC detectors: Time-to-charge waveform capture, local clocks, dynamic networks, September 2018.

[48] C. Adams, et al. Ionization electron signal processing in single phase LArTPCs. Part I. Algorithm description and quantitative evaluation with MicroBooNE simulation. *JINST*, 13(07):P07006, 2018.

[49] C. Adams, et al. Ionization electron signal processing in single phase LArTPCs. Part II. Data/simulation comparison and performance in MicroBooNE. *JINST*, 13(07):P07007, 2018.

[50] G. De Geronimo, et al. Front-end ASIC for a liquid argon TPC. *IEEE Transactions on Nuclear Science*, 58:1376–1385, 2011.

[51] Noise dependence on temperature and LAr fill level in the MicroBooNE time projection chamber, August 2015.

[52] M. Kuźniak and A. M. Szelc. Wavelength shifters for applications in liquid argon detectors. *Instruments*, 5(1):4, 2020.

[53] R. Acciarri, et al. Effects of nitrogen contamination in liquid argon. *JINST*, 5:P06003, 2010.

[54] A. A. Machado and E. Segreto. ARAPUCA a new device for liquid argon scintillation light detection. *JINST*, 11(02):C02004, 2016.

[55] A. A. Machado, E. Segreto, D. Warner, A. Fauth, B. Gelli, R. Maximo, A. Pizolatti, L. Paulucci, and F. Marinho. The X-ARAPUCA: An improvement of the ARAPUCA device, April 2018.

[56] L. Paulucci. The DUNE vertical drift photon detection system. *JINST*, 17(01):C01067, 2022.

[57] M. Auger, et al. A novel cosmic ray tagger system for liquid argon TPC neutrino detectors. *Instruments*, 1(1):2, 2017.

[58] C. Adams, et al. Design and construction of the MicroBooNE cosmic ray tagger system. *JINST*, 14(04):P04004, 2019.

[59] S. Amoruso, et al. Analysis of the liquid argon purity in the ICARUS T600 TPC. *Nuclear Instruments and Methods in Physics Research Section A: Accelerators, Spectrometers, Detectors and Associated Equipment*, 516:68–79, 2004.

[60] R. Acciarri, et al. Noise characterization and filtering in the MicroBooNE liquid argon TPC. *JINST*, 12(08):P08003, 2017.

[61] A. Abed Abud, et al. Reconstruction of interactions in the ProtoDUNE-SP detector with Pandora. June 2022.

[62] R. Acciarri, et al. The Pandora multi-algorithm approach to automated pattern recognition of cosmic-ray muon and neutrino events in the MicroBooNE detector. *European Physical Journal C*, 78(1):82, 2018.

[63] P. Abratenko, et al. Cosmic ray background rejection with wire-cell LArTPC event reconstruction in the MicroBooNE detector. *Physical Review Applied*, 15(6):064071, 2021.

[64] X. Qian, C. Zhang, B. Viren, and M. Diwan. Three-dimensional imaging for large LArTPCs. *JINST*, 13(05):P05032, 2018.

[65] A. Abed Abud, *et al*. Separation of track- and shower-like energy deposits in ProtoDUNE-SP using a convolutional neural network. *European Physical Journal C*, 82(10):903, 2022.

[66] R. Acciarri, *et al*. Convolutional neural networks applied to neutrino events in a liquid argon time projection chamber. *JINST*, 12(03):P03011, 2017.

[67] B. Abi, *et al*. Neutrino interaction classification with a convolutional neural network in the DUNE far detector. *Physical Review D*, 102(9):092003, 2020.

[68] P. Abratenko, *et al*. Convolutional neural network for multiple particle identification in the MicroBooNE liquid argon time projection chamber. *Physical Review D*, 103(9):092003, 2021.

[69] C. Adams, *et al*. Deep neural network for pixel-level electromagnetic particle identification in the MicroBooNE liquid argon time projection chamber. *Physical Review D*, 99(9):092001, 2019.

[70] P. Abratenko, *et al*. Semantic segmentation with a sparse convolutional neural network for event reconstruction in MicroBooNE. *Physical Review D*, 103(5):052012, 2021.

[71] R. Acciarri, *et al*. A deep-learning based raw waveform region-of-interest finder for the liquid argon time projection chamber. *JINST*, 17(01):P01018, 2022.

[72] P. Abratenko, *et al*. Neutrino event selection in the MicroBooNE liquid argon time projection chamber using Wire-Cell 3D imaging, clustering, and charge-light matching. *JINST*, 16(06):P06043, 2021.

[73] P. Abratenko, *et al*. Measurement of differential cross sections for ν_μ - Ar charged-current interactions with protons and no pions in the final state with the MicroBooNE detector. *Physical Review D*, 102(11):112013, 2020.

[74] M. Antonello, *et al*. Muon momentum measurement in ICARUS-T600 LArTPC via multiple scattering in few-GeV range. *JINST*, 12(04):P04010, 2017.

[75] P. Abratenko, *et al*. Determination of muon momentum in the MicroBooNE LArTPC using an improved model of multiple Coulomb scattering. *JINST*, 12(10):P10010, 2017.

[76] S. Amoruso, *et al*. Measurement of the mu decay spectrum with the ICARUS liquid argon TPC. *European Physical Journal C*, 33:233–241, 2004.

[77] C. Adams, *et al*. Reconstruction and measurement of $\mathcal{O}(100)$ MeV energy electromagnetic activity from $\pi^0 \to \gamma\gamma$ decays in the MicroBooNE LArTPC. *JINST*, 15(02):P02007, 2020.

[78] R. Acciarri, *et al*. Michel electron reconstruction using cosmic-ray data from the MicroBooNE LArTPC. *JINST*, 12(09):P09014, 2017.

[79] P. Abratenko, *et al*. Electromagnetic shower reconstruction and energy validation with Michel electrons and π^0 samples for the deep-learning-based analyses in MicroBooNE. *JINST*, 16(12):T12017, 2021.

[80] R. Acciarri, *et al*. Convolutional neural networks applied to neutrino events in a liquid argon time projection chamber. *JINST*, 12(03):P03011, 2017.

[81] B. Abi, et al. Neutrino interaction classification with a convolutional neural network in the DUNE far detector. *Physical Review D*, 102(9):092003, 2020.

[82] P. Abratenko, et al. Semantic segmentation with a sparse convolutional neural network for event reconstruction in MicroBooNE. *Physical Review D*, 103(5):052012, 2021.

[83] C. Adams, et al. Deep neural network for pixel-level electromagnetic particle identification in the MicroBooNE liquid argon time projection chamber. *Physical Review D*, 99(9):092001, 2019.

[84] P. Abratenko, et al. Convolutional neural network for multiple particle identification in the MicroBooNE liquid argon time projection chamber. *Physical Review D*, 103(9):092003, 2021.

[85] A. Ereditato, I. Kreslo, M. Lüthi, C. Rudolf von Rohr, M. Schenk, T. Strauss, M. Weber, and M. Zeller. A steerable UV laser system for the calibration of liquid argon time projection chambers. *JINST*, 9(11):T11007, 2014.

[86] J. Conrad, B. J. P. Jones, Z. Moss, T. Strauss, and M. Toups. The photomultiplier tube calibration system of the MicroBooNE experiment. *JINST*, 10(06):T06001, 2015.

[87] W. Foreman, et al. Calorimetry for low-energy electrons using charge and light in liquid argon. *Physical Review D*, 101(1):012010, 2020.

[88] MicroBooNE Collaboration. Measuring light yield with isolated protons in MicroBooNE, August 2022. https://microboone.fnal.gov/wp-content/uploads/MICROBOONE-NOTE-1119-PUB.pdf.

[89] MicroBooNE Collaboration. Light yield calibration in MicroBooNE, August 2022. https://microboone.fnal.gov/wp-content/uploads/MICROBOONE-NOTE-1120-TECH.pdf.

[90] C. Adams, et al. Calibration of the charge and energy loss per unit length of the MicroBooNE liquid argon time projection chamber using muons and protons. *JINST*, 15(03):P03022, 2020.

[91] T. Yang. Calibration of calorimetric measurement in a liquid argon time projection chamber. *Instruments*, 5(1):2, 2020.

Chapter 8

The Super-Kamiokande and Other Detectors: A Case Study of Large Volume Cherenkov Neutrino Detectors

Shunichi Mine

Kamioka Observatory, ICRR, The Univ. of Tokyo, Higashi-Mozumi 456, Kamioka-cho, Hida-city Gifu 506-1205, Japan
mine@km.icrr.u-tokyo.ac.jp

8.1 Introduction

This chapter describes the Cherenkov neutrino detectors used in high-energy "high-E" (above about 0.1 GeV) physics experiments. These experiments include atmospheric neutrino oscillation measurements, nucleon decay searches (atmospheric neutrinos are the background), long baseline beam neutrino oscillation measurements, astrophysical neutrino searches, and so on. Each section also briefly mentions differences from low-energy "Low-E" ($\ll 100\,\mathrm{MeV}$) physics measurements such as solar and supernova neutrinos.

We can only detect the presence of a neutrino if it interacts with the detector material. Neutrinos interact in two ways: charged-current (CC) interactions and neutral-current (NC) interactions. In CC interactions, the neutrino converts into the equivalent charged lepton (e.g., $\nu_e + n \to e^- + p$) and the experiments detect the charged lepton, as well as any other charged particles which emit Cherenkov light. In NC interactions, the neutrino remains a neutrino and we detect the other particles produced by the interaction.

In principle, CC interactions are easier to work with because electrons and muons have characteristic signatures in the Cherenkov detectors and are thus fairly easy to identify. Tau neutrinos are more difficult to identify for two reasons. Since the tau is more massive, the energy thresholds for the CC tau production (3.5 GeV) and the Cherenkov radiation of the tau lepton (2.7 GeV) are much higher. Also, the tau is extremely short-lived and therefore does not travel far enough to emit much Cherenkov light ($c\tau \sim 9 \times 10^{-5}$ m).

Since neutrinos only weakly interact with matter, neutrino detectors must be generally very large to detect a significant number of events. The combination of a transparent medium such as water with photomultiplier tubes (PMTs) as light sensors is beneficial for achieving a large effective volume at a low cost. Such a medium serves as a target for neutrino interactions and is well suited for propagating Cherenkov light from charged final-state particles and energetic photons. PMTs are chosen with wavelength sensitivity overlapping with the produced Cherenkov spectrum. A water purification system can remove radioactive substances and impurities and constantly circulate ultrapure water in the tank. The detectors are often built underground to isolate the detector from cosmic rays. In addition, detectors outside the main detector absorb or discriminate against incoming background particles from outside.

In order to extract as much information from each event as possible, the detector should have the following: good angular resolution so that the direction of the particle can be accurately reconstructed; good particle identification, allowing particle discrimination; good energy measurement so that the energy can be reconstructed; good time resolution so that the time evolution of transient signals can be studied; and so on.

In Section 8.2, the principle of the Cherenkov detector is explained using the Super-Kamiokande (Super-K or SK) detector [1–4]. SK is the world's largest water tank Cherenkov neutrino detector. This is the second-generation detector, with more mature experimental techniques than the first-generation IMB/Kamiokande detectors [5–7]. Since water Cherenkov detectors have a relatively sparsely instrumented readout per detector mass, using only PMTs to view a large active water volume, software for event reconstruction is very important. For this reason, the Appendix details how to reconstruct the events by software programs in SK. Section 8.3 summarizes past and present Cherenkov detectors. Section 8.4 presents the instrumentation and techniques of the hardware used in SK. Calibrating the detector is necessary and important for good physics results, so we explain this in Section 8.5. Some future prospects

8.2 Principle of Cherenkov Detectors

8.2.1 *Cherenkov radiation*

Water has a refractive index of 1.33 for visible light wavelengths, so light in water travels at $0.75c$. Particles aren't affected by the refractive index, so a particle traveling at almost the speed of light through water will be traveling faster than the local speed of light in that medium. As an aircraft traveling faster than the speed of sound emits a sonic boom, similarly, a particle traveling through a transparent medium faster than the speed of light in that medium emits a kind of "light boom," a coherent cone of blue light known as Cherenkov radiation [8].

Figure 8.1 shows the geometry of Cherenkov radiation. The charged particle is traveling left to right at speed βc through a medium with

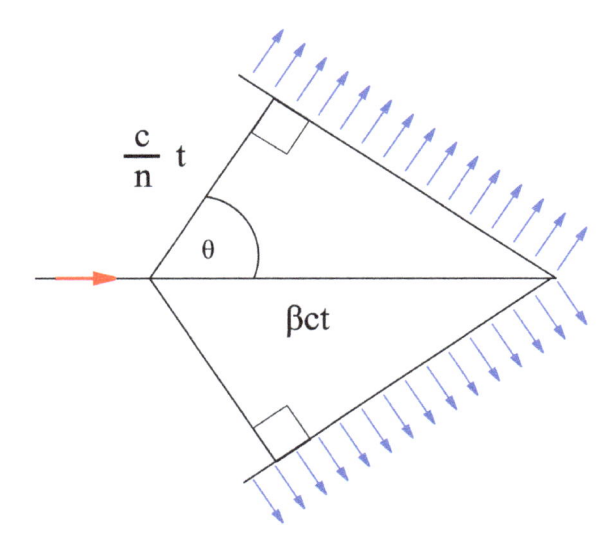

Figure 8.1. The geometry of Cherenkov radiation. From the left, a charged particle propagates, polarizing atoms and molecules around it. These atoms and molecules emit electromagnetic radiation, which propagates outward from each point along the track as a spherical wave. The wavefronts form a forward-propagating cone. The horizontal arrow is the velocity (v) of the charged particle, β is v/c, and n is the refractive index of the medium. The diagonal arrows show the direction of the radiation.

Source: Wikipedia.

refractive index n. The Cherenkov cone has half-angle θ given by

$$\cos\theta = \frac{1}{n\beta} \qquad (8.1)$$

To emit Cherenkov light, β must be above $1/n$ ($\cos\theta$ must be less than 1). The energy threshold above which Cherenkov radiation is emitted is 0.8 MeV for an electron, 160 MeV for a muon, and 2.7 GeV for a tau. At the highest speeds ($\beta \sim 1$), the "Cherenkov opening angle" is about 41° in water and the Cherenkov light is radiated almost diagonally.

The spectrum of the Cherenkov light as a function of the wavelength λ is

$$\frac{dN}{d\lambda} = \frac{2\pi\alpha x}{c}\left(1 - \frac{1}{n^2\beta^2}\right)\frac{1}{\lambda^2}, \qquad (8.2)$$

where $\alpha \approx 1/137$ (fine structure constant) and x is the length of the charged particle trajectory. A charged particle emits about 400 photons per centimeter pathlength in water in the wavelength region, 300–700 nm, where photomultiplier tubes (PMTs) are sensitive. Most visible Cherenkov radiation is in the violet range because of Eq. (8.2).

8.2.2 Cherenkov light detection (overview)

The detector water is contained in a tank lined with PMTs. For each event, the number of photons and time information for each PMT is recorded

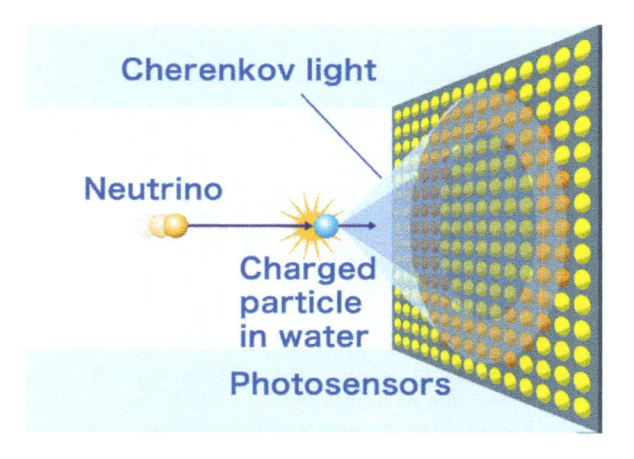

Figure 8.2. Cherenkov light is emitted in a cone shape along the direction of the charged particle. A tank wall is lined with photosensors (PMTs).
Source: Kamioka Observatory, ICRR, The University of Tokyo.

(Section 8.5.1.1). The Cherenkov light produced by a charged particle is reconstructed as a ring of hit PMTs (Fig. 8.2). Cherenkov light is emitted in a cone shape, surrounding the direction of the charged particle, and therefore the energy, direction, particle type, and so on are determined using information obtained from the PMTs, such as the amount of light detected and the ring shape (see Appendix). By assuming a type of neutrino interaction (e.g., quasi-elastic scattering), measurements of each charged particle in the final state can be used to reconstruct the original neutrino direction, energy, and so on.

Cherenkov light in the medium undergoes repeated wavelength-dependent absorption and scattering until it reaches the PMT. The attenuation of light due to absorption and scattering in water is measured as a function of wavelength (Sections 8.5.1.2 and 8.5.1.3) and is taken into account in the detector simulation and the event reconstruction for the physics analyses. For example, as discussed in Section 8.5.1.2, symmetric Rayleigh scattering is dominant at shorter wavelengths and absorption is dominant at longer wavelengths. The light transmittance is highest at around 400 nm. SK's PMT has the maximum quantum efficiency of the photocathode at about 400 nm [1]. The water transparency in SK is $\sim$100 m at around 400 nm, while in ice (Section 8.3.1.2) it is a couple of tens of meters.

Muons are leptons with a rest mass of 106 MeV/c^2 and a mean lifetime of 2.2×10^{-6} s. They have a relatively large mass with respect to that of the electron (0.511 MeV/c^2), and these particles do not participate in the strong interaction. Muons can penetrate longer distances in water ($\sim$0.5 cm/MeV) and are less susceptible to radiative energy losses compared to electrons. Over a broad energy range, the dominant energy loss mechanism is that of ionization. For example, the track length of a 1 GeV/c muon is about 5 m in water.

High-energy electrons and positrons primarily emit photons as they traverse the detector medium, a process called bremsstrahlung. Photons above a few MeV interact with matter primarily via pair production (an electron–positron pair). These processes (electromagnetic showers) continue, leading to a cascade of particles of decreasing energy until photons fall below the pair-production threshold, and electron (positron) energy losses become dominated by processes other than bremsstrahlung. Therefore, primary electrons have shorter track lengths than muons of the initial same energy.

As illustrated in Fig. 8.2, if a charged particle stops before reaching the surface of the water tank, it emits Cherenkov light at an angle determined by Eqs. (8.1) and (8.2) and is imaged on the tank surface as a Cherenkov light ring. The size and thickness of the Cherenkov ring image on the tank surface are determined by the distance of the particle vertex to the surface, particle type, direction, and momentum of the charged particles. In contrast, if the charged particle penetrates the tank before stopping, it will result in a Cherenkov light pattern with no holes. See Section 8.4.11 for the classification of events according to where charged particles are created and stop.

Muons are basically single particles and make sharp rings, whereas electrons, positrons, and gamma ray photons initiate electromagnetic showers and the nearly parallel electrons and positrons in the shower combine to make a fuzzy ring, as shown in Fig. 8.3. Coulomb scattering of the electron also contributes to the fuzziness of the ring.

Neutrino interactions can produce mesons such as pions, as well as the charged lepton. In general, charged pions (a mass of $140 \,\text{MeV}/c^2$ and a lifetime of 2.6×10^{-8} s) are difficult to distinguish from muons, but they may be able to be identified by using the properties of hadronic scattering (A.1.8).

When multiple Cherenkov rings are observed, the direction, particle type, and momentum are obtained for each ring, as in the case of the single-ring event. A neutral pion π^0 (a mass of $135 \,\text{MeV}/c^2$ and a lifetime of 8.4×10^{-17} s) immediately decays into two gammas and is detected as two "electron-like" rings. Figure 8.4 shows a typical reconstructed invariant mass distribution of the neutral pion events. See Section A.1 for details of reconstructing the event in the high-E physics analyses.

Tau leptons (a mass of $1.78 \,\text{GeV}/c^2$ and a lifetime of 2.9×10^{-13} s) produced in CC ν_τ interactions decay quickly to secondary particles. Due to the short lifetime of the tau lepton, it is impossible to directly detect taus in SK. The decay modes of the tau lepton are classified into leptonic and hadronic decay, based on the secondary particles in the decay. The leptonic decays produce neutrinos and an electron or a muon. These events look quite similar to the atmospheric CC ν_e or ν_μ background. The hadronic decays of the tau are dominant and produce one or more pions, plus a neutrino. The existence of extra pions in the hadronic decays of tau allows for the separation of the CC ν_τ signal from CC ν_μ, CC ν_e, and NC background. Multiple-ring events are relatively easy to separate from single-ring atmospheric neutrino events. However, the multi-ring background events, resulting from multi-pion/DIS atmospheric neutrino interactions, are difficult to distinguish from the tau signal. Simple selection criteria based on kinematic variables do not efficiently identify CC ν_τ events.

The Super-Kamiokande and Other Detectors   257

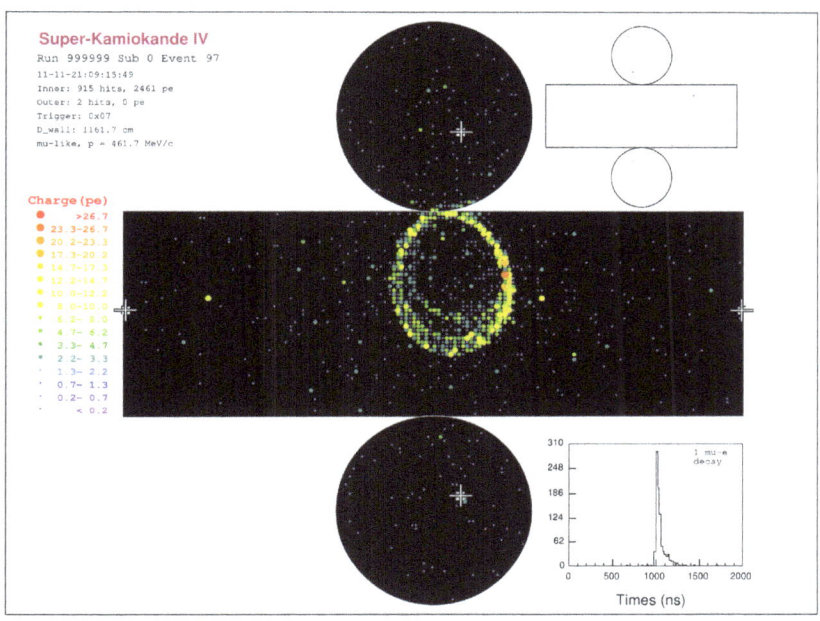

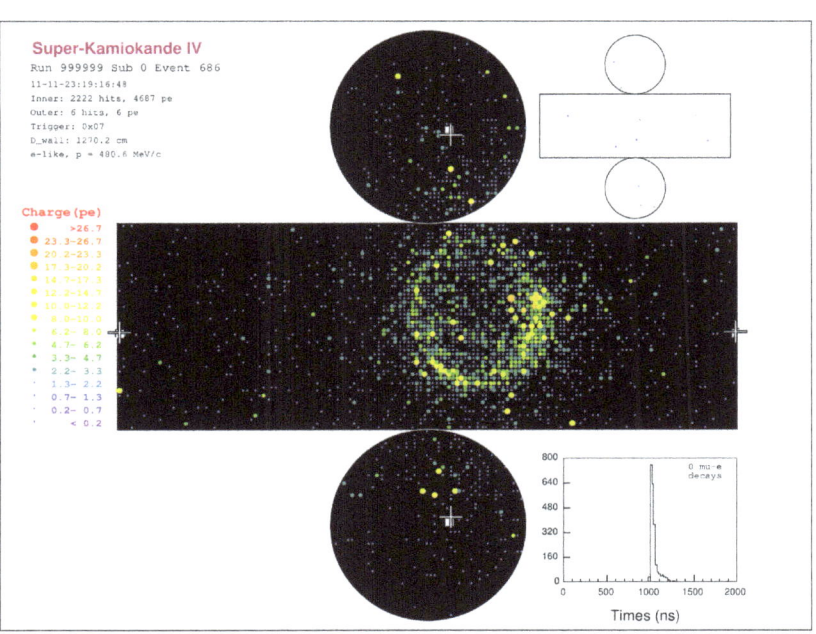

Figure 8.3 (*Continued*). The upper and lower displays are a typical muon and an electron-simulated event, respectively, in SK. In this example, a muon (electron) of about 500 MeV/c was created near the center of the tank, which headed toward the barrel wall. The display shows the expansion of the vertical cylindrical detector ($\sim$40 m diameter and $\sim$40 m height). The larger and smaller figures are the inner detector (ID) and the outer detector (OD), respectively. See Section 8.4 for each detector component. The color dots visualize which PMTs were hit in the event. The color correspond to the number of photoelectrons registered at that particular PMT. The timing distribution of the hit PMTs is shown at the bottom right.

Source: Kamioka Observatory, ICRR, The University of Tokyo.

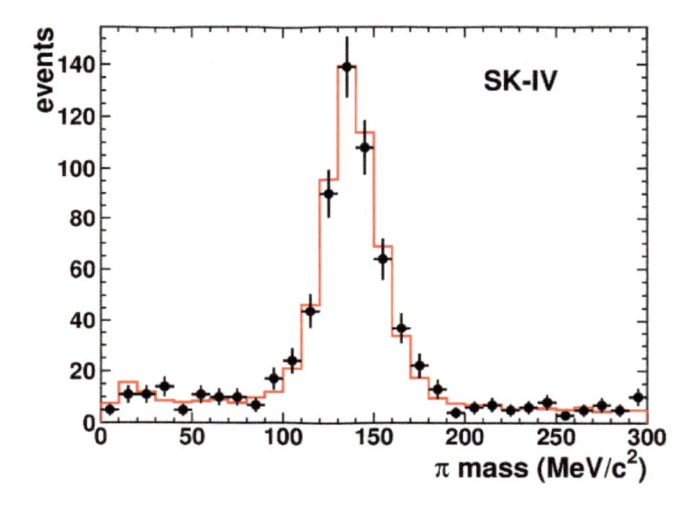

Figure 8.4. Reconstructed invariant mass distribution of the atmospheric-neutrino-induced π^0 events in the observed data (dot) and simulated atmospheric neutrino samples (histogram) in SK [9]. Two "electron-like" ring events are selected without any tagged "Michel electron" (an electron from a muon decay) in the fiducial volume (defined as a distance from the reconstructed vertex to the nearest inner detector wall is greater than 2 m). See Section A.1 for the meaning of each cut parameter.

In order to statistically identify events with the expected characteristics that differentiate signal and background, a multivariate method is applied in the SK analysis, based on the reconstructed parameters [10].

The final-state charged particles used in lower-energy "low-E" ($\ll$100 MeV) neutrino data analysis are almost exclusively electrons or gamma rays. For example, for electrons at SK, approximately several photoelectrons per MeV (pe/MeV) are detected at the PMT. Since more than 10,000 PMTs (Section 8.4.3) are mounted on the inner detector

(ID) wall of SK, the number of photoelectrons per PMT for each low-E neutrino event is typically one. Therefore, unlike for high-E events, low-E event reconstruction basically does not use PMT charge information but only PMT hit and time information. The high-E and low-E event reconstructions, respectively, are detailed in Appendix A. In high-E physics data analyses, we use events above the energy threshold of several tens of MeV, so a few MeV radioactivity is not a background.

8.3 Past and Present Cherenkov Detectors
8.3.1 *Water Cherenkov experiments*
8.3.1.1 *Densely instrumented artificial tanks*

The pioneering Irvine–Michigan–Brookhaven [5, 7] (IMB, 1982–1991) and Kamioka Nucleon Decay Experiment [6] (Kamiokande, 1983–1996) experiments, made famous by their observations of SN 1987A, were originally conceived as detectors for nucleon decay. Atmospheric neutrinos are a background for the nucleon decay search as the neutrino interactions produce charged particles, sometimes kinetically indistinguishable from the nucleon decay signal.

Super-Kamiokande [1–4] (1996–present) was designed to extend and improve upon the experience gained by its scientific predecessors, the Kamiokande and IMB experiments, as well as to detect high-E neutrinos, search for nucleon decay, study solar and atmospheric neutrinos, and watch for supernovae in the Milky Way galaxy. As an example, the results of the discovery of atmospheric neutrino oscillations published in 1998 [11] are shown in Fig. 8.5. SK has also been used as a far detector for long-baseline beam neutrino oscillation experiments: K2K [12] (1999–2004) and T2K [13] (2010–present).

The K2K experiment used a 1-kiloton (1 KT) water Cherenkov neutrino detector located at about 300 m from an aluminum neutrino production target to determine the neutrino beam characteristics and to measure neutrino cross-sections [12,14,15]. This 1KT "near detector" was a scaled-down version of the 50-kiloton SK "far detector." The vertical cylindrical tank ($\sim$11 m in diameter and $\sim$11 m in height), the PMT type, the photocoverage (40%), the water purification system, the readout electronics, the event reconstruction algorithms, and the detector calibration methods were basically the same as those in SK. Having the same detector technology along the beam line allowed for a significant reduction of

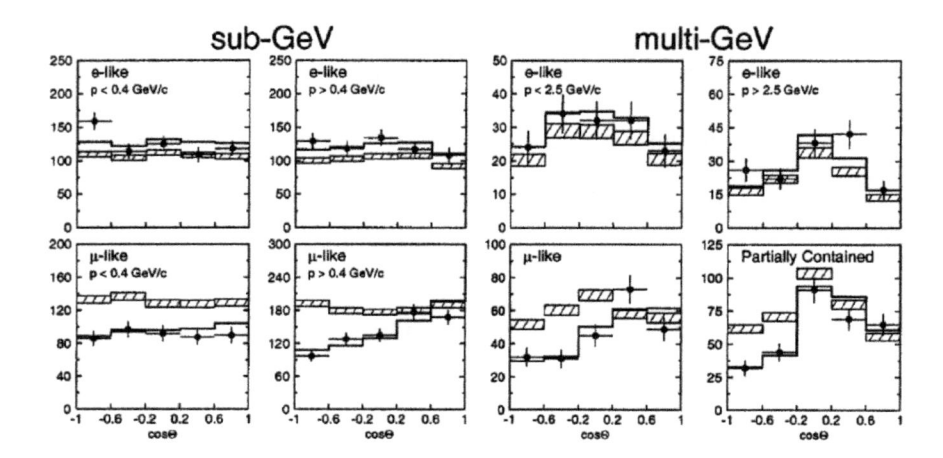

Figure 8.5. Zenith angle distributions of muon-like and electron-like events for sub-GeV and multi-GeV datasets [11]. Upward-going particles (longer atmospheric neutrino travel length) have $\cos\Theta < 0$ and downward-going particles (shorter travel length) have $\cos\Theta > 0$. The hatched histogram shows the Monte Carlo expectation. The bold line is the best-fit expectation for $\nu_\mu \leftrightarrow \nu_\tau$ oscillations. "Partially Contained" is explained in Section 8.4.11.

detector uncertainties related to the prediction of the neutrino interaction rate and energy spectrum in SK before the neutrino oscillations. The detector performance, including the energy reconstruction and particle identification, was similar to that achieved in high-E analyses at SK. The 1KT detector also provided high statistics measurements of neutrino–water interactions.

8.3.1.2 *Sparsely instrumented natural water (neutrino telescopes)*

A very large volume of natural water can be instrumented with a sparse array of PMTs dispersed throughout the volume [16–20]. The cone geometry is not visually apparent but can be reconstructed using the time at which each hit PMT records its pulse, where the Cherenkov opening angle is known because these detectors observe only high-E particles. The threshold of these detectors depends on the spacing of the PMTs and is typically high (tens or hundreds of GeV). These detectors typically reconstruct muons, which make a long straight track, much better than electrons, which deposit all their energy in a fairly small volume and are thus seen by fewer PMTs.

The pioneering project for the construction of an underwater neutrino telescope was developed by the DUMAND collaboration [16], which attempted to deploy a detector off the coast of Hawaii in the 1980s. In parallel, the BAIKAL collaboration [17] began to build a similar detector system in the Baikal Lake. These experiments have been continued in the Mediterranean Sea by the ANTARES collaboration [18], which completed in 2008 the construction of the largest neutrino telescope ($\sim 0.1\,\text{km}^2$) in the northern hemisphere.

Regarding deep ice, a major step toward the construction of a large neutrino detector was made by the AMANDA collaboration [19], which completed their detector in 2000, and then IceCube, completed in 2011, with a total of 86 scheduled strings [20].

The detection principle to search for high-E neutrinos of astrophysical origin relies mainly on the detection of Cherenkov light emitted from an up-going muon induced by a ν_μ that penetrated the Earth. The detection strategy is based on the measurement of the intensity and arrival time of Cherenkov light produced along the muon track on a three-dimensional array of PMTs. A typical event display which shows an upgoing muon event is shown in Fig. 8.6.

The main problem in track reconstruction is that, at the energy of interest in neutrino telescopes (namely, from about 100 GeV to about 1 PeV), high-energy muons can produce electromagnetic radiation in the interaction with the Coulomb field of the nuclei in the material. These radiative energy loss processes generate secondary charged particles along the muon trajectory, which also produce Cherenkov radiation that arrives after the ideal Cherenkov cone. These track-correlated processes are stochastic, and their relative photon yield fluctuates with distribution around the true track position, depending on the physics of the underlying secondary process and on the properties of the detector. In addition, the arrival time of the photons on the PMTs has poor resolution because of the optical properties of the material in which the tracks propagate. The dominant effect is the scattering that depends strongly on the distance of the PMT from the track. Moreover, environmental background contributes to a large noise counting rate. In an undersea neutrino detector, the decay of radioactive elements in the water, mainly the β-decay of potassium isotope ^{40}K, generates electrons that produce Cherenkov light, resulting in an isotropic background of photons. These photons may degrade the hit pattern of a neutrino-induced event and consequently the event reconstruction, even when coincidence methods significantly reduce the contamination.

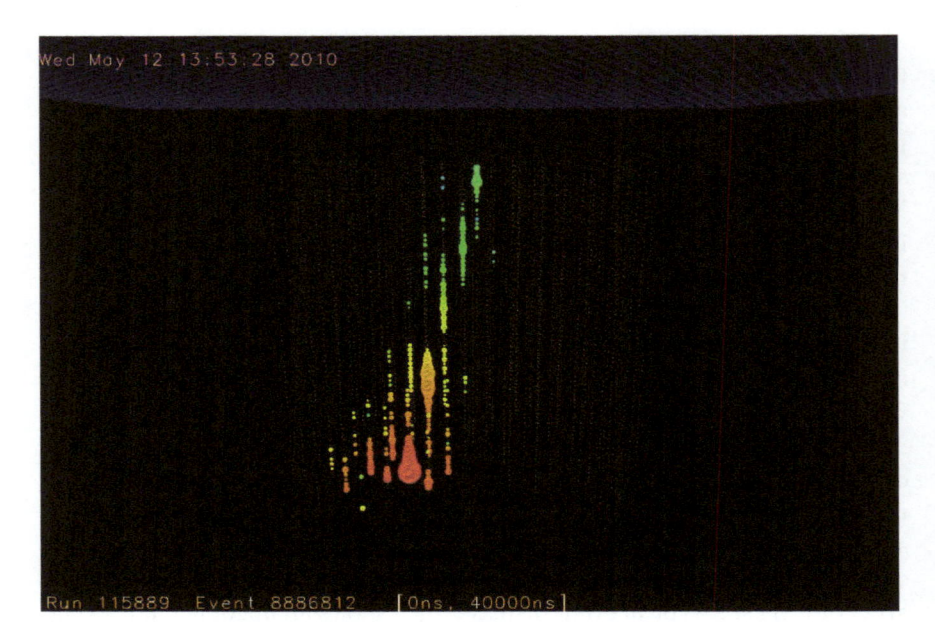

Figure 8.6. An upgoing muon event from the 59-string configuration. The light collected by each sensor is shown with a gray sphere. The color shows the time sequence. The size reflects the number of photons detected.
Source: IceCube Collaboration.

8.3.2 Other Cherenkov experiments

The Sudbury Neutrino Observatory [21] (SNO, 1999–2006) used 1,000 tons of ultrapure heavy water contained in a 12 m diameter vessel made of 5 cm thick acrylic plastic. This target was observed by 9,456 PMTs on a 17.8 m diameter support structure contained within a cylindrical cavern that was 22 m in diameter and 34 m height and filled with ordinary ultrapure water. The experiment was located 2 km underground in an active nickel mine near Sudbury, Ontario. Heavy water, D_2O, replaces normal hydrogen with its heavier isotope deuterium (^{2}H or D), whose nucleus contains a neutron in addition to the proton of normal hydrogen. Deuterium is extremely weakly bound and therefore easily broken up when struck. Therefore, in addition to the neutrino interactions visible in a regular water detector, a neutrino can break up the deuterium in heavy water in two different ways: CC reactions result in an electron signal and are sensitive to only electron neutrinos, while NC reactions can break apart a deuteron, resulting in a neutron signal and are sensitive to all active neutrino flavors.

SNO used three different techniques to detect neutrons, all viable due to the presence of heavy water. The first of these techniques [22] used pure D_2O in which neutron capture on D resulted in a 6.25 MeV gamma ray. With the addition of 2 tons of NaCl to the D_2O [23], neutrons captured by ^{35}Cl with a higher capture cross-section resulted in a gamma cascade of 8.6 MeV and higher neutron detection efficiency. These events can be statistically separated from the electron signal making use of parameters sensitive to the event isotropy. In the final phase of the experiment, the salt was removed and ^3He proportional counters were deployed [24] to measure the neutron signal independently of the Cherenkov signal.

8.4 Detector Components in SK

We introduce the instrumentation and techniques of the hardware used in SK [1–4].

8.4.1 *Detector location*

SK is located in the Mozumi mine of the Kamioka Mining and Smelting Company, Gifu, Japan. The detector cavity lies under the peak of Mt. Ikenoyama, with 1,000 m of rock, or 2,700 m-water-equivalent (m.w.e.) mean overburden (Fig. 8.7).

Cosmic ray muons with energy of less than 1.3 TeV cannot penetrate to a depth of 2,700 m.w.e. The observed muon flux, which does not pose a significant background for the experiment, is $6 \times 10^{-8}\,\mathrm{cm^{-2}\,s^{-1}\,sr^{-1}}$.

8.4.2 *Water tank*

The SK detector consists of a welded stainless-steel vertical cylindrical tank, 39 m in diameter and 42 m in height, with a total nominal water capacity of 50,000 tons. The tank top itself is used as a platform to support electronics huts, equipment for calibration, water quality monitoring, and other facilities.

On the top of the tank, there are multiple feedthroughs with an inner diameter of ∼20 cm at intervals of several meters along the x and y axes (z is defined vertically upward) and along the outer wall for detector calibration (Section 8.5) of the inner detector (ID) and the outer detector (OD), respectively. For calibrations relevant to low-E physics (solar neutrinos, supernova neutrinos, and so on), an electron linear accelerator "LINAC" [25] is installed at the top of the tank toward the $+x$ direction.

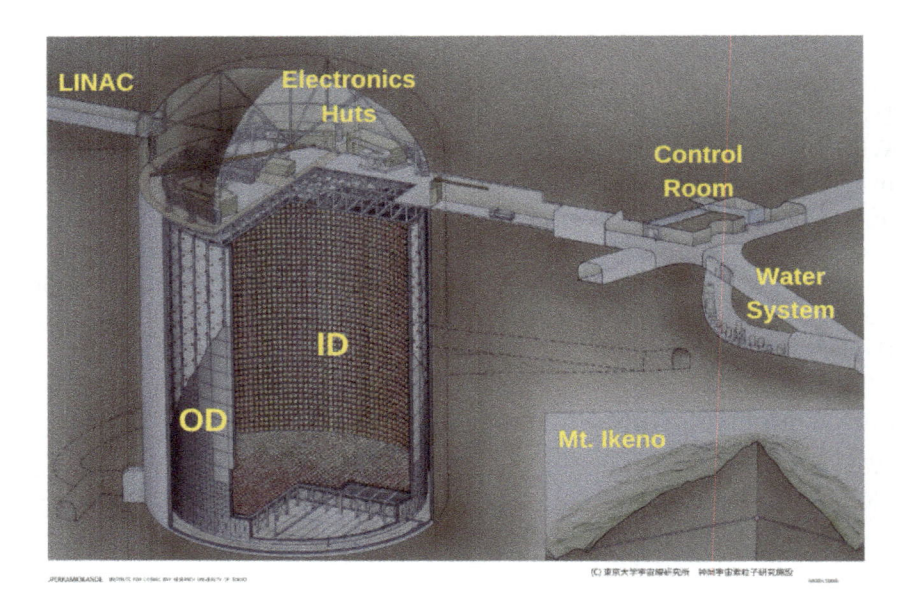

Figure 8.7. A sketch of the SK detector site, under Mt. Ikenoyama.
Source: Kamioka Observatory, ICRR, The University of Tokyo.

The electron beam is bent 90° toward the $-z$ direction and delivered into the tank through the ID calibration holes.

8.4.3 ID PMTs

Within the tank, a stainless-steel framework of thickness 55 cm, spaced approximately 2–2.5 m inside the tank walls on all sides, supports separate arrays of inward-facing and outward-facing PMTs. The inward-facing array consists of 11,146 Hamamatsu Type R3600 50 cm diameter hemispherical PMTs in SK-I (1996–2001). These PMTs have a photocathode made of bialkali (Sb-K-Cs), with a maximal photon conversion probability in the wavelength range of Cherenkov light (Eq. (8.2)). The inward-facing PMTs, and the volume of water they view, are referred to as the inner detector (ID).

The experiment began data-taking in 1996 and was shut down for maintenance in 2001. This initial phase is called SK-I. Due to an accident during the ensuing upgrade work, the experiment resumed in 2002 with only about half of its original number of ID-PMTs (SK-II). To prevent similar accidents, all ID-PMTs were encased in fiber-reinforced plastic (FRP) cases with ultraviolet transparent acrylic front windows starting from SK-II. Such protective covers are needed to avoid any cascade implosion of the PMTs.

Figure 8.8. The inner detector during the full reconstruction in 2006. The light-colored figures in the lower left are three people.
Source: Kamioka Observatory, ICRR, The University of Tokyo.

Figure 8.8 shows the ID during the full reconstruction before filling water toward SK-III (2006–2008) in 2006.

The density of PMTs in the ID was such that effectively about 40% (20%) of the ID surface area was covered by a photocathode in SK-I, SK III-VIII (SK-II). Although the photo coverage was half that of other eras, SK-II (2002–2005) has obtained a detector performance that is almost the same as other detector phases in high-energy physics analyses. See Refs. [26, 27], for example.

8.4.4 OD PMTs

Optically isolated from the ID is an array of 1,885 outward-facing Hamamatsu R1408 20 cm diameter hemispherical PMTs in SK-I. These PMTs, and the water volume they view, are referred to as the outer detector (OD). The optical isolation of ID and OD is made by black polyethylene terephthalate sheets ("black sheets") on the inside of the barrier and highly reflective Tyvek® sheets on its outside. Tank walls are also lined with Tyvek® sheets. Each OD PMT is attached to a 50 cm × 50 cm acrylic wavelength shifting (WS) plate. These features improve light collection efficiency for the OD, compensating for the relatively sparse PMT array.

8.4.5 Compensation coils

Twenty-six sets of horizontal and vertical coils are arranged around the inner surface of the tank to neutralize the geomagnetic field that would otherwise affect photoelectron trajectories in the PMTs. Before and during filling of the tank with water for SK-III, and with the coils carrying their design currents, the residual fields at 458 PMT locations around the detector were measured using a device that can simultaneously measure the magnetic field vector along three orthogonal axes.

During the SK refurbishment tank open work between SK-IV (2008–2018) and SK-V (2019–2020), we remeasured the residual magnetic field for all the PMTs with a tilted-corrected and waterproof 3-axis magnetometer, providing a magnetic field map of the whole tank. These results confirm that the geomagnetic field is successfully reduced by the coils. Dynode directions are also recorded for all the PMTs.

The geomagnetic compensation coil cables failed in three places at the end of 2023. In 2024, SK-VIII (2024 - present) was started after we successfully installed new coils to restore geomagnetic field cancellation.

8.4.6 PMT HVs and electronics

The cables from each PMT for both ID and OD are brought up to the top of the tank and connected to the signal digitizers and the HV supplies in four quadrant electronics huts. The PMT cables have the same length to maintain the delay, and the signal qualities are the same. The ID (OD) PMTs are operated with a gain of about 10^7 at a supply HV ranging from 1,600 to 2,000 V (1,300–1,800 V). See Section 8.5.1.1 for the PMT HV determinations.

At the start of the SK-IV period in 2008, the front-end electronics were upgraded to a system with an ASIC based on a high-speed charge-to-time converter (QTC), the QBEE (QTC-based electronics with Ethernet) [28] module for both ID and OD. This allows us to record all hits above the discriminator threshold, including the dark noise hits. All the hit data are sent to the front-end readout computers without filtering, and the events are defined using the software trigger. Each ID PMT has a dynamic range from 1 photoelectron (pe) to about 300 (1,000) pe in SK-I-III (from SK-IV).

An electronics hut, the "Central Hut," contains electronics and associated computers for GPS time stamping. Also, various detector monitoring systems are located there.

8.4.7 Online software trigger system

The Cherenkov photon signal for a given event is emitted promptly along particle tracks and clusters within a few hundred ns, but dark noise is random and does not form a timing cluster. Pulses on PMTs exceeding a charge threshold corresponding to roughly 0.25 pe are registered as hits, all of which are processed by a software trigger system. The number of hits (~ 6 hits/MeV except SK-II) within 200 ns is counted, and if the number of hits exceeds a certain value (equivalent to a few MeV), a trigger is applied. The online trigger efficiency for the high-E ($> \sim 100$ MeV) physics events is 100%.

T2K-SK beam events (Section 8.3.1.1) are selected by requiring the time of the event in SK to coincide with the expected time of arrival of the neutrino beam. For T2K analyses [13], all hits occurring in the 1 ms windows centered on each beam spill arrival are written to disk. Beam spills are excluded from the analysis if they coincide with problems in the data acquisition system or the GPS system used to synchronize SK with the accelerator at J-PARC. Additionally, spills that occur within 100 μs of a beam-unrelated event are rejected to reduce the contamination of T2K data with cosmic-ray muon Michel electrons.

8.4.8 GPS system

The K2K experiment (Section 8.3.1.1) required synchronization of clocks with ~ 100 ns accuracy at the near and far detector sites (KEK and SK, respectively), which are separated by 250 km [29]. GPS provides a means for satisfying this requirement at a very low cost. Commercial GPS receivers output a 1 pulse per sec (1 PPS) signal whose leading edge is synchronized with GPS seconds rollovers to well within the required accuracy. For each beam spill trigger at KEK, and each event trigger at SK, 50 MHz free-running Local Time Clock (LTC) modules at each site provide fractional second data with 20 ns ticks. At each site, two GPS clocks run in parallel, providing hardware backup as well as data quality checks.

Basically, the same GPS system is used for the T2K experiment with some updates. The standard GPS receivers installed in Kamioka and Tokai (the near detector site) cannot verify whether their outputs are well synchronized with each other. To overcome this problem, special GPS receivers, capable of providing information for the "Common View" method, were installed at both sites. The Common View method uses information from the same GPS satellites, which are visible from different locations simultaneously, and provides the data to compare timing signals from standard GPS receivers. These devices confirm that the reference clock signals from two GPS receivers in Kamioka and Tokai are synchronized and stable within the order of $\mathcal{O}(100\,\mathrm{ns})$.

8.4.9 Water purification system

The water used in SK is sourced from natural water in the Kamioka mine. To maximize water transparency and minimize backgrounds due to natural radioactivity, the water used to fill the SK tank is highly purified by a multi-step system including deionization (DI) resins, filtration, reverse osmosis (RO), and degasification. Water purity is maintained by recirculation through the purification system.

The water circulation pattern in the detector can be manipulated by changing the temperature of the water that is fed into the bottom of the detector from the recirculation system. Since the temperature difference between the top of the tank and the bottom of the tank is only about 0.1° C, the water temperature should be controlled with better than 0.01° precision. To achieve temperature control, a thermometer with 0.0001° precision is a component of the recirculation system and its output is fed back to the water-cooling system. The same thermometers are placed at eight positions in both the ID and the OD to monitor the water temperature and their time dependence.

In order to improve SK's neutron detection efficiency and to thereby increase its sensitivity to the diffuse supernova neutrino background flux, 13 tons of $Gd_2(SO_4)_3 \cdot 8H_2O$ (gadolinium sulfate octahydrate) was dissolved into the SK tank in 2020, marking the start of the SK-Gd phase of operations, SK-VI [3] (2020–2022). In 2022, an additional 26 tons of $Gd_2(SO_4)_3 \cdot 8H_2O$ was dissolved, and SK-VII [4] (2022–2024) was started yielding a 75% neutron capture efficiency on Gd. In the recirculation system for the gadolinium sulfate water, DI and RO are omitted, and instead, special anion and cation resins that keep Gd^{3+} and SO_4^{2-} are deployed. Nevertheless, the attenuation length of the Cherenkov light in the SK tank (Section 8.5.1.3) is sufficiently long and stable for physics analyses.

8.4.10 Fresh air system

To mitigate the relatively high radon background present in the air of the mine, the tank area is supplied with fresh air pumped in from a site outside and well away from the mine entrance. The flow rate and temperature of the air sent from a prefabricated hut ("Radon Hut") outside the mine entrance to the SK are a few $100\,\mathrm{m}^3/\mathrm{min}$ and about $10°$ C, respectively. It is necessary to lower the temperature in order to reduce the condensation in the pipe that sends air into the mine. Air is cooled with source water in the mine and a chiller. Radon concentration is checked by monitors installed in the Radon Hut and at several locations in the SK.

This Rn-reduced air is supplied to the gap between the water surface and the top of the SK tank and is kept at a slight overpressure to help prevent ambient radon-laden air from entering the detector. The air purification system consists of three compressors, a buffer tank, dryers, filters, and activated charcoal filters. The SK Rn-reduced air system uses 50 L of chilled activated charcoal to remove Rn from throughput air. Since it is known that the Rn-trapping efficiency of activated charcoal increases significantly below temperatures around $-60°$ C, the system was upgraded in 2013 to bring its cooling power down from $-40°$ C in the previous design to below this threshold. In order to do so, the system's coolant was changed to 3 MNOVEC^{TM} 7100 and its refrigerator upgraded [30]

8.4.11 *Event classifications*

Neutrino interactions are detected via the Cherenkov light emitted by the charged particles produced (Section 8.2). Events due to entering charged particles can be identified using the OD PMTs. Neutrino interaction candidates produced in the ID are defined as events producing Cherenkov light in the ID, with no evidence of entering particles in the OD. Neutrino events produced in the ID are termed "fully contained" if there is no activity in the OD indicating exiting or entering particles.

Events originating in the ID with OD light patterns consistent with exiting particles are termed "partially contained." For these events, the energy deposited in the detector is only a lower limit on the neutrino energy. Upward-going muons, which are assumed to be products of neutrino interactions in the rock below SK, are also recorded. For upward muons, Cherenkov light patterns consistent with an entering muon are required.

Finally, downward-going muons (cosmic-ray muons), products of meson decay in the atmosphere, are observed at a net rate of about 2 Hz. These events provide useful housekeeping and calibration data. For example, through-going muons are used to monitor the attenuation length of Cherenkov light in water, and stopping muons are used to check the energy scale for physics analyses (Section 8.5.1.3).

8.5 Detector Calibrations

We describe the calibration methods in SK-IV [2], in which new readout QBEE electronics (Section 8.4.6) have been operating.

8.5.1 *ID calibration*

8.5.1.1 *PMT and electronics calibrations*

For the PMT charge calibration, it is necessary to measure the "gain" and "QE" of each PMT. "Gain" is a conversion factor from the number of photoelectrons to charge (pC), and "QE" is defined as the product of the quantum efficiency and the photoelectron collection efficiency on the first dynode of the PMT.

An appropriate HV value for each PMT was determined so that all PMTs output approximately the same charge. A scintillator ball connected to an Xe lamp was placed in the center of the SK tank to emit isotropic light. The number of photons reaching each PMT is affected not only by the difference in geometrical distance within the large cylinder tank (Section 8.4.2) but also by the position dependence of light propagation due to water quality and light reflection on the tank surface (Section 8.5.1.2). To avoid this problem, 420 "standard PMTs" with HV pre-determined using a pre-calibration system were installed in the tank. PMTs at the same distance from the light source are grouped and the standard PMT placed in each group is used as a reference. The scintillator ball connected to the Xe lamp remains in the center of the tank as a monitor of PMT gain as well as other measurements.

The PMT gain measurement is performed in two stages. First, the relative gain is measured and then the absolute gain is calibrated. A diffuser ball connected to a nitrogen-laser-driven dye laser through an optical fiber was placed in the tank for the relative gain measurements. High light intensity (sufficient number of photons reaching each PMT) and low light intensity (~ 1 photon reaching each PMT) measurements were made. The former measures the charge $Q_{\text{obs}}(i)$ and the latter measures the number of hits $N_{\text{obs}}(i)$ in each PMT i:

$$Q_{\text{obs}}(i) \propto I_s \times a(i) \times \varepsilon_{\text{qe}}(i) \times G(i) \tag{8.3}$$

$$N_{\text{obs}}(i) \propto I_w \times a(i) \times \varepsilon_{\text{qe}}(i) \tag{8.4}$$

where I_s and I_w are the average intensities of the high- and low-intensity flashes, respectively, $a(i)$ is the acceptance of PMT, $\varepsilon_{\text{qe}}(i)$ denotes its QE, and $G(i)$ is its gain. The gain of each PMT can be derived by taking the ratio of Eqs. (8.3) and (8.4), except for a factor common to all PMTs:

$$G(i) \propto \frac{Q_{\text{obs}}(i)}{N_{\text{obs}}(i)} \qquad (8.5)$$

Then the relative gain of each PMT is derived by normalization with the average gain over all PMTs. These relative gains are used as corrections in the conversion of each PMT's output charge to an observed number of photoelectrons.

The absolute gain is calibrated from the cumulative single-photoelectron charge distribution for all PMTs. A Ni-Cf ("nickel source") deployed in the center of the tank is used for this purpose. It produces isotropic gamma rays of about 9 MeV when thermal neutrons from the spontaneous fission of the californium capture on the nickel. To evaluate the distribution below the usual threshold of 0.25 pe, data with double the usual PMT gain and half the usual discrimination threshold were taken (Fig. 8.9). The value averaged over the whole pC region was defined as the conversion factor from pC to single photoelectron. The single-photoelectron charge distribution is implemented in the SK detector simulation ("SK-MC").

The nickel source data are also used to measure the relative QE. As shown in Eq. (8.4), the number of hits observed at each PMT is proportional to each QE value when a source with sufficiently low light intensity is used. The nickel source data were taken after achieving uniform water quality by causing water convection in the tank. The uniform water quality can be identified by measuring the water temperature profile in the tank (Section 8.4.9). The ratio of the number of hits between data and MC is implemented in the SK-MC as the relative QE for each PMT.

The charge linearity of the whole system combined with PMT and readout electronics was measured using the laser system. It is necessary to cover all possible charge ranges including the saturation region (above a few 100 pe). For this purpose, data were taken with the diffuser ball placed at an off-center location in the tank. The average nonlinearity measured for all PMTs is taken into account in the SK-MC.

The time response of each readout channel, including PMTs and readout electronics, must be calibrated for precise reconstruction of the event (Appendix A). The response time of readout channels can vary due to differences in the transit time of PMTs, lengths of PMT signal cables, and processing time of readout electronics. In addition, the timing of discriminator output depends on the pulse heights of PMTs since the rise

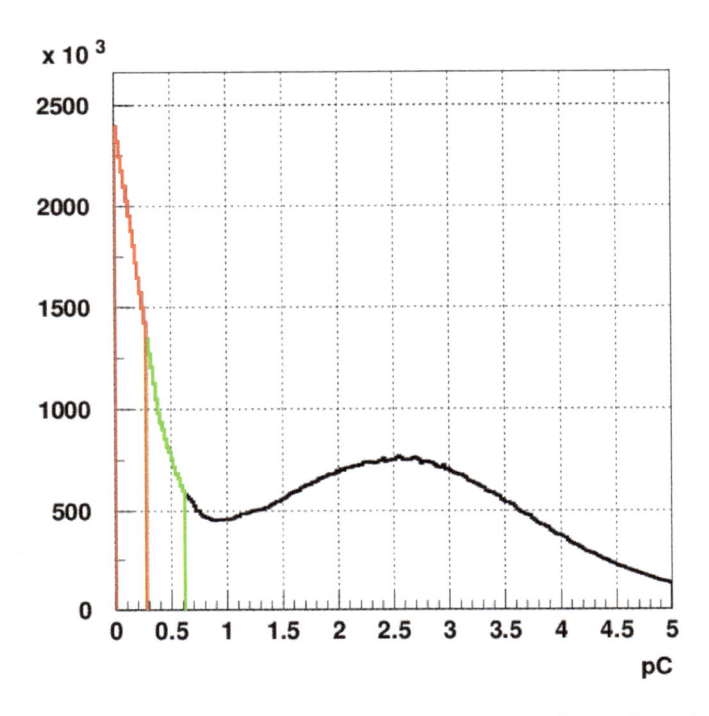

Figure 8.9. The single-photoelectron charge distributions in pC unit for nickel source data [2]. The black line ($>\sim 0.6$ pC) shows the data with normal PMT gain, the green line (~ 0.3–0.6 pC) shows the data with double gain and half threshold, and the red line below ~ 0.3 pC is linear extrapolation.

time of a large pulse is shorter than that of a smaller one. This is known as the "time-walk" effect.

The overall processing time and the time-walk effect were measured by injecting fast light pulses from a gas flow nitrogen laser into PMTs in the tank. The wavelength of the laser light is shifted to 398 nm by a dye, where the convoluted response with Cherenkov spectrum (Eq. (8.2)), light absorption spectrum (Fig. 8.11), and quantum efficiency of the PMTs is almost maximum. The fast light is guided into the tank by an optical fiber and injected into a diffuser ball located near the center of the tank to produce an isotropic light. Figure 8.10 shows a typical scatter plot of the time-charge "TQ" distribution for one readout channel. The calibration constants, called the "TQ-map," are derived by fitting the TQ distribution to polynomial functions. Each readout channel has its own TQ-map used to correct the time response.

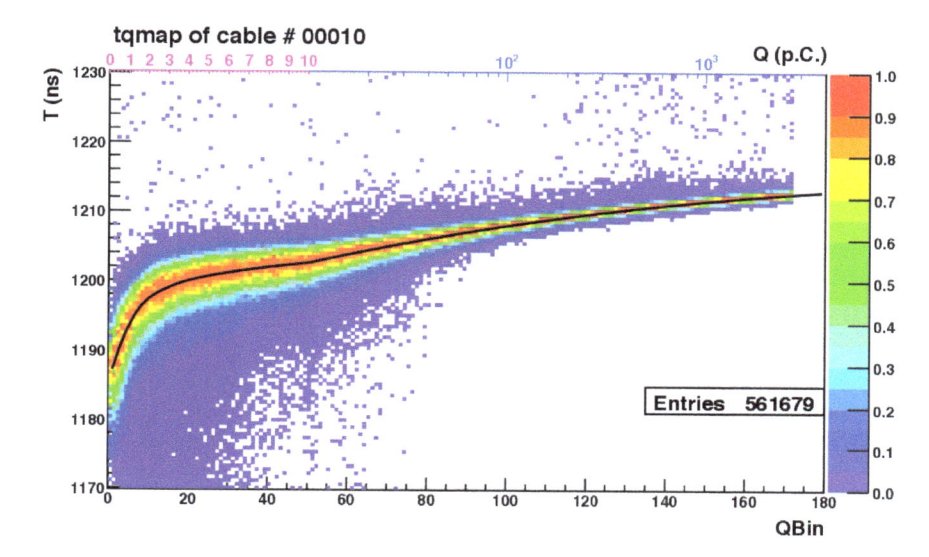

Figure 8.10. Typical TQ distribution for a readout channel [2]. The horizontal axis is the charge (QBin) of each hit, and the vertical axis is the time-of-flight-corrected timing (T) of the hits. Larger (smaller) T corresponds to earlier (later) hits in this figure.

The timing resolution is evaluated using the same dataset used for the TQ-map evaluation. All PMT timing distributions corrected by their TQ-maps are fitted by an asymmetric Gaussian to evaluate the timing resolution in each charge region. The results are implemented in the SK-MC.

For the real-time monitoring of the time response, SK employs a nitrogen laser that uses sealed nitrogen as a gain medium and is better suited for continuous operation. The light output of the dye system is injected into the same diffuser ball used for the TQ-map measurement.

8.5.1.2 *Calibration for photon tracking*

By injecting a collimated laser beam into the SK tank and comparing the timing and spatial distributions of the light with MC, we can extract light absorption and scattering coefficients as functions of wavelength. The earlier timing PMT hits come from scattered photons, while the later timing hits represent photons reflected by the PMTs and black sheets. The total number of scattered photons and the shape of the time distribution are used to tune the symmetric and asymmetric scattering and the absorption parameters for the SK-MC in each SK phase. Figure 8.11 shows each water

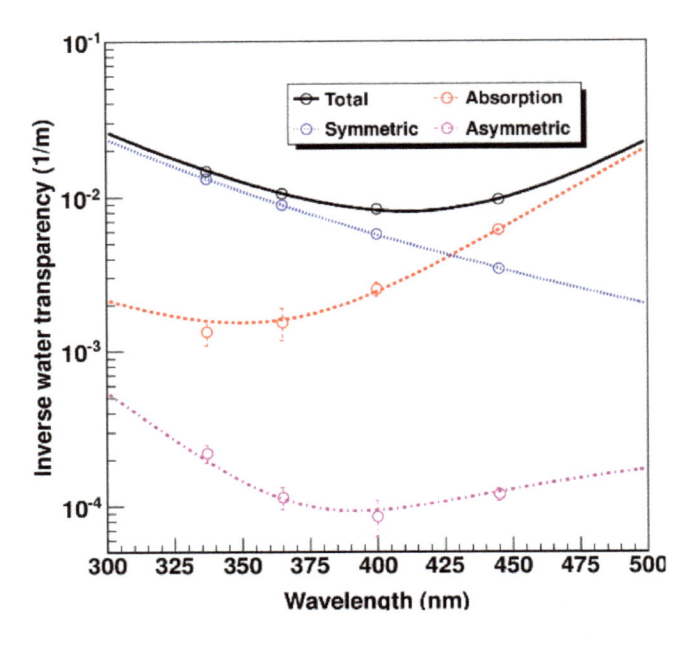

Figure 8.11. Typical water quality parameters implemented in the SK-MC [2]. The points are the data obtained in April 2009. Each line through the points for absorption (dashed), symmetric scattering (dotted), and asymmetric scattering (dash-dotted) shows the fitted functions while the top line shows the total of all fitted functions added together.

quality parameter as a function of wavelength. Time variation of each parameter is monitored at several wavelengths using several light injectors directed vertically and horizontally in the tank.

It has been found that water quality depends on position in the tank, especially vertically, and mainly comes from differences in absorption. This is because the pure water enters from the bottom area of the OD in the tank (Section 8.4.9). The vertical dependence of absorption in the SK-MC is determined from the monthly nickel source data and monitored with the real-time Xe system (Section 8.5.1.1).

Light reflection at the PMT surface in the SK-MC is tuned using the same laser data used to determine the water quality parameters. Four layers of material (refractive index) from the surface to the inside of the PMT are taken into account: water (1.33), glass ($1.472 + 3670/\lambda^2$, where λ is the wavelength in nm), bialkali ($n_{\text{real}} + i \cdot n_{\text{img}}$), and vacuum (1.0). Here, n_{real} and n_{img} are the real and imaginary parts of the complex refractive index.

Cherenkov photons are reflected or absorbed on the black sheet (Section 8.4.4). The reflectivity of the black sheet in the SK-MC is measured by a light injector set in the tank. The reflected charge was measured at three incident angles with three wavelengths.

8.5.1.3 *Energy scale calibrations in physics analyses*

The energy scale is one of the most important detector parameters. This is because the reconstructed energy is used in all physics analyses, and its uncertainty is the source of the most dominant detector systematic uncertainties in most physics analyses.

We describe how to calibrate the energy scale and estimate its uncertainty in the high-E physics analyses first. Following the various basic detector calibrations (Sections 8.5.1.1 and 8.5.1.2), the SK-MC's global photon yield is tuned using cosmic-ray through-going muon data. The Cherenkov light attenuation length used in the event reconstructions (A.1) is also determined by using cosmic-ray through-going muon MC. The electron and muon particle gun MC events are generated using the SK-MC, and the relationship between the corrected charge (A.1.2) and true momentum is obtained. The time variation of PMT gain is monitored by using the charge distribution of dark noise for each PMT. The time variation of the attenuation length is monitored by the cosmic-ray through-going muon data. Details on the attenuation length measurement analysis are available elsewhere [1]. Both time variations are taken into account in the corrected charge as a function of time.

After the above energy scale calibrations, the energy scale uncertainty is estimated using control data samples, cosmic-ray stopping muons, π^0s produced by the atmospheric neutrino interactions (Fig. 8.4), and Michel electrons from the cosmic-ray stopping muons [26, 27]. These samples are independent of the control data samples used in the prior calibrations. The uncertainty of the energy scale used in the physics analyses is estimated from the quadratic sum of the uncertainty of the absolute scale (difference between data and MC) in the calibration reference time period and the residual time variation in each SK phase. The directional dependence of the energy scale obtained using Michel electron data is also considered in the physics analyses as another energy scale error source.

For the low-E analyses, the same initial calibrations (Sections 8.5.1.1 and 8.5.1.2) are used. The energy scale is of particular importance to the solar neutrino analysis since the recoil electron spectrum from the elastic scattering of ^{8}B solar neutrinos on electrons is steeply falling with energy. A precise calibration is performed with the LINAC (Section 8.4.2), which also checks angular resolution. To crosscheck it and to study the directional dependence of the energy scale, ^{16}N is produced via (n, p) reactions on ^{16}O using a deuterium-tritium ("DT") fusion neutron generator which makes $\sim$14 MeV neutrons [31]. ^{16}N decay is dominated by an electron whose maximum energy of 4.3 MeV coincides with gamma radiation at 6.1 MeV. Spallation from cosmic-ray muons produces neutrons that can also be used for calibration [32].

8.5.2 OD calibration

Requirements for OD calibration are not as stringent as for the ID in the current SK physics analyses. The readout electronics and the PMT HV power supply are similar to those for the ID (Section 8.4.6). The calibration and simulation parameter tuning for the OD-PMTs are similar but done separately from that for the ID. The PMT timing and gain calibrations are done using cosmic-ray muon and dark noise data, respectively. Cosmic-ray muon data are also used to tune the light reflection on the Tyvek® sheets (Section 8.4.4) in the SK-MC.

From the same laser used in the ID PMT timing monitor, 52 optical fibers extend through OD calibration holes (Section 8.4.2) to each OD region in the tank. A diffuser is attached to the end of each fiber. The charge nonlinearity of each readout channel is measured with all the light injector data and implemented in the SK-MC.

8.6 Future Prospects

Hyper-Kamiokande (Hyper-K or HK) [33] and KM3Net [34] are used as examples to discuss the prospects for future plans for high-E physics experiments.

8.6.1 Densely instrumented artificial tanks

The planned HK detector will consist of an order of magnitude larger tank than the predecessor, SK, and will be equipped with ultrahigh sensitivity photosensors. A larger HK tank will be used in order to obtain in only 10 years an amount of data corresponding to 100 years of data collection time using SK. This allows for the observation of previously unrevealed rare phenomena and to probe CP violation in neutrino oscillation. We expect an order of magnitude higher sensitivity to nucleon decay searches. We have been developing the world's largest 50 cm PMTs, which exhibit a photodetection efficiency two times greater than that of the SK PMTs. These new PMTs are able to perform light intensity and timing measurements with a much higher precision.

Thanks to its vertical cylindrical tank which is similar to SK, there is no major concern which makes the expected energy scale error significantly larger in HK. However, it is necessary to reduce the systematic error of the energy scale in HK to reach the target physics analysis sensitivities. For this purpose, improved measurements of low-level detector parameters, such as scattering and absorption in the water and PMT performance will be pursued and linked to the uncertainties on reconstructed particle parameters. To improve physics sensitivities, a more extensive OD calibration will be performed, including the effect of water quality and Tyvek® reflectivity.

In addition to the 50 cm PMT from Hamamatsu, use of Multi-PMT ("mPMT") is currently considered in HK. The mPMT module consists of 19 PMTs of 7.5 cm diameter, contained in a vessel with a diameter of 50 cm. The mPMT module has the following advantages over the conventional 50 cm PMT:

- It has high granularity with multiple 7.5 cm PMTs, which provides higher resolution of Cherenkov ring images, especially for charged particles created at a short distance from the wall.
- Its directionality improves the reconstruction of the edge of the Cherenkov ring image, which is not recognized with 50 cm PMTs if the vertex is close to the wall. This also helps to identify the position of light emission.
- It has better timing resolution with small PMTs.

The mPMT will be also used for calibration. As the mPMT is an independent photon detection device from the 50 cm PMT system, comparing the two readout systems can investigate possible biases in one of the photo-sensor systems. Due to the large difference in the acceptance (the photo-cathode area of a 50 cm PMT is a factor 40 larger than that of the 7.5 cm PMT), the two photo-sensor systems measure different numbers of photoelectrons from the same light source. Therefore, nonlinear bias, if it exists, appears differently in 50 cm PMTs and mPMTs. The directionality and improved timing resolution of the mPMTs, combined with data from the 50 cm PMTs, can also enable improved measurements of low-level detector parameters from calibration sources.

For large detectors, measurements necessary for calibration must be automated as much as possible. It is also necessary to find ways to further reduce detector downtime in order to not miss important physics events like supernovae.

Attempts to reconstruct events using new software techniques such as machine learning have also begun in both SK and HK. These new techniques are expected to improve the performance of event reconstruction.

8.6.2 *Sparsely instrumented natural water*

The Cubic Kilometer Neutrino Telescope, or KM3NeT, is a future European research infrastructure that will be located at the bottom of the Mediterranean Sea. It will host the next-generation neutrino telescope in the form of a water Cherenkov detector with an instrumented volume of several cubic kilometers distributed over three locations in the Mediterranean.

KM3NeT will search for neutrinos from distant astrophysical sources like Core-Collapse Supernovae (CCSN), the dramatic explosions of giant stars at the end of their evolution, or their remnants (SNR), gamma-ray bursts, or colliding stars and will be a powerful tool in the search for dark matter in the universe. Its prime objective is to detect neutrinos from sources in our galaxy. Arrays of thousands of optical sensor modules will detect the faint light in the deep sea from charged particles originating from collisions of the neutrinos and the water or rock in the vicinity of the detector.

The KM3NeT telescope will consist of two sites, realizing two neutrino telescopes: ARCA and ORCA. With the ARCA telescope, which is more sparsely instrumented but has a larger volume, KM3NeT will be sensitive to higher-energy, lower-rate distant astrophysical sources, such as gamma-ray bursts or colliding stars. The ORCA telescope is more densely instrumented and will be sensitive to lower energy neutrinos, and is the instrument for KM3NeT scientists studying neutrino properties by exploiting neutrinos generated in the Earth's atmosphere.

ARCA will be installed at the KM3NeT-It site, about 100 km offshore of the small town of Portopalo di Capo Passero in Sicily, Italy. The detection units of the ARCA telescope will be anchored at a depth of about 3,500 m. With an appropriate granularity of the three-dimensional arrays of sensor modules in the detector of the ARCA telescope, cosmic neutrinos with energies between several tens of GeV and 100 PeV can be observed [35].

Deployed at the KM3NeT-Fr installation site about 40 km offshore from Toulon, France, the ORCA neutrino detector will take advantage of the excellent optical properties of deep-sea water to achieve the angular and energy resolutions required for resolving the neutrino mass hierarchy. A multi-megaton scale array of KM3NeT light sensor modules will be used with a granularity optimized for studying reactions of atmospheric neutrinos with the seawater. The sensor modules will be arranged on vertical detection units with a height of about 150 m and in the dense configuration required for detection of neutrinos with energies as low as about a GeV, three orders of magnitude lower than the typical energy scale probed by the detector of the ARCA telescope for neutrino astroparticle physics.

In the future and pending funding, the full neutrino telescope will contain on the order of 12,000 pressure-resistant glass spheres attached to about 600 strings. The strings hold sensor spheres, each anchored to the sea floor and supported by floats. Each sphere, called a "digital optical module" (DOM), is about 43 cm in diameter, contains 7.6 cm PMTs with supporting electronics, and is connected to shore via a high-bandwidth optical network. At the shore of each KM3NeT installation site, a farm of computers will perform the first data filter in the search for the signal of cosmic neutrinos, prior to streaming the data to a central KM3NeT data center for storage and further analysis.

8.6.3 *Contrasts and complementarities with DUNE*

A future detector with overlapping physics aims with respect to next-generation water Cherenkov detectors is the Deep Underground Neutrino Experiment (DUNE). DUNE is a 40-kton liquid argon time-projection chamber (LArTPC) planned for an underground site in South Dakota, in conjunction with a neutrino beam from Fermilab. LArTPC technology has advantages and disadvantages with respect to water Cherenkov technology; in particular, LArTPCs provide fine-grained tracking capability without a Cherenkov threshold so that hadronic components of final-state products of neutrino interactions are not lost. The disadvantage of LArTPCs is greater cost and complexity; water detectors can be made very large for a relatively low cost per kiloton and so often win in statistics. There are many complementarities between the detector types, depending on the physics topic. An especially notable complementarity is that for supernova neutrinos; water is primarily sensitive to electron antineutrinos via inverse beta decay on free protons, whereas LAr is primarily sensitive to electron neutrinos via charged-current absorption on ^{40}Ar. Having both types of detectors active during the next supernova burst will greatly enhance the worldwide scientific reach.

8.7 Achievement of Cherenkov Neutrino Detectors

Large water Cherenkov detectors have accumulated many very high-impact scientific achievements over the past decades. The first-generation Kamiokande-II experiment made the first directional solar neutrino measurement; Kamiokande-II and IMB were the first to observe hints of atmospheric neutrino disappearance. Both Kamiokande-II and IMB made the historic observation of a neutrino burst from SN 1987A in the Large Magellanic Cloud. "For pioneering contributions to astrophysics, in particular for the detection of cosmic neutrinos," Masatoshi Koshiba was awarded the Nobel Prize in Physics in 2002 [36]. The next generation of water Cherenkov detectors brought precision solar-neutrino oscillation information by Super-K and SNO, along with clear evidence of atmospheric neutrino oscillation by Super-K. These measurements resulted in the Nobel Prize to Takaaki Kajita and Arthur B. McDonald in 2015 [37]. Furthermore, Super-K was and is the far detector for the K2K and T2K long-baseline experiments, which confirmed atmospheric neutrino oscillations with artificial neutrino beams and improved knowledge of oscillation parameters. In addition, IceCube has made the first observations of cosmic neutrinos, both as a diffuse flux and from specific sources. Super-K has also produced stringent limits on baryon number violation via many channels — while baryon number violation has not been observed, this lack of observation has been highly influential for constraining grand unified theories. These achievements represent only a few highlights. Super-K, SNO, and long-string neutrino telescopes have produced decades of additional science, including indirect dark matter constraints, cosmic ray measurements, astrophysical neutrino source searches, and diverse searches for beyond-the-standard-model physics.

Appendix A: Event Reconstruction Algorithm in SK

This appendix describes the reconstructions of "fully contained" events (Section 8.4.11) in the current SK physics analyses. For comparison with high-E physics event reconstruction, low-E physics event reconstruction is also described. Some future prospects are mentioned in Section 8.6.1.

A.1 High-E Physics Analyses

In the high-E neutrino interactions ($>\sim 100$ MeV), we reconstruct the physics quantities of an event, such as vertex position, the number of Cherenkov rings, momentum, particle type, and the number of Michel electrons. The primary event fitter is called "APfit" [38]. We start from the vertex fitter program to obtain the vertex position of events. With the knowledge of the reconstructed vertex, the ring fitter identifies each ring. After that, the particle identification program identifies the particle type for each ring. The momentum for each ring is determined and Michel electrons are identified.

A.1.1 *Vertex*

The principle of the vertex fitting is that the timing residual (i.e., (photon arrival time) − (time of flight)) distribution should be most peaked with the correct vertex position. In the timing distribution, the time resolution of each hit PMT as a function of detected photoelectrons (Section 8.5.1.1) and the track length of the charged particle (Section 8.2.2) are taken into account.

The timing fit has worse vertex resolution in the direction of Cherenkov ring, compared to that in the perpendicular direction. This is because the time residual does not change along the line on which the distance from all hit PMTs is the same. For single-ring events, the Cherenkov ring charge pattern information in addition to the timing information is used to improve the vertex reconstruction in the longitudinal direction. The expected and observed photoelectrons are compared for each hit PMT to optimize the reconstructed vertex. The result of the particle identification (A.1.4) is used in the expectation.

A.1.2 *Ring fitting*

Possible Cherenkov rings are looked for and probable rings are reconstructed in the ring fitter. We use a known technique for pattern recognition, the Hough transformation [39], to search for possible ring candidates. The Hough transformation transforms rings of an assumed radius to peaks in Hough space (see Fig. 8.12). In the ring fitter, the Hough space is made up of two-dimensional arrays divided by polar angle Θ and azimuthal angle Φ. These angles are measured from the reconstructed vertex. Detected photoelectrons in each hit PMT are mapped to the (Θ, Φ)

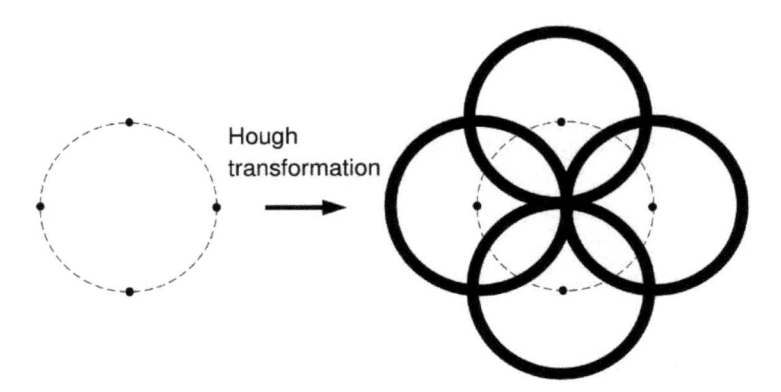

Figure 8.12. Principle of the ring fitting [38] Suppose there are four hit PMTs on the unknown ring (radius r) and we want to find the center of the ring (left figure). By Hough transformation, the detected photoelectrons are mapped to a circle with radius r centered on the PMT (right figure). By accumulating the mapped circles, we can find the peak in the Hough space, giving the center of the unknown ring.

pixels. The number of photoelectrons is corrected ("corrected charge") by taking into account the distance from the vertex, the attenuation length of Cherenkov light in water, time variation of the PMT gain (Section 8.5.1.3), and acceptance of the PMT as a function of photon incident angle. Then, we draw a circle of a fixed radius R corresponding to a 42° Cherenkov cone from each hit PMT. By accumulating the mapped circles, we can find the peak in the Hough space, giving the center of the unknown ring.

If the momentum of one of the reconstructed rings is too low, or if the direction of the ring is too close to the other, this ring is discarded as a fake ring by the ring correction program for the multi-ring candidate events. This is to avoid double counting of the number of rings due to an additional Cherenkov ring from scattering of a single charged particle, for example.

A.1.3 *Ring edge finding*

The Cherenkov opening angle (Eq. (8.1)) is reconstructed for each ring by using the photoelectron distribution as a function of the opening angle, as shown in Fig. 8.13. As shown in the lower plot, the edge of the ring is reconstructed at the second derivative $= 0$ above the peak angle in the upper plot.

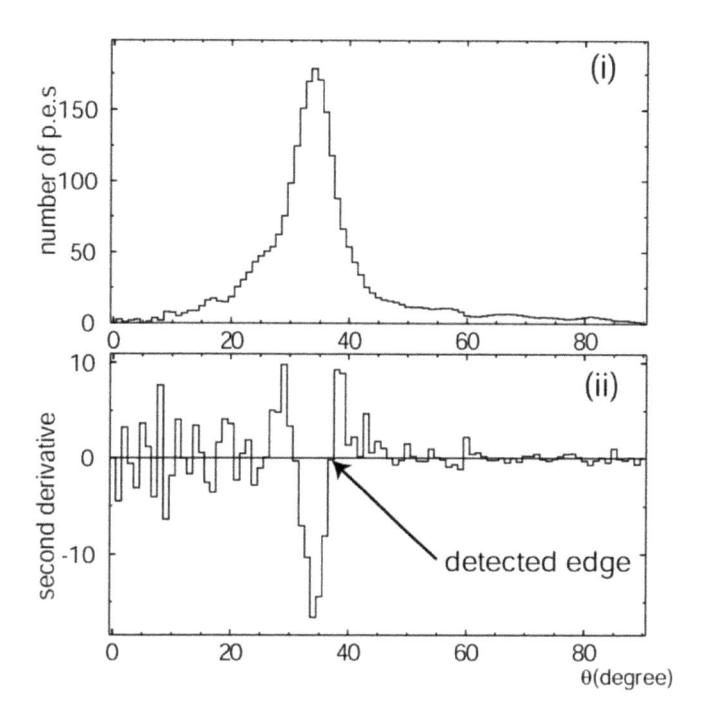

Figure 8.13. Typical observed photoelectrons as a function of the opening angle in the upper plot [38]. The second derivative from the upper distribution in the lower plot.

A.1.4 *Particle identification*

The primary particle identification (PID) classifies a particle as a showering particle ($e^{\pm}$, γ) or a non-showering particle ($\mu^{\pm}$, $\pi^{\pm}$, etc.), using the photon distribution of its Cherenkov ring (Fig. 8.3, for example). Showering (non-showering) particles are called "electron-like" ("muon-like"). At first, the expected photoelectron distributions are calculated with an assumption of the particle type. Using the expected photoelectron distributions, the likelihood is calculated for the electron and muon assumptions, respectively. The calculation of the probability is performed for all of the PMTs for which the opening angles toward the ring direction are within 1.5 times the Cherenkov opening angle. The particle type (electron-like or muon-like) with greater likelihood is chosen. Optimizations of the expected charges are performed by changing the direction and the Cherenkov opening angle of the ring, in order to get maximum-likelihood values.

Additionally, another probability is defined using only the Cherenkov opening angle to utilize the β dependence (Eq. (8.1)). The Cherenkov angle of muons with lower momentum ($<\sim 500$ MeV/c) is smaller than $41°$. Depending on the needs of the physics analysis, the second likelihood can be combined with the first one, or not. For example, the selection of the π^0 events (Fig. 8.4) does not use the second likelihood. This is because the gamma radiation length in water is ~ 36 cm (no Cherenkov light emission), and the event vertex is reconstructed at the electron–positron pair-production point instead of the π^0 decay point.

The excellent PID performance (e/π^0 separation) was experimentally confirmed using a 1-kiloton water Cherenkov detector with electron and muon beams from the 12 GeV proton synchrotron at KEK [40].

Some physics analyses use recoil protons (a mass of 938 MeV/c^2) with energies above the Cherenkov threshold (1.4 GeV) [41, 42]. Muons and protons both have sharp ring edges, whereas electrons produce fuzzy rings and are relatively easy to distinguish. To distinguish protons from muons, a proton fitter using the light pattern and ring topology has been developed (see Ref. [41] for details). A characteristic of protons is that they tend to have hadronic interactions in water and lose energy by producing secondary particles. If both the secondary particles and the scattered proton are below the Cherenkov threshold, the Cherenkov light emission is truncated and leaves a narrow proton Cherenkov ring, for example.

A.1.5 *Ring separation*

The ring separation program determines the fraction of photoelectrons in each PMT for each ring for multi-ring events. We need this separation for the ring fitting and the particle identification of each ring as well as the momentum reconstruction.

Given a vertex position and each ring direction, the expected photoelectron distribution for each ring as a function of opening angle is calculated assuming the flat distribution in azimuthal angle. The PID result is used to calculate the expected charge. The charge fraction for each ring is determined by comparing the expected and observed charges for different fractions.

A.1.6 *Momentum*

The momentum is estimated from the total sum of the corrected charge within a $70°$ half opening angle from the reconstructed ring direction. To reconstruct momentum from the corrected charge, conversion tables are prepared based on the electron and muon particle gun MC events. The PID results define electron-like or muon-like momentum, respectively. For multi-ring events, the ring separation program is applied with each particle-type assumption to determine the fraction of photoelectrons in each PMT due to each ring. Then, the momentum for each ring is determined by the same method used for single-ring events.

The energy (momentum) scale calibrations and checks dedicated for each physics analysis using several control data samples are described in Section 8.5.1.3.

A.1.7 Michel electron tagging

We identify Michel electrons, namely, the electrons produced when muons decay, following the primary events. Michel electron candidates are detected when the number of hits in the sliding timing window of a few 10 ns after the primary event exceeds a certain threshold, the vertex is well reconstructed, and the total number of photoelectrons does not exceed a certain threshold.

A.1.8 Alternative fitter

Depending on the physics analyses, the other fitter called "fiTQun" can be applied [13, 43, 44]. FiTQun employs a maximum likelihood method to reconstruct particle types and determine kinematics in the detector simultaneously. The algorithm is based on methods developed for the MiniBooNE experiment [45] but has been developed from scratch for SK with additional features, such as multi-ring reconstruction for events with multiple final-state particles. Compared to APFit, fiTQun uses more information, including information from PMT hits outside of the expected Cherenkov cone and hit timing information, during the fitting procedure. For a given event, fiTQun's fit procedure will run multiple times to determine the best kinematic parameters for each possible particle configuration hypothesis, while APFit basically fits those parameters once. Since fiTQun uses more information, the fitting process time is slower and a better understanding of the detector parameters implemented in the detector simulator (Section 8.5.1) is required with respect to APfit.

Unlike APfit, a charged pion hypothesis with a single track compatible with a charged pion undergoing a hard scatter can be used [46].

A.2 Low-E Physics Analyses

There are several differences in the reconstruction of "low-E" neutrino interactions ($\ll 100$ MeV): particle identification is less important since almost all resulting charged particles are electrons and gamma rays, and remaining muons or pions are barely above the Cherenkov threshold, so the ring opening angle is significantly smaller than $41°$. Also, events with multiple rings typically cannot be separated, so the ring fitter is replaced by cuts designed to suppress multi-ring events. Due to multiple Coulomb scattering of electrons due to neutrino interactions or Compton scattering of gamma rays, ring patterns are less clear than those resulting from high-E interactions. Also, individual PMTs typically record only single-photoelectrons. As a consequence, vertex reconstruction is based almost exclusively on PMT timing. Recorded pulse heights only reflect the single-photoelectron distribution, and the PMT hit pattern does not help much.

The "BONSAI" (Branch Optimization Navigating Successive Annealing Iterations) event reconstruction [47] compares the hit time residual distribution (assuming direct photon travel from that point to the PMT) viewed from a single point in space-time (the 3D vertex combined with the assumed time of light emission) with a likelihood. The point with the largest likelihood is chosen as the reconstructed vertex. To find the maximum, BONSAI uses a simple search tree, combined with an annealing algorithm designed to avoid local maxima: instead of selecting a single, best branch to continue, all branches within a range of likelihoods are continued. This is particularly important for the lowest electron energies since dark noise fluctuations combined with radioactive background from the detector boundaries can result in several local maxima deep inside the detector. The tree is started from multiple points calculated from sets of "four-hit combinations." A four-hit combination results in an analytical exact solution to the four equations corresponding to the four-hit times. The four-hit combinations are drawn from a subset of PMT hits. That subset is constructed from the requirement that every hit pair has a time difference smaller than their spatial separation, divided by the speed of light in water.

Once a vertex is determined, another likelihood fit determines the event direction after hits are selected within 20 ns in the time-of-flight subtracted hit times. This likelihood is energy dependent. At first, an initial energy estimate is used; after the full energy reconstruction is finished, the direction likelihood fit is repeated.

In addition to the direction reconstruction, three-hit combinations are formed from the 20 ns hit selection and the Cherenkov angle distribution is made from these (each three-hit combination gives an exact solution of the opening angle). Multiple photon events result in a peak of this distribution that is larger than 41°, while muons and pions have a peak below this angle. Electrons always have velocities close to the speed of light if they emit Cherenkov light at all, so their opening angle is always near 41°. The amount of multiple Coulomb scattering is estimated from the "sharpness" of the Cherenkov cone by forming event direction candidates from hit pairs and comparing the length of the vector sum of these candidates to the maximal possible length (number of such pairs). Electrons undergoing a larger than usual amount of multiple Coulomb scattering are typically lower in energy and are observed only due to an up-fluctuation in light yield. An example is ^{210}Bi beta decay (with an endpoint energy of $\sim$3 MeV). Also, beta-gamma decays such as ^{208}Tl (endpoint $\sim$5 MeV) look like single-photoelectrons with a larger than usual amount of multiple Coulomb scattering.

The energy is reconstructed by counting PMT hits within 50 ns after time-of-flight subtraction. Multiple photon hits are corrected by PMT occupancy. Additional corrections are made to account for water transparency, late arrival photons, dark noise, number of dead PMTs, effective cathode area of the PMT, and PMT gain variation.

Neutrons may be generated by neutrino interactions. To detect the neutron signal, two independent approaches have been implemented. The first approach is to detect the single 2.2 MeV gamma released from neutron capture on hydrogen. This approach requires a 500-μs forced trigger scheme following a normal trigger, in order to identify the 2.2 MeV gamma offline [48]. The second approach involves doping the water with a water-soluble chemical compound of gadolinium (Section 8.4.9), neutron capture on which yields a gamma cascade with a total energy of about 8 MeV. These relatively high-energy gamma rays should be readily seen in SK-VI [3], SK-VII [4], and SK-VIII.

Acknowledgments

The author would like to thank the following people (in alphabetical order): Hans-Gerd Berns, Yoshinari Hayato, Edward Kearns, Yusuke Koshio, Fabio Lacob, Shigenobu Matsuno, Neil Kevin McCauley, Makoto Miura, Masayuki Nakahata, Yuichi Oyama, Gianfranca de Rosa, Seiya Sakai, Kate

Scholberg, Hiroyuki Sekiya, Masato Shiozawa, Michael Smy, Henry W. Sobel, Yasuo Takeuchi, and Roger A. Wendell.

References

[1] Y. Fukuda et al. The super-Kamiokande detector. *Nuclear Instruments and Methods in Physics Research Section A: Accelerators, Spectrometers, Detectors and Associated Equipment*, 501:418–462, 2003.

[2] K. Abe et al. Calibration of the super-Kamiokande detector. *Nuclear Instruments and Methods in Physics Research Section A: Accelerators, Spectrometers, Detectors and Associated Equipment*, 737:253–272, 2014.

[3] K. Abe et al. First gadolinium loading to Super-Kamiokande. *Nuclear Instruments and Methods in Physics Research Section A: Accelerators, Spectrometers, Detectors and Associated Equipment*, 1027:166248, 2022.

[4] K. Abe et al. Second gadolinium loading to Super-Kamiokande, *Nuclear Instruments and Methods in Physics Research A* 1065 (2024) 169480.

[5] R. M. Bionta et al. Observation of a neutrino burst in coincidence with supernova 1987a in the large Magellanic cloud. *Physical Review Letters*, 58:1494, 1987.

[6] K. Hirata et al. Observation of a neutrino burst from the supernova SN1987A. *Physical Review Letters*, 58:1490, 1987.

[7] L. A. Cole. *Chasing the Ghost, Nobelist Fred Reines and the Neutrino*. Rutgers University, USA, March 2021. https://doi.org/10.1142/12125, https://www.worldscientific.com/worldscibooks/10.1142/12125.

[8] J. D. Jackson. *Classical Electrodynamics*, 3rd edn. Wiley, New York, 1999, pp. 637–638.

[9] K. Abe et al. Search for nucleon decay into charged antilepton plus meson in 0.316 megaton · years exposure of the super-Kamiokande water Cherenkov detector. *Physical Review D*, 96:012003, 2017.

[10] Z. Li et al. Measurement of the tau neutrino cross section in atmospheric neutrino oscillations with super-Kamiokande. *Physical Review D*, 98:052006, 2018.

[11] Y. Fukuda et al. Evidence for oscillation of atmospheric neutrinos. *Physical Review Letters*, 81:1562, 1998.

[12] M. H. Ahn et al. Measurement of neutrino oscillation by the K2K experiment. *Physical Review D*, 74:072003, 2006.

[13] K. Abe et al. Improved constraints on neutrino mixing from the T2K experiment with 3.13×10^{21} protons on target. *Physical Review D*, 103:112008, 2021.

[14] S. Nakayama et al. Measurement of single π^0 production in neutral current neutrino interactions with water by a 1.3 GeV wide band muon neutrino beam. *Physics Letters B* 619:255–262, 2005.

[15] S. Mine et al. Experimental study of the atmospheric neutrino backgrounds for $p \to e^+\pi^0$ searches in water Cherenkov detectors. *Physical Review D*, 77:032003, 2008.

[16] K. K. Young et al. DUMAND-II (deep underwater muon and neutrino detector) progress report. *AIP Conference Proceedings*, 338:886, 1995.
[17] I. A. Belolaptikov et al. The Baikal underwater neutrino telescope: Design, performance, and first results. *Astroparticle Physics*, 7:263–282, 1997.
[18] S. Adrián-Martínez et al. Measurement of atmospheric neutrino oscillations with the ANTARES neutrino telescope. *Physics Letters B*, 714:224–230, 2012.
[19] J. Ahrens et al. Observation of high energy atmospheric with the Antarctic muon and neutrino detector array. *Physical Review D*, 66:012005, 2002.
[20] M. G. Aartsen et al. Evidence for high-energy extraterrestrial neutrinos at the IceCube detector. *Science*, 342, 2013.
[21] J. Boger et al. The Sudbury neutrino observatory. *Nuclear Instruments and Methods in Physics Research Section A: Accelerators, Spectrometers, Detectors and Associated Equipment*, 449:172–207, 2000.
[22] Q. R. Ahmad et al. Direct evidence for neutrino flavor transformation from neutral-current interactions in the Sudbury neutrino observatory. *Physical Review Letters*, 89:011301, 2002.
[23] S. N. Ahmed et al. Measurement of the total active ^{8}B solar neutrino flux at the Sudbury neutrino observatory with enhanced neutral current sensitivity. *Physical Review Letters*, 92:181301, 2004.
[24] B. Aharmim et al. Independent measurement of the total active ^{8}B solar neutrino flux using an array of ^{3}He proportional counters at the Sudbury neutrino observatory. *Physical Review Letters*, 101:111301, 2008.
[25] M. Nakahata et al. Calibration of super-Kamiokande using an electron LINAC. *Nuclear Instruments and Methods in Physics Research Section A: Accelerators, Spectrometers, Detectors and Associated Equipment*, 421:113–129, 1999.
[26] K. Abe et al. Atmospheric neutrino oscillation analysis with external constraints in super-Kamiokande I-IV. *Physical Review D*, 97:072001, 2018.
[27] A. Takenaka et al. Search for proton decay via $p \to e + \pi^0$ and $p \to \mu + \pi^0$ with an enlarged fiducial volume in super-Kamiokande I-IV. *Physical Review D*, 102:112011, 2020.
[28] H. Nishino et al. High-speed charge-to-time converter ASIC for the super-Kamiokande detector. *Nuclear Instruments and Methods in Physics Research Section A: Accelerators, Spectrometers, Detectors and Associated Equipment*, 610:710–717, 2009.
[29] H. G. Berns et al. GPS time synchronization system for K2K. In: *11th IEEE NPSS*, 1999, p. 480.
[30] Y. Nakano et al. Measurement of radon concentration in super-Kamiokande's buffer gas. *Nuclear Instruments and Methods in Physics Research Section A: Accelerators, Spectrometers, Detectors and Associated Equipment*, 867:108–114, 2017.
[31] E. Blaufuss et al. ^{16}N as a calibration source for super-Kamiokande. *Nuclear Instruments and Methods in Physics Research Section A: Accelerators, Spectrometers, Detectors and Associated Equipment*, 458:638–649, 2001.

[32] S. Locke et al. New methods and simulations for cosmogenic induced spallation removal in super-Kamiokande-IV. arXiv:2112.00092.
[33] K. Abe et al. Hyper-Kamiokande design report. arXiv: 2101.05269.
[34] S. Adrián-Martínez et al. Letter of intent for KM3NeT 2.0. arXiv: 1601.07459.
[35] F. Benfenati et al. First scientific results of the KM3NeT neutrino telescope. *EPJ Web of Conferences*, 283:04009, 2023.
[36] The Nobel Prize in Physics 2002. https://www.nobelprize.org/prizes/physics/2002/summary/.
[37] The Nobel Prize in Physics 2015. https://www.nobelprize.org/prizes/physics/2015/summary/.
[38] M. Shiozawa, Reconstruction algorithms in the super-Kamiokande large water Cherenkov detector. *Nuclear Instruments and Methods in Physics Research Section A: Accelerators, Spectrometers, Detectors and Associated Equipment*, 433:240–246, 1999.
[39] E. R. Davies. *Machine Vision: Theory, Algorithms, Practicalities*. Academic Press, San Diego, 1997.
[40] S. Kasuga et al. A study on the e/μ identification capability of a water Cerenkov detector and the atmospheric neutrino problem. *Physics Letters B*, 374:238–242, 1996.
[41] M. Fechner et al. Kinematic reconstruction of atmospheric neutrino events in a large water Cherenkov detector with proton identification. *Physical Review D*, 79:112010, 2009.
[42] K. Abe et al. Search for cosmic-ray boosted sub-GeV dark matter using recoil protons at super-Kamiokande. *Physical Review Letters*, 130:031802, 2023.
[43] K. Abe et al. Evidence of electron neutrino appearance in a muon neutrino beam. *Physical Review D*, 88:032002, 2013.
[44] M. Jiang et al. Atmospheric neutrino oscillation analysis with improved event reconstruction in super-Kamiokande IV. *Progress of Theoretical and Experimental Physics*, 053F01, 2019, 39 p.
[45] R. B. Patterson et al. The extended-track event reconstruction for Mini-BooNE. Nuclear Instruments and Methods in Physics Research Section A: Accelerators, Spectrometers, Detectors and Associated Equipment, 608:206, 2009.
[46] N. F. Calabria. Search for proton decay in super-Kamiokande and perspectives in the hyper-Kamiokande experiments. PhD Thesis, The University of Naples Federico II, 2022.
[47] K. Abe et al. Solar neutrino measurements in super-Kamiokande-IV. *Physical Review D*, 94:052010, 2016.
[48] H. Watanabe et al. First study of neutron tagging with a water Cherenkov detector. *Astroparticle Physics*, 31:320–328, 2009.

Index

AC coupling, 51
activation function, 153
AdaBoost, 151
Adam, 141
aerogel, 101, 105, 114, 116, 117
AMANDA, 261
Amplifier (silicon), 55
annealing, 74, 76, 77, 85, 87, 90
anomaly detection, 165
ANTARES, 261
anti-k_t algorithm, 195
anti-oxidants, 79
APfit, 281
ARCA, 278
aromatic, 79
aromatic compound, 78
Arthur B. McDonald, 280
ASICs, 163
astrophysical neutrino searches, 251
atmospheric neutrino oscillations, 259
atmospheric neutrinos, 251
attenuation length, 75
attenuation length measurement, 275
available energy, 19

BaBar, 114–116
backpropagation, 155
bagging, 150
BAIKAL, 261
batch normalization, 160
beam, 45

BELLE, 114
benzene, 78, 81, 87
beyond-the-standard-model, 280
bias, 131
bias-variance decomposition, 136
bimolecular reactions, 85, 90
binary cross-entropy, 130
Birks, 80, 83
Birks' constant, 84
BONSAI, 286
boosting, 150
Bragg peak, 224
bremsstrahlung, 2, 255
bubble chamber, 46

C_4F_{10}, 109
calibration, 243
CALICE, 26
CAlorimeter for LInear Collider
    Experiment, 26
calorimetry, 237
Cambridge/Aachen algorithm, 194
CF_4, 116
charged pions, 256
Cherenkov, 7, 97–122
Cherenkov cone, 254
Cherenkov light, 251
Cherenkov neutrino detectors, 251
Cherenkov opening angle, 254
Cherenkov radiation, 252
Cherenkov rings, 256

Cherenkov spectrum, 272
classification, 127
color center, 74, 75
common view, 267
compensation coils, 266
compression, 144
Compton scattering, 4
Cone algorithms, 190
convolutional neural networks, 157
corrected charge, 275, 282
cosmic-ray muons, 263, 269
cosmic-ray stopping muons, 275
cosmic-ray through-going muon, 275
Coulomb scattering, 256
crazing, 77
critical energy, 4
crosslinking, 87, 89
Cubic Kilometer Neutrino Telescope, 278

dark matter, 280
DC-DC converters, 69
decision trees, 147
deep underground neutrino experiment, 279
degradation, 87
depletion region, 50
detector calibrations, 270
deuterium, 262
deuterium-tritium ("DT") fusion neutron generator, 276
diffuse supernova neutrino background, 268
diffusion, 231
diffusion coefficient, 90
digital optical module, 279
discoloration, 74, 76
dispersion, 103, 104, 106, 116
DOM, 279
dose constant, 72
dose rate, 72, 75–77, 89, 90
doublet, 63
downward-going muons, 269
DREAM, 26
drift chamber, 46

dropout, 143
dual-readout approach, 27
Dual-REAdout Module, 26
DUMAND, 261
DUNE, 279
Dynode directions, 266

EGS4, 5
electromagnetic calorimeters, 7
electromagnetic fraction (f_{em}), 13
electromagnetic showers, 255
electrons, 255
Eljen, 73, 74
EM showers, 221
energy resolution, 7
energy scale, 275
Equivalent Noise Charge, 57
event reconstruction algorithm, 280

f_{em}, 14
Förster mechanism, 82
fake (track), 62
far detector, 259
fast timing, 102, 106, 111, 112, 121
FCC-hh, 72
feedback (amplifier), 56
fiTQun, 285
fluorophore, 79
Fluors, 82
FPGAs, 163
Franck-Rabinowtich, 86
fresh air system, 268
Front End (silicon), 55
fully contained, 269
function, 129

g-value, 79
gadolinium sulfate octahydrate, 268
gamma ray photons, 256
gas flow nitrogen laser, 272
GEANT4, 35
gel, 87
generalization, 135
generalized k_t algorithm, 193

geomagnetic field, 266
GPS, 266
gradient descent, 138
graph neural networks, 163
grooming and tagging, 199

hadronic scattering, 256
hadronic showers, 11
hadronic vertices, 33
Hamamatsu, 264
heavy water, 262
Henry's law, 88
Higgs boson, 160
HK, 276
hls4ml, 164
homogeneous, 7
Hough transform, 119
Hough transformation, 281
Hund's rule, 82
hybrid photodiodes (HPD), 109–111, 117
Hyper-K, 276
Hyper-Kamiokande, 276

IceCube, 261
IMB, 252
impurity measure, 150
index of refraction, 76
inductive bias, 156
information entropy, 52
InfraRed and Collinear (IRC) safety, 185
integrated circuit, 47
invariant mass, 256
"invisible" energy, 12
ion recombination, 227
ionization, 255
ionization statistics, 47
Irvine–Michigan–Brookhaven, 259

J-PARC, 267
JADE algorithm, 192
jet algorithm, 182
jet definition, 182
jet substructure, 198

jet tagging, 161
jets, 180

k_t algorithm, 194
K2K, 259
Kalman filter, 64
Kamioka Nucleon Decay Experiment, 259
Kamiokande, 252
KEK, 267
Keras, 165
KM3NeT, 276, 278

L_1 regularization, 142
L_2 regularization, 142
leakage current, 51
learning objective, 129
LHCb, 99, 109, 116–121
light absorption spectrum, 272
LINAC, 263
linear models, 131
liquid scintillators, 71
Local Time Clock, 267
log likelihood, 120
long baseline beam neutrino oscillation measurements, 251
loss, 129
LTC, 267

machine learning, 125, 278
Masatoshi Koshiba, 280
Michel electrons, 275
microchannel plate (MCP), 109, 110, 115
microscopy, 45
mini-batch, 141
MiniBooNE, 285
mirrors, 102, 103, 106, 107, 117, 119
module, 48
Molière radius, 5
momentum resolution, 61
monomer, 78
Moore's Law, 53
multi-anode photomultiplier tube (MaPMT), 109, 113, 117, 118

Multi-PMT, 277
multiple scattering, 53
Muons, 255

near detector, 259
neural network, 152
neutral pion, 256
neutrino interactions, 256
neutrino telescopes, 260
neutron capture, 268
Ni-Cf, 271
nickel source, 271
nitrogen-laser-driven dye laser, 270
Nobel Prize in Physics, 280
Noise, 57, 235
noise, 238
nuclear interaction length, 14
nucleon decay searches, 251

1-kiloton (1 KT) water Cherenkov neutrino detector, 259
objective function, 129
ORCA, 278
organic scintillators, 71
overfitting, 135
oxygen, 74, 76, 77, 88–90

pair production, 255
partially contained, 269
particle flow approach, 30
particle identification, 97, 99, 100, 121, 122, 283
pattern recognition, 62
PbWO, 8
permanent damage, 77
peroxides, 89
phosphorescence, 82
photo-coverage, 259
photocathode, 255, 264
photoelectric effect, 4, 17
photoelectrons, 258
photomultiplier tubes, 252
photon detectors, 98, 103, 106–109, 112, 117–122
PID, 283

pixels (hybrid), 49
pixels (monolithic), 49
plastic scintillator, 71
PMTs, 252
polymer, 78
polymers, 72, 77
polystyrene (PS), 72
polyvinyl toluene (PVT), 72
pooling, 159
positrons, 255
post-training quantization, 145
protons, 284
proximity focusing, 108
pruning, 144
PS, 74, 87
PVT, 87
PyTorch, 165

QBEE, 266
QTC-based electronics with Ethernet, 266
quantization, 144
quantization-aware training, 145
quantum efficiency, 255, 272
quenching, 229

radiation damage, 65, 71, 74
radiation length, 2, 61, 284
radiators, 101, 103–105, 107–109, 113, 114, 116–119, 121
radical–radical interactions, 84
radicals, 78, 84, 85, 89, 90
Radon concentration, 269
random forests, 151
Rayleigh scattering, 232, 255
recombination scheme, 182
recurrent neural networks, 162
refractive index, 104, 105, 109, 118, 253
regression, 127
regularization, 141
resolution, 102, 106–108, 110, 111, 113, 115
RICH, 97–122
ring edge finding, 282

ring fitting, 281
ring separation, 284
Rn-reduced air system, 269

sampling calorimeter, 10
sampling fraction, 17
scintillating tiles, 76
scintillation, 80
sensor (silicon), 50
sequential recombination algorithms, 191
serial power, 69
shot noise, 57
shower maximum, 5
shower timing, 34
signal processing, 237
signal-to-noise, 235
silicon detector, 47
silicon photomultiplier (SiPM), 108, 109, 111, 112, 118
Single Event Upsets, 65
single-photoelectron, 271
SK, 252
SK-MC, 271
SN 1987A, 259, 280
SNO, 262
Snowmass accord, 183
SoftDrop, 201
software trigger system, 267
solar neutrino, 275
solubility, 88
space charge, 231
space points, 62
sparsity, 147
stochastic gradient descent, 139
Stokes' shift, 82
strips (silicon), 48
substrate, 79
Sudbury Neutrino Observatory, 262
Super-K, 252

Super-Kamiokande, 252
supernova neutrinos, 251
supernovae, 259
supervised learning, 127

T2K, 259
Takaaki Kajita, 280
Tau leptons, 256
Tau neutrinos, 252
template, 253
temporary damage, 77
TensorFlow, 165
the e/h ratio, 16
threshold, 53, 100, 101, 112–114, 121
time over threshold, 53
"time-walk" effect, 272
timing, 35
track fitting, 63
tracking, 46
trigger, 163
Trigger (tracker), 54
Tyvek®, 265

underfitting, 135
unimolecular reactions, 85, 90
unsupervised learning, 127
upward-going muons, 269

water purification system, 268
water quality parameters, 274
water transparency, 255, 268
wavelength shifters, 235
wavelength shifting fiber, 76
wavelength shifting fluor, 79
weight, 131

Xe lamp, 270
XGBoost, 152

yellowed, 74

www.ingramcontent.com/pod-product-compliance
Ingram Content Group UK Ltd.
Pitfield, Milton Keynes, MK11 3LW, UK
UKHW051141171224
452492UK00002B/10